Promoting Responsible Tourism With Digital Platforms

Youssef El Archi
Abdelmalek Essaadi University, Morocco & Hungarian University of Agriculture and Life Sciences, Hungary & Centre for Tourism Research, Development, and Innovation (CITUR), Portugal

Brahim Benbba
Abdelmalek Essaadi University, Morocco

Lóránt Dénes Dávid
John von Neumann University, Hungary & Hungarian University of Agriculture and Life Sciences, Hungary

Lucília Cardoso
Centre for Tourism Research, Development, and Innovation (CITUR), Portugal

A volume in the Advances in Hospitality, Tourism, and the Services Industry (AHTSI) Book Series

IGI Global
PUBLISHER of TIMELY KNOWLEDGE

Published in the United States of America by
IGI Global
Business Science Reference (an imprint of IGI Global)
701 E. Chocolate Avenue
Hershey PA, USA 17033
Tel: 717-533-8845
Fax: 717-533-8661
E-mail: cust@igi-global.com
Web site: http://www.igi-global.com

Library of Congress Cataloging-in-Publication Data

CIP Data Pending
ISBN: 979-8-3693-3286-3
eISBN: 979-8-3693-3287-0

This book is published in the IGI Global book series Advances in Hospitality, Tourism, and the Services Industry (AHTSI) (ISSN: 2475-6547; eISSN: 2475-6555)

British Cataloguing in Publication Data
A Cataloguing in Publication record for this book is available from the British Library.

For electronic access to this publication, please contact: eresources@igi-global.com.

Advances in Hospitality, Tourism, and the Services Industry (AHTSI) Book Series

Maximiliano Korstanje
University of Palermo, Argentina

ISSN:2475-6547
EISSN:2475-6555

Mission

Globally, the hospitality, travel, tourism, and services industries generate a significant percentage of revenue and represent a large portion of the business world. Even in tough economic times, these industries thrive as individuals continue to spend on leisure and recreation activities as well as services.

The Advances in Hospitality, Tourism, and the Services Industry (AHTSI) book series offers diverse publications relating to the management, promotion, and profitability of the leisure, recreation, and services industries. Highlighting current research pertaining to various topics within the realm of hospitality, travel, tourism, and services management, the titles found within the AHTSI book series are pertinent to the research and professional needs of managers, business practitioners, researchers, and upper-level students studying in the field.

Coverage

- Hotel Management
- Travel Agency Management
- Destination Marketing and Management
- Service Management
- Health and Wellness Tourism
- Service Design
- Leisure & Business Travel
- Food and Beverage Management
- Sustainable Tourism
- Service Training

IGI Global is currently accepting manuscripts for publication within this series. To submit a proposal for a volume in this series, please contact our Acquisition Editors at Acquisitions@igi-global.com or visit: http://www.igi-global.com/publish/.

The Advances in Hospitality, Tourism, and the Services Industry (AHTSI) Book Series (ISSN 2475-6547) is published by IGI Global, 701 E. Chocolate Avenue, Hershey, PA 17033-1240, USA, www.igi-global.com. This series is composed of titles available for purchase individually; each title is edited to be contextually exclusive from any other title within the series. For pricing and ordering information please visit http://www.igi-global.com/book-series/advances-hospitality-tourism-services-industry/121014. Postmaster: Send all address changes to above address.

Titles in this Series

For a list of additional titles in this series, please visit:
www.igi-global.com/book-series/advances-hospitality-tourism-services-industry/121014

Managing Tourism and Hospitality Sectors for Sustainable Global Transformation
Pankaj Kumar (Mizoram University, India) S.K. Gupta (HNB Garhwal University, India) Maximiliano E. Korstanje (University of Palermo, Argentina) Prakash Chandra Rout (PSS Central Institute of Vocational Education, India) and Madhurima (Department of Environmental Science, Mizoram University, ndia)
Business Science Reference • © 2024 • 323pp • H/C (ISBN: 9798369362600) • US $345.00

Revolutionizing the Service Industry With OpenAI Models
Mahmut Demir (Isparta University of Applied Sciences, Turkey) and Ali Dalgic (Isparta University of Applied Sciences, Trkey)
Business Science Reference • © 2024 • 315pp • H/C (ISBN: 9798369312391) • US $290.00

Achieving Sustainable Transformation in Tourism and Hospitality Sectors
Pankaj Kumar (Mizoram University, India) S.K. Gupta (HNB Garhwal University, India) Maximiliano E. Korstanje (University of Palermo, Argentina) Prakash Chandra Rout (PSS Central Institute of Vocational Education, India) and Madhurima (Department of Environmental Science, Mizoram University, India)
Business Science Reference • © 2024 • 461pp • H/C (ISBN: 9798369333907) • US $285.00

Social Media Strategies for Tourism Interactivity
Célia M.Q. Ramos (University of Algarve, Portugal) Teresa Costa (Instituto Politécnico de Setúbal, Portugal) Filipe Segurado Severino (Estoril Higher Institute for Tourism and Hotel Studies, Portugal) and Maria de Lurdes Calisto (Estoril Higher Institute for Tourism and Hotel Studies, Portugal)
Business Science Reference • © 2024 • 345pp • H/C (ISBN: 9798369309605) • US $285.00

Gastronomic Sustainability Solutions for Community and Tourism Resilience
Shivam Bhartiya (Jain (Deemed-to-be) University, India) Vaibhav Bhatt (Central University of Tamil Nadu, India) and Andrea Edurne Jimenez Ruiz (Universidad Autonoma del Estado de Mexico, Mexico)
Business Science Reference • © 2024 • 315pp • H/C (ISBN: 9798369341353) • US $295.00

Shifts in Knowledge Sharing and Creativity for Business Tourism
Muhammad Waseem Bari (Government College University, Faisalabad, Pakistan) Mohammad Soliman (Research and Consultation Department, University of Technology and Applied Sciences, Salalah, Oman) and Emilia Ala-verdov (Georgian Technical University, Georgia)
Business Science Reference • © 2024 • 336pp • H/C (ISBN: 9798369326190) • US $295.00

IGI Global
PUBLISHER of TIMELY KNOWLEDGE

701 East Chocolate Avenue, Hershey, PA 17033, USA
Tel: 717-533-8845 x100 • Fax: 717-533-8661
E-Mail: cust@igi-global.com • www.igi-global.com

Table of Contents

Detailed Table of Contents

in Fez. The main findings reveal that digital interventions significantly improve visitor engagement, facilitate appreciation of cultural heritage, and support sustainable tourism practices. Specifically, tools like AR and AI-powered guides offer personalized and immersive experiences that connect visitors more closely with the Medina's rich history, contributing to the conservation of the site.

This chapter discusses the potential impact of blockchain technology on sustainable tourism, highlighting the increasing importance of sustainability in the tourism industry. Blockchain's decentralized and transparent nature offers transformative solutions to address environmental, social, and economic challenges. The chapter explores how blockchain can contribute to sustainable practices, including resource management, community empowerment, and responsible supply chains. It emphasizes the role of blockchain in enhancing sustainability through transparency, traceability, and accountability. Furthermore, blockchain-based systems streamline the measurement and verification of sustainability indicators, promoting responsible practices and providing reliable information to travelers. Overall, the chapter provides valuable insights on utilizing blockchain to advance sustainability in tourism destinations, attracting environmentally conscious travelers, and fostering sustained socio-economic development.

The introduction of geotagging has transformed the tourism industry, impacting environmental conservation, social engagement, and economic development. Through GPS-enabled devices, augmented reality applications, social media networks, and more, tourists and stakeholders can enhance experiences and support responsible tourism practices. Geotagging can contribute significantly to biodiversity conservation, sustainable management, and environmental education, fostering a deeper appreciation for nature. Sociologically, it can promote community connection, cultural preservation, and tourist safety. Economically, it can boost revenue generation, marketing, and infrastructure development. Challenges include privacy concerns, data security risks, and the digital divide, which need addressing through prioritizing data protection and closing the digital gap. Future research should focus on enhancing geotagging tools, exploring blockchain and AI, and ensuring ethical use of geotagged data, thus propelling the tourism industry towards sustainable growth and innovation.

The research on smart tourism experiences that take into account the responsible behavior of travelers is booming. The relationship between the development of memorial consumer experiences and the

This research seeks to analyze how the use of smart data analytics solutions by tourism stakeholders can promote the prediction of responsible customer behavior. After a theoretical and conceptual framework, the theoretical study is based on an exploratory literature review. It revealed the adoption of a range of smart data analytics solutions by businesses and tourist destinations in order to better predict attitudes and concrete responsible actions reflecting the eco-responsible behavior of tourists. Despite the inherent limitations of the approach, the results show multiple good practices that can be adopted by industry players to better understand, anticipate, and even guide the responsible behavior of their customers and prospects.

Digital technology has become an essential tool for the tourism sector (comparing destinations, choosing hotels, booking flights, etc.). As a result, the majority of tourism bookings are now made using online travel platforms. This revolution in the use of digital technologies and artificial intelligence by travel companies and tourists has been amplified by the spread of the COVID-19 pandemic, prompting managers to switch to e-marketing strategy. Internet use has certainly contributed to accelerated growth in the tourism sector, but the question is how do tourists and companies use social media to make their choices? Using the PRISMA method, the authors attempt to explore the state of the art on the use of social networks and artificial intelligence by professionals and users in the tourism sector. The results show a positive impact of the use of digital technology on the evolution of the tourism sector.

In this research, the authors estimate the revenue loss due to COVID-19 on homestays in 13 districts of Uttarakhand, India. They use novel techniques and unique variables to calculate revenue and expenditure. They also attempted to uncover resilience strategies adopted by homestay owners to recover their respective businesses. They found six themes from the eleven semi-structured interviews with these homestay hosts. They defined these themes as untact hospitality, upgrading homestay services with digital technologies, digitally transforming booking policies, sustainability with legal registration, supporting intrapreneurship, and social entrepreneurship model. This study contributes to advancing the crisis management literature within the tourism sector by evaluating the impact of the pandemic on homestays of Uttarakhand and thereafter elaborating on resilience strategies adopted. Practitioners and policymakers can use the study outcomes to re-evaluate the existing guidelines concerning homestay operations according to the new normal solitary concept.

Preface

In the age of digital transformation, the tourism industry faces a pressing challenge: balancing the growing demand for travel with the imperative to protect the environment and preserve local cultures. The rise of digital platforms has revolutionized how people plan, book, and experience travel, but it has also intensified concerns about overtourism, cultural commodification, and environmental degradation. Without thoughtful intervention, these trends threaten to undermine the destinations travelers seek to explore.

Promoting Responsible Tourism With Digital Platforms offers a comprehensive solution by exploring how digital platforms can be leveraged to promote responsible travel practices. By examining case studies, theoretical frameworks, and the latest technological advancements, the book provides actionable insights for policymakers, industry professionals, and travelers alike. It serves as a roadmap for integrating responsible tourism principles into the digital landscape, ensuring that tourism remains sustainable and beneficial for all stakeholders.

The book explores sustainable destination management, community engagement through social media, and blockchain technology for ethical tourism practices. It equips readers with the knowledge and tools needed to navigate the complex intersection of responsible tourism and digital platforms. Promoting a deeper understanding of these issues, this book aims to foster a more conscientious and sustainable approach to travel, ensuring that future generations can continue exploring the world's wonders without compromising the integrity of local communities and ecosystems.

The first chapter, titled "Role of Virtual Reality Technology in Sustainable Travel Behaviour and Engagement Among Millennials," involves around minding the gap across different generation's traits and behavior, the tech-savvy millennials identified as potential visitors seeking immersive experiences, higher sensory stimulation, more engagement, and higher behavioral intention towards the sustainability of destinations. Thus, this study aimed to analyze the effectiveness of virtual reality technologies among millennial travel behavioral intention and engagement towards tourism destinations providing an insight into technological and sustainable capacity and capability building for Malaysian destination marketers. The research framework integrates the technology acceptance model (TAM) and the stimulus–organism–response (SOR) to conduct an empirical analysis of 263 valid participants using a partial-least-squares based structural equation modeling approach, identifying several positive consequences. Expanding the millennial's travel panorama through virtual reality technologies at tourism destinations can significantly help the marketers position themselves uniquely and attract more potential customers in the future.

The second chapter, titled "From Pixels to Preservation: AI and Big Data Redefining Responsible Tourism," fills the gap regarding the AI and big data analytics as it relates to responsible travel behavior. The objective of this chapter is to investigate the growing role of AI and big data in the context of responsible tourism practices, prospects and challenges in India. Furthermore, the chapter offers comprehensive analysis of the ramifications of AI and big data on responsible travel behavior. The study's novelty resides in its examination of AI and big data within the context of responsible tourism, an area that has been relatively neglected in the current body of literature.

The third chapter, titled "The Role of Digital Technology and Sustainable Practices in Tourists' Decision Making," involves around the process of choosing a tourist destination considering technology and sustainability as a key factors. This study delves into prominent models and theories within the literature, such as the destination choice set model, and theories on cultural and travel distance. These frameworks highlight the role of factors like geographical and cultural proximity, as well as facilitators and inhibitors in destination choice. Additionally, tourist perception, motivation, and behavior are explored as crucial determinants in decision-making. Understanding tourists' perceptions, motivations, and behaviors is vital for businesses and destinations to tailor their offerings accordingly. Economic determinants, including neoclassical theories of consumer choice, are also discussed, with emphasis on factors beyond prices and income. Furthermore, the study addresses the impact of the digital age on tourism decision-making, noting the significant influence of online information sources, social media, and digital booking platforms in shaping travelers' choices regarding responsible practices.

The fourth chapter,ntitled "Digital Technologies for Sustainable Tourism: The Case of Ecobnb," explores the role of digital technologies in favouring the match of the demand and the supply of sustainable tourism activities. The evolution of the concept of tourism and the main traits of sustainable tourism are described. The chapter also sheds light on how digital technologies affod smart travellers to satisfy their needs of torism activities more respectful of the environment and the local communities. Lastly, the chapter analyses the case of Ecobnb, an Italian digital plaform that gathers all the sustainable accommodations in Italy and proposes travel itineraries that respect the environment.

The fifth chapter, "Digital Journey Through Fez's Historic Medina: Embracing the Core of Responsible Tourism," aims to explore the role of digital technologies in enhancing sustainable tourism within the historic Medina of Fez in Morocco, with a focus on preserving its cultural heritage while fostering responsible tourism practices. Through a mixed-methods approach that combines a comprehensive literature review with qualitative analysis—comprising interviews with key tourism stakeholders and analysis of online reviews —this research investigates the adoption and impact of online platforms, interactive digital guides, Augmented Reality (AR) applications, and Artificial Intelligence (AI) systems on the tourism experience in Fez. The main findings reveal that digital interventions significantly improve visitor engagement, facilitate appreciation of cultural heritage, and support sustainable tourism practices. Specifically, tools like AR and AI-powered guides offer personalized and immersive experiences that connect visitors more closely with the Medina's rich history, contributing to the conservation of the site.

The sixth chapter, titled "Breaking Chains, Building Futures: Blockchain's Impact on Sustainable Tourism," discusses the potential impact of blockchain technology on sustainable tourism, highlighting the increasing importance of sustainability in the tourism industry. Blockchain's decentralized and transparent nature offers transformative solutions to address environmental, social, and economic challenges. The chapter explores how blockchain can contribute to sustainable practices, including resource management, community empowerment, and responsible supply chains. It emphasizes the role of blockchain in enhancing sustainability through transparency, traceability, and accountability. Furthermore, blockchain-based systems streamline the measurement and verification of sustainability indicators, promoting responsible practices and providing reliable information to travelers. The chapter provides valuable insights on utilizing blockchain to advance sustainability in tourism destinations, attract environmentally conscious travelers, and foster sustained socio-economic development.

The seventh chapter, titled "Revolutionizing Responsible Tourism Cutting-Edge Technological Advances in Geotagging," focuses on the introduction of geotagging which has transformed the tourism industry, impacting environmental conservation, social engagement, and economic development. Through GPS-enabled devices, augmented reality applications, social media networks, and more, tourists and stakeholders can enhance experiences and support responsible tourism practices. Geotagging can contribute significantly to biodiversity conservation, sustainable management, and environmental education, fostering a deeper appreciation for nature. Sociologically, it can promote community connection, cultural preservation, and tourist safety. Economically, it can boost revenue generation, marketing, and infrastructure development. Challenges include privacy concerns, data security risks, and the digital divide, which need addressing through prioritizing data protection and closing the digital gap. Future research should focus on enhancing geotagging tools, exploring blockchain and AI, and ensuring ethical use of geotagged data, thus propelling the tourism industry towards sustainable growth and innovation.

The eighth chapter, titled "Smart Tourism Experience and Responsible Travelers Behavior: A Systematic Literature Review," presents a systematic review of the literature on the interaction between tourism experience and responsible traveler behavior in the context of the transformation of responsible tourism through digital platforms. In today's digital world, it is crucial to reconsider our model with sustainability in mind. The research on smart tourism experiences that take into account the responsible behavior of travelers is booming. The relationship between the development of memorial consumer experiences and the promotion of responsible consumer behavior is a highly sought-after area of research. This is especially true regarding the potential use of emerging technology solutions to enhance tourism services, improve customer smartness, predict responsible behavior, and increase customer satisfaction and loyalty. The purpose of this research is to investigate how the advancement of smart tourism experiences can promote responsible behavior among customers of tourism businesses and destinations.

The ninth chapter, titled "Leveraging Digital Platforms for Responsible Sports Tourism: Budapest's Role in the 2020 European Football Championship," aims to present the safety and security systems for sporting events, with a special focus on the theoretical and practical tools used during the 2020 European Football Championship, which are of particular importance for ensuring security. The authors have analysed the policing statistics regarding the 2020 European Football Championship and described the latest sustainable technological solutions in use, with a particular focus on artificial intelligence, drone technology, and digitalization. It was pointed out how technological development, innovation and digitization affect the sustainable operation and activity of law, sports law, sport tourism, law enforcement and sports police.

The tenth chapter, titled "Web of Wonders: Sustainable Digital Marketing Strategies for Unforgettable African Tourism Destinations," investigates sustainable digital marketing strategies tailored for African tourism destinations, with a focus on promoting unforgettable experiences while preserving the natural and cultural heritage of these destinations. It also highlights the importance of aligning digital marketing efforts with sustainable tourism principles, such as promoting responsible travel behavior and supporting local communities. The chapter emphasizes the need to showcase the unique natural and cultural heritage of African destinations through visually appealing content and immersive digital experiences. Furthermore, the potential of emerging technologies, such as blockchain and artificial intelligence, is explored to enhance the effectiveness and sustainability of digital marketing campaigns. The outcomes of this study aim to provide practical recommendations for African tourism destinations to enhance their digital marketing efforts while ensuring long-term sustainability and success.

The eleventh chapter, titled "Smart Data Analysis and Prediction of Responsible Customer Behaviour in Tourism: An Exploratory Review of the Literature," seeks to analyze how the use of smart data analytics solutions by tourism stakeholders can promote the prediction of responsible customer behavior. With the massification of data both emitted, collected, and shared by the customer, tourism players are more than ever invited, particularly through intelligent solutions for analyzing customer data, to better understand the mobiles and determinants of its behavior, here responsible, and thus better satisfy, retain. After a theoretical and conceptual framework, our theoretical study is based on an exploratory literature review, revealing the adoption of a range of smart data analytics solutions by businesses and tourist destinations to better predict attitudes and concrete responsible actions reflecting the eco-responsible behavior of tourists. Despite the inherent limitations of our approach, the results show multiple good practices that can be adopted by industry players to better understand, anticipate, and even guide the responsible behavior of their customers and prospects.

The twelfth chapter, "Exploring Sustainability in Tourism Marketing Through Digital and Social Networks: A Literature Review," explores the state of the art on the sustainable use of social networks and digital technologies by stakeholders and tourists in the tourism sector. Digital technology has become an essential tool for the tourism sector (comparing destinations, choosing hotels, booking flights…etc.). As a result, the majority of tourism bookings are now made using online travel platforms. This revolution in the use of digital technologies and artificial intelligence by travel companies and tourists has been amplified by the spread of the COVID-19 pandemic, prompting managers to switch to e-marketing strategy. Internet use has certainly contributed to accelerated growth in the tourism sector, but the question is how do tourists and companies use social media to make their choices?

The thirteenth chapter, "Assessing Loss Due to COVID-19 in the Homestay Sector and Comprehending Resilience Strategies," estimates the revenue loss due to COVID-19 on homestays in 13 districts of Uttarakhand, India. The authors use novel techniques and unique variables to calculate revenue and expenditure. The authors also attempted to uncover resilience strategies adopted by homestay owners to recover their respective businesses. They found six themes from the eleven semi-structured interviews with these homestay hosts. They defined these themes as intact hospitality, upgrading homestay services with digital technologies, digitally transforming booking policies, sustainability with legal registration, supporting intrapreneurship, and social entrepreneurship model. This study contributes to advancing the crisis management literature within the tourism sector by evaluating the impact of the pandemic on homestays of Uttarakhand and thereafter elaborating on resilience strategies adopted. Practitioners and policymakers can use the study outcomes to re-evaluate the existing guidelines concerning homestay operations according to the new normal solitary concept.

In an era marked by the rapid evolution of digital technologies and the urgent need for sustainable practices in tourism, this edited volume emerges as a pivotal contribution to the field. This book not only identifies the pressing challenges faced by the tourism industry, such as overtourism, cultural commodification, and environmental degradation but also provides a comprehensive roadmap for leveraging digital platforms to address these issues. Each chapter delves into diverse aspects of the intersection between digital innovation and responsible tourism, offering valuable insights and practical strategies for industry stakeholders, policymakers, and travelers. From exploring the role of virtual reality in shaping millennial travel behavior to examining the potential of blockchain technology in fostering sustainable practices, the book navigates through the complex landscape of digital transformation in tourism.

By emphasizing concepts such as sustainable destination management, community engagement through social media, and the ethical implications of AI and big data analytics, the book bridges the gap between theory and practice. It underscores the importance of integrating responsible tourism principles into the digital ecosystem to ensure that tourism remains not only economically viable but also socially and environmentally responsible. Furthermore, the book sheds light on innovative case studies, such as the utilization of geotagging for environmental conservation and the application of smart data analytics to predict responsible customer behavior. These examples demonstrate the transformative power of digital technologies in redefining the tourism experience while preserving cultural heritage and natural ecosystems.

Moreover, the book's exploration of the impact of the COVID-19 pandemic on tourism, particularly in terms of loss assessment and resilience strategies in the homestay sector, offers timely insights for navigating future crises and promoting sustainable recovery. Additionally, this volume serves as a beacon of guidance in an increasingly digitized and environmentally conscious world. By equipping stakeholders with the knowledge and tools needed to navigate the complex landscape of responsible tourism, this book not only contributes to academic discourse but also paves the way for a more sustainable and inclusive future for global tourism.

Youssef El Archi
Abdelmalek Essaadi University, Morocco & Hungarian University of Agriculture and Life Sciences, Hungary & Centre for Tourism Research, Development, and Innovation (CITUR), Portugal

Brahim Benbba
Abdelmalek Essaadi University, Morocco

Lóránt Dénes Dávid
John von Neumann University, Hungary & Hungarian University of Agriculture and Life Sciences, Hungary

Lucília Cardoso
Centre for Tourism Research, Development, and Innovation (CITUR), Portugal

Chapter 1
Role of Virtual Reality Technology in Sustainable Travel Behaviour and Engagement Among Millennials

Rupam Konar
https://orcid.org/0000-0002-3235-3842
Taylor's University, Malaysia

Lazey Doma Bhutia
https://orcid.org/0009-0009-5705-0547
Asia Pacific University of Technology and Innovation, Malaysia

Kevin Fuchs
https://orcid.org/0000-0003-3253-5133
Prince of Songkla University, Thailand

Kandappan Balasubramanian
https://orcid.org/0000-0001-7634-4676
Taylor's University, Malaysia

ABSTRACT

This study aimed to analyze the effectiveness of virtual reality technologies among millennial travel behavioral intention and engagement towards tourism destinations providing an insight into technological and sustainable capacity and capability building for Malaysian destination marketers. The research framework integrates the technology acceptance model (TAM) and the stimulus–organism–response (SOR) to conduct an empirical analysis of 263 valid participants using a partial-least-squares-based structural equation modeling approach, identifying several positive consequences. Expanding the millennial's travel panorama through virtual reality technologies at tourism destinations can significantly help the marketers position themselves uniquely and attract more potential customers in the future.

DOI: 10.4018/979-8-3693-3286-3.ch001

INTRODUCTION

Tourism, as a global sector, has two mandates: economic expansion and sustainable development. These goals, which are consistent with the UN's Sustainable Development Goals (UN, 2015), have economic, social, and environmental components. In the digital era, technological innovation is critical for simplifying destination administration and supporting sustainable practices (Gretzel et al., 2020). Malaysia's established tourist industry has the ability to harness new technology for long-term success. The COVID-19 pandemic significantly impacted worldwide tourism, forcing harsh precautions such as physical distance and travel restrictions, resulting in a decrease in activity (GNFI, 2020). In response, the industry shifted to technology-driven solutions. Virtual Reality (VR), together with Augmented Reality (AR) and Mixed Reality (MR), offers an exciting potential to reinvent visitor interaction and destination marketing, providing new solutions in the face of bigger problems (Wang et al., 2023).

Malaysia's early use of VR technology in tourism illustrates its understanding of the technology's revolutionary potential. A recent study by Hassan et al. (2023) highlights the positive impact of VR on tourist interest in Malaysian cultural heritage sites. While the pandemic's impacts persist, VR continues to demonstrate its adaptability in industries such as amusement parks, cruise lines, museums, and destination marketing (Morgan et al., 2023). This embrace of innovation demonstrates Malaysia's commitment to using technology to drive tourism growth, improve tourist interaction, and protect its natural treasures. Although previous studies have looked at how VR affects travel plans (Liu et al., 2018), the intricate sensory impressions provided by VR experiences demand more investigation. VR utilizes immersive, multi-sensory experiences to change customer views of locations, making it a strong marketing tool (Huang et al., 2022). Its key components - visualization, immersion, and interaction - provide a dynamic platform for exhibiting locales and creating unique experiences that go beyond traditional media (Choi et al., 2023). Malaysia is committed to sustainable tourism. Recognizing the environmental impact of traditional travel, the government explores new solutions to reduce its footprint. Virtual reality has the ability to replicate destination experiences without the requirement for actual travel. By replacing certain physical travels with interesting VR tours, Malaysia can reduce overtourism (Lee & Gretzel, 2023). Furthermore, including educational components about environmental protection into VR experiences might raise tourists' environmental consciousness, resulting in a healthy interaction between tourism and nature (Chen et al., 2022). Malaysia's tourist resilience during the COVID-19 pandemic was dependent on its capacity to adopt disruptive technology like VR. VR provided a safe alternative to traditional travel, allowing for virtual excursions while also protecting public health during epidemics (Kim et al., 2022).

VR's ability to expedite information-gathering operations while also offering a more immersive and complete experience highlights its potential to improve travel itineraries (Gretzel et al., 2020). Recognizing its long-term relevance, especially in appealing to the millennial generation (Xiang & Gretzel, 2020), this study seeks to evaluate the impact of virtual reality on millennial travel engagement and behavior in Malaysia. While there is an increasing corpus of research on VR in tourism, the current literature provides an imperfect picture of visitors' decision-making and behavioural processes. A significant study vacuum exists about how virtual stimuli influence visitor perceptions and travel intentions (Morgan et al., 2023). Furthermore, a thorough, theoretically grounded model of behavior in this particular situation has yet to be fully created. Existing research on VR's impact on sustainable travel behaviour and participation among Malaysian millennials finds major gaps (Lee & Gretzel, 2023). To begin, although current research investigates VR's impact on visiting intentions (Liu et al., 2018), in-person sensory experiences during VR interactions are mostly unknown (Morgan et al., 2023). Furthermore, research on VR's potential contribution to sustainable tourism, particularly its function in reducing travel's environmental effect, is limited (Chen et al., 2022). Additionally, there is a major gap in empirical research concentrating just on the Malaysian setting, necessitating targeted investigations adapted to the country's tourist industry's distinct dynamics (Hassan et al., 2023). Addressing these gaps is critical to developing a thorough knowledge of VR's diverse role in promoting sustainable tourism in Malaysia. By diving into the fundamental features of VR experiences, this chapter hopes to give practical insights for tourism stakeholders, particularly destination managers, who want to include VR into their strategic plans as both a promotional tool and a tourist attraction. As a result, the potential of VR technology to promote sustainable travel behaviour and participation among Malaysian millennials is quite promising. As Malaysia navigates the changing tourism market, adopting VR emerges as a strategic priority, ushering in a new age of immersive and sustainable travel experiences. To demonstrate the growing adoption of VR technology in Malaysia, consider mentioning VR arcades like VR Lab (Kuala Lumpur) with its focus on physical movement, or VAR LIVE Malaysia known for its immersive experiences. As of early 2024, these facilities offer a glimpse into the exciting VR landscape emerging within the country.

Theoretical Background

Virtual Reality in Tourism

The early vision for virtual reality (VR) in tourist marketing was as a game-changing instrument for the business (Cheong, 1995; Williams and Hobson, 1995). This original assumption was based on the known function of travel intermediaries, such as tour operators and travel agencies, in selling tourist products to prospective clients (Holloway, 2009). Traditionally, printed brochures were the major means of conveying information and advertisements (Holloway 2004). However, as the internet evolved from web 1.0 to web 5.0, opportunities arose for using immersive platforms such as VR to connect with prospective visitors (Tavakoli and Wijesinghe, 2019). While VR's first usage in tourist marketing was centred on increasing brand recognition, making relationships, and sustaining brand presence, its applications have grown beyond these initial aims. According to recent studies, virtual reality can also be a useful tool for market research. Its immersive capabilities provide a cost-effective and informative approach to understanding client behaviour and preferences in the tourist business. Experiential stimuli are important in tourist marketing, especially considering the intangible character of tourism products (Goossens, 2000; Hyun & O'Keeffe, 2012). Tourists' destination pictures, which strongly impact their decision-making processes, are frequently formed via experiences (Cai, 2002). Prior to the introduction of VR, experience information was mostly confined to comments from past visitors via text, images, and videos. Today's visitors, however, are increasingly documenting their trip experiences in immersive VR forms, thanks to readily available technology (Cooper & Macneil, 2005). According to research, the amount of VR immersion determines its impact, with VR headsets allowing for more emotive and cognitive assessment than desktop-based 360-degree movies (Adachi et al., 2020). Tourism providers are increasingly using user-generated VR content, particularly 360-degree movies, for social media marketing because of its perceived authenticity among visitors (Buhalis & O'Connor, 2005; Navío-Marco, Ruiz-Gómez, & Sevilla-Sevilla, 2018). Indeed, millennials substantially rely on such material for trip planning (UNWTO, 2017).

According to Williams and Hobson (1995), while VR's capacity to adapt virtual experiences is important, its usefulness as a complete replacement for tourism is determined by individual incentives (Cooper & MacNeil, 2005; Guttentag, 2010). Authenticity, particularly how sensory motions in VR affect users' cognitive and emotional reactions, is critical for acceptance (Mura et al., 2017; Kim, Lee, & Jung, 2020). VR has the potential to replace for certain activities like theme parks, museums, and cultural places. This appeals to both budget-conscious travellers who lack the resources to travel physically (Cooper & Macneil, 2005; Dewailly, 1999) and affluent persons wanting in-depth location information. VR's capacity to improve user perception and mental picture construction of places adds to its attractiveness (Tussyadiah et al., 2018; Bogicevic et al., 2019). However, Musil and Pigel's (1994) argument against VR replacing tourism is consistent with travellers' overall rejection to VR replacing actual experiences (Prideaux 2002). Cheong (1995) underlined the fundamental social component of travel, which remains an issue despite VR's communication capabilities. The absence of a true human touch in VR social interactions casts doubt on its completeness as a solution. Furthermore, worries about income losses for tourism-dependent regions and possible infrastructure development challenges, particularly in distant locations of developing nations (Cheong, 1995; Jude & Ukekwe, 2020), urge for additional investigation. As a potential alternative, VR provides both benefits and problems, necessitating context-specific evaluations based on the distinct characteristics and development goals of each location. Currently, virtual tourism research focuses mostly on idea formulation, virtual technology implementation, on VR tourism, subjective enjoyment and perceived immersion are essential parts of VR consumers' senses as the benefits and drawbacks of virtual tourism, and virtual tourism-based destination marketing. There has been few research on virtual tourism that look at travellers' sustainable behaviour.

Technology Acceptance Model

The Technology Acceptance Model (TAM) is based on the foundation of behaviour theory, introduced by Davis et al. (1989) this model focuses on final adaptation behaviour in respect to certain technology which determines perceived usefulness and perceived ease of use explaining people's acceptance and their final adaptation behavioural attitude. Perceived usefulness is explained as "the degree to which a person believes that using a particular system would enhance his or her job performance", and perceived ease of use is defined as "the degree to which a person believes that using a particular system would be free of effort". The so-called attitude is the user's perception of the progressive and depraved aspects of technology, such as perceived utility and simplicity of use. Because the original TAM primarily focused on the users' perceptions of technology and desire to use it, following research have used the expanded TAM as its theoretical basis when analysing the consequences of users' senses in terms of technical characteristics on users' behaviours and attitudes. The TAM model is universally applicable and has been tested in a variety of situations, including mobile technologies, virtual communities, and online gaming. Similarly, the Technology Acceptance Theory has been frequently used to the study of tourism-related behavioural intentions. It is thought that perceived utility and perceived simplicity of use are major elements influencing customers' online travel reservation intentions, tourism app use, and tourism website use.

In the past literature of VR research, TAM is the most widely accepted and adopted model which assisted in identifying the antecedents to VR consumers experiences. As a result, when consumers utilise VR technology, they appear to expect to experience subjective well-being, such as satisfaction and happiness, which influences their behavioural intentions. In tourism, TAM is employed in several situations, such as cultural heritage, historic visitor attractions, and wine tourism. Attitudes are impacted by perceived usefulness and simplicity of usage. Perceived usefulness, attitude, and perceived simplicity of use all have an impact on intention to use. The TAM sees information technologies, including the Internet, and VR headset as a tool for enhancing user performance. Toros et al. (2024), Bano and Siddiqui (2022) Tawafak et al. (2023) confirmed that perceived utility and perceived simplicity of use are the two most important elements in understanding the desire to utilise technology. Huang et al. (2013) and Yang and Han (2021) found that perceived utility and perceived ease of use influence destination choice in the setting of 3D virtual environments. As a result, perceived utility and perceived ease of use are two critical TAM components examined in this study.

Stimuli Organism Response Framework (S-O-R)

The stimuli-organism-response (SOR) framework aims to "describe the individual behaviour through the stimuli creating cognitive and emotional states, which, in turn, lead to responses". The SOR framework consist of three parts comprising of stimulus (S), organism (O), and response (R). Extended from the environmental psychology, the SOR has been dominantly used to understand the consumer behaviours in the domain of technology applications in different contexts. The SOR model is also used widely across the tourism research, such as mobile social tourism shopping, environmental responsibility behaviours in tourism using the eco-friendly repute perception of a destination as the stimulus, the organism was evaluated using consumption emotion, and response was based on tourism satisfaction and tourist's environmental responsibility behaviour. The Stimulus-Organism-Response (S-O-R) framework offers a valuable lens for understanding user behavior in virtual reality (VR) tourism. This framework, rooted in the work of Mehrabian and Russell (1974), posits that environmental stimuli influence an individual's internal state (organism), ultimately leading to behavioral responses. Marketing researchers have successfully employed S-O-R to comprehend the impact of environmental elements on consumer behavior (Xu et al., 2014).

There is a growing recognition of S-O-R's potential in VR tourism research. The framework sheds light on how users perceive and respond to VR stimuli. Studies by Kim et al. (2021, 2020) exemplify this application. Their research employed S-O-R to explore how VR experiences influence user attitudes and visit intentions. Kim et al. (2020) specifically investigated the role of perceived authenticity in VR experiences, demonstrating its influence on cognitive and affective responses, which in turn, impact visit intention. In this vein, the present study adopts the S-O-R framework to examine the interplay between users' sensory experiences (organism) while engaging with VR's technological features (stimuli) and how this translates into changes in attitude and intention to visit a destination (responses).

Sustainable Travel through Virtual Reality

There's a critical gap in tourism research: understanding if people with strong environmental concerns are more likely to adopt virtual tourism due to its immersive nature. Public awareness of tourism's environmental footprint is on the rise (Gössling, Scott, & Hall, 2019), prompting consumers to make eco-conscious choices (Lenzen et al., 2023). While research supports virtual consumption as a substitute for physical goods (Wang et al., 2022), it's unclear if this concept translates directly to tourism. This study proposes that heightened environmental consciousness could motivate individuals to seek virtual travel alternatives. The recent pandemic serves as a powerful example: travel restrictions demonstrably led to a significant reduction in carbon emissions (Scott, Hall, & Gössling, 2021). VR technology holds immense potential to make tourism more sustainable overall (Gretzel, Hwang, & Gibson, 2022; Ioannides & Hajibabaei, 2020). However, a hurdle exists. Some studies suggest a preference for the authenticity and sensory richness of real travel experiences (Huang et al., 2023). Xiang, Gretzel, Gong, & Li (2020) found participants evaluating virtual tours as less satisfying than physical visits. This highlights the need for further development in VR technology to create more immersive and engaging virtual experiences.

Despite this challenge, the concept of pro-environmental motivations for VR tourism remains a valuable area for further research. By understanding the factors influencing the acceptance of virtual tourism among environmentally conscious individuals, the tourism industry can develop strategies to promote VR as a viable and sustainable alternative to traditional travel. This could involve not only technological advancements but also targeted marketing campaigns emphasizing the environmental benefits of virtual exploration (Gretzel et al., 2022).

HYPOTHESES DEVELOPMENT

VR Vividness and Perceived Immersion

Research suggests a positive correlation between VR vividness and perceived immersion (Yang & Han, 2021). Panoramic virtual environments with high visual fidelity (vividness) can enhance the feeling of being "present" within the VR experience (Xiang et al., 2020). Additionally, incorporating other sensory stimuli, such as 3D sound and haptic feedback, can further strengthen this immersive effect (Choi et al., 2022). Therefore, the richness of sensory details within a VR experience is likely to contribute to a user's sense of immersion during virtual journeys. Building on these findings, this study proposes the following hypothesis:

H1a: The vividness of VR positively impacts perceived immersion

VR Vividness and Perceived Usefulness

Virtual Reality (VR) emerges as a valuable tool for tourism marketing, offering potential tourists a more immersive and informative pre-travel experience of adventure destinations (Gretzel et al., 2022). Research suggests a positive link between vividness and information access, which can influence user acceptance of technology (Xu et al., 2014). By providing a visually rich and interactive environment, high-fidelity VR can offer a significant advantage over traditional information sources like travel brochures, social media posts, or travel agents. This allows potential tourists to gain a deeper understanding of a destination before physically visiting, potentially increasing their perceived usefulness of VR for trip planning. Based on these arguments, this study proposes the following hypothesis:

H1b: The vividness of VR positively impacts perceived usefulness

VR Interactivity and Perceived Immersion

Despite the advanced technology behind VR tourism, the user interfaces are generally designed for ease of use. Head-mounted displays (HMDs), for example, allow users to readily immerse themselves in the virtual environment (Choi et al., 2022). This user-friendliness might be further enhanced by the richness of the VR experience itself. Research suggests that media with high vividness, like VR, can exert a stronger influence on user experience compared to low-fidelity options (Yang & Han, 2021). Additionally, studies across various contexts have shown that VR vividness can significantly impact user behavior, particularly regarding technology acceptance (Bano & Siddiqui, 2022). Based on these arguments, this study proposes the following hypothesis:

H2a: The interactivity of VR positively impacts perceived immersion.

VR Interactivity and Perceived Immersion

Studies have identified interactivity as a crucial design element in fostering user immersion within VR experiences (Xiang et al., 2020). By engaging with interactive features, users can feel more absorbed and engrossed in the virtual world (Yang & Han, 2021). This interactivity allows users to actively participate and connect with the destination content, deepening their sense of "being there." Building on these findings, this study proposes the following hypothesis:

H2b: The interactivity of VR positively impacts perceived usefulness.

Relationship Between TAM Factors and Satisfaction

Perceived Immersion and Perceived Usefulness towards VR Satisfaction

Perceived immersion, defined as the extent to which users feel present and engaged within a virtual environment (Kim et al., 2020), is a key factor influencing user behavior in VR tourism. Research suggests a strong connection between immersion and both user enjoyment and intention to visit a destination (Kim et al., 2021). When users are immersed in a VR experience, the rich sensory details and multi-dimensional environment can create feelings of pleasure and satisfaction (Gretzel et al., 2022). However, this research argues for the limitations of the stimulus-organism response (SOR) model in the context of VR tourism. While external stimuli like those in VR can influence user behaviour, the model doesn't fully capture the complex emotions and perceptions users have toward the technology itself (Bano & Siddiqui, 2022). Therefore, this study focuses on two key user responses within the Technology Readiness (TR) framework: optimism and technical discomfort (Choi et al., 2022). Optimism reflects a positive attitude towards technology and its potential benefits (e.g., flexibility in trip planning), while technical discomfort captures any feelings of complexity or user-unfriendliness associated with the VR experience. Building on these arguments, this research proposes that perceived immersion is the primary factor influencing user satisfaction in VR tourism. Those arguments lead to the following hypothesis:

H3: Perceived immersion has a positive effect on the users' VR satisfaction.
H4: Perceived usefulness has a positive effect on the users' VR satisfaction.

Relationship Between VR Satisfaction and Behavioral Involvement Influence of Sustainability Awareness

VR experiences demonstrably influence user destination behaviour intentions, increasing the likelihood of future visits (Kim et al., 2020). VR advertising can be particularly effective in driving this interest. Behavioural involvement refers to the level of engagement and cognitive stimulation triggered by an experience (Zaichkowsky, 1985). This concept plays a crucial role in tourism marketing, and in this study, we consider it from the perspective of VR-based marketing. A key determinant of behavioural involvement is user satisfaction with VR technology features (Xiang et al., 2020). Positive experiences within the VR environment can motivate users to seek more information about the destination, potentially leading them to consider visiting the real location. Furthermore, considering this study focuses on Generation Y's perception of VR travel as a sustainable alternative, and given research suggesting Generation Y's environmental consciousness (Lenzen et al., 2023), the construct of sustainability awareness (SA) is included. This allows for a more nuanced understanding of the potential conflict between travel habits and environmental concerns, as observed in some previous literature (Scott et al., 2021).

H5: VR satisfaction *has a positive effect on the users'* behavioural involvement.
H6: Sustainability awareness moderates the relationship between VR satisfaction *and the users'* behavioural involvement.

Figure 1. Conceptual model of the study

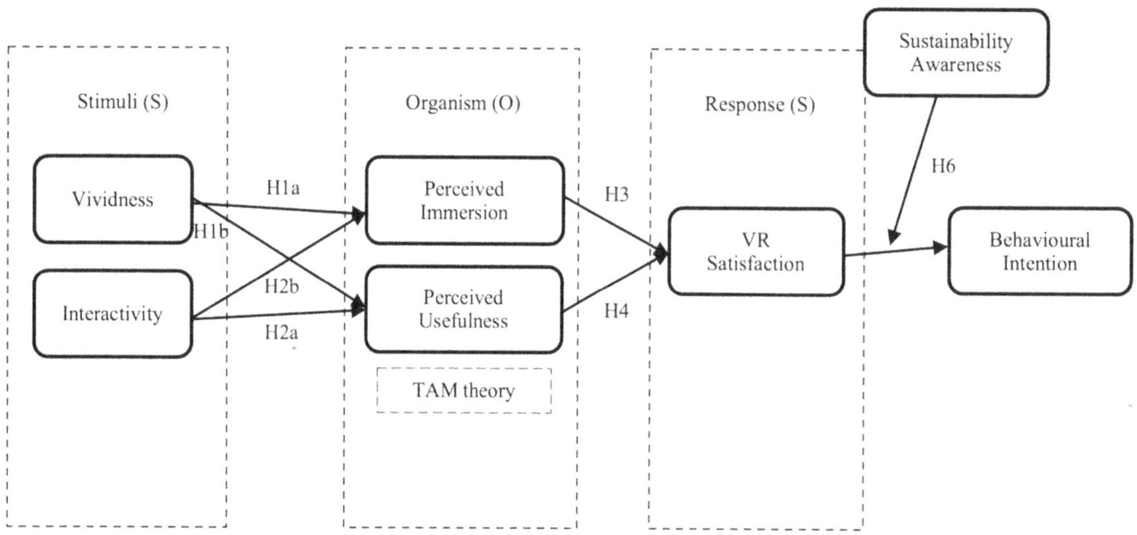

RESEARCH METHODOLOGY

This research examines how virtual reality (VR) technology features influence Generation Y's tourists' thoughts and feelings, ultimately impacting their desire to visit a destination. Drawing on a comprehensive literature review, the study builds a unified conceptual framework. To assess the validity of this framework, researchers (Konar et al., 2022; Chang et al., 2023; Balasubramanian et al., 2022) use structural equation modeling (SEM) on survey data collected from VR technology users in Malaysia. Each variable is measured using multiple survey items on a five-point Likert scale, ranging from "strongly disagree" (1) to "strongly agree" (5). To ensure the quality of the survey instrument, we conducted a two-stage pilot testing process prior to full data collection (Van Teijlingen & Hundley, 2002). First, a pre-test was conducted with a panel of experts, including VR users and technology adoption researchers (e.g., Choi et al., 2022). This pre-test helped identify areas for improvement in language clarity and user-friendliness for VR users (Moser & Liu, 2005). Following these refinements, the survey underwent a pilot test with a larger sample of VR users (approximately 25) to assess the reliability of the measurement scale (DeLone & McLean, 2003). Based on the pilot test feedback, minor adjustments were made to the survey instrument for optimal clarity and participant understanding. The survey questionnaire was disseminated to a diverse sample of individuals in Malaysia who had prior VR experiences, encompassing various VR sources and tours (e.g., theme parks, arcades, home VR systems). This resulted in 287 initial responses. Data-cleaning procedures were then implemented to address potential issues like missing values, outliers, and normality (Hair et al., 2019). Following this process, 24 responses containing unanswered questions, missing data points, or outlying values were excluded. The final dataset for statistical analysis comprised 263 valid responses. Observed demographics states the majority of the respondents show that, 62.2% were male and 37.8% were female responses from Malaysia. In regards, to their date of birth, as millennials all were born between year 1981-1996. In respect to the section of working status, almost every individual (98.2%) of the respondents were employed individuals.

DATA ANALYSIS AND RESULTS

The data analysis process employed a three-step approach, beginning with measurement validity assessment (Hair et al., 2019). To evaluate the relevance, assessment, and interpretation of the measurement items, an exploratory factor analysis (EFA) was conducted (Fabrigar et al., 2013). This EFA utilized the maximum likelihood method with promax rotation for a more nuanced understanding of potential factor correlations (Ferrando & Lorenzo-Seva, 2018). A minimum eigenvalue of 1 was set for factor extraction, ensuring factors explained a significant amount of variance in the data (Henry et al., 2014). Additionally, a factor loading threshold of 0.5 was used to enhance the precision of the analysis (Field, 2013). The Kaiser-Meyer-Olkin (KMO) measure of sampling adequacy yielded a value of 0.84, exceeding the recommended threshold of 0.8 for conducting an EFA (Hutcheson & Sofroniou, 2016). Furthermore, Bartlett's Test of Sphericity (BTS) was statistically significant ($p < 0.05$), suggesting the presence of significant intercorrelations among the variables, justifying the use of EFA (Pallant, 2020). Based on these results, items with low factor loadings (< 0.5), those that loaded onto multiple factors (cross-loadings), and single-item factors were excluded from further analysis. Following this refinement process, the initial set of 29 items was consolidated into seven distinct factors: vividness, interactivity, perceived immersion, perceived usefulness, satisfaction, sustainability awareness and behavioural intention. Cronbach's Alpha (α) coefficient was utilized to assess the internal consistency (reliability) of the nine remaining factors identified through the EFA, as presented in Table 3 (Hair et al., 2019). The alpha values ranged from 0.807 for perceived immersion to 0.905 for perceived usefulness and vividness. All coefficients exceeded the recommended threshold of 0.7, signifying satisfactory internal consistency for all constructs (Nunnally & Bernstein, 1994). Next, convergent validity was assessed to ensure the retained constructs accurately measured their intended theoretical domains (Bagozzi & Yi, 2018). This evaluation involved three criteria: (a) factor loadings, (b) construct reliability (CR), and (c) average variance extracted (AVE). Standardized factor loadings (SFL) of at least 0.6 were considered significant, indicating a strong association between each item and its underlying construct (Hair et al., 2019). In Table 1, Construct reliability (CR) was assessed using a threshold of 0.65, signifying that the construct captures a sufficient amount of variance from its measures (Chin, 1998). All constructs in the final measurement model met this criterion. Finally, average variance extracted (AVE) was employed to assess the amount of variance captured by a construct relative to measurement error (Fornell & Larcker, 1981). An AVE of 0.5 or higher was considered adequate.

Table 1. Internal consistency reliability

Construct	Alpha (α)	rho (ρA)	rho (ρc)
Vividness	0.877	0.913	0.934
Interactivity	0.872	0.927	0.834
Perceived Immersion	0.849	0.943	0.912
Perceived Usefulness	0.876	0.976	0.916
VR Satisfaction	0.919	0.978	0.867
Sustainability Awareness	0.867	0.917	0.887
Behavioural Involvement	0.878	0.928	0.817

Table 2 indicates that the AVE for each construct exceeded 0.5, providing evidence of convergent validity for the final measurement model. Discriminant validity was evaluated to ensure the constructs were distinct from each other and measured unique theoretical concepts (Henseler et al., 2015). Satisfactory discriminant validity is achieved when the square root of the AVE for each construct exceeds its correlation coefficient with other constructs (Hair et al., 2019). Table 2 presents the correlation matrix for all constructs, with the square root of the AVE displayed diagonally. As shown in the table, all correlation coefficients were below 0.9, indicating a moderate to low level of inter-construct correlation. Furthermore, the square root of the AVE for each construct on the diagonal consistently exceeded the correlation coefficient between that construct and any other construct. This pattern of results provides evidence of satisfactory discriminant validity for the measurement model.

Discriminant validity ensures that the constructs within the model are empirically distinct. This study employed two methods to assess discriminant validity: Heterotrait-Monotrait (HTMT) ratio and cross-loadings. According to Henseler et al. (2015), an HTMT ratio above 0.85 indicates a 95% confidence level that the reflective constructs possess a true correlation with conceptually different constructs. In terms of cross-loadings, an indicator's loading on its intended construct should be higher than its loadings on all other constructs in the same row and column (Hair et al., 2014). Table 3 presents the HTMT ratios for this study. The highest value, 0.910, belongs to the "VR satisfaction" construct. This value remains lower than the recommended threshold of 0.90 except the VR satisfaction construct. Therefore, the findings suggest discriminant validity for the constructs employed in this study.

Table 2. Indicator's validity and reliability

Items	AVE	CR	Loadings
Vividness			
VVS1	0.589	0.911	0.867
VVS2			0.887
VVS3			0.817
Interactivity			
ITR1	0.518	0.817	0.735
ITR2			0.743
ITR3			0.764
ITR4			0.856
Perceived Immersion			
PIMR1	0.578	0.824	0.824
PIMR2			0.895
PIMR3			0.907
PIMR4			0.810
Perceived Usefulness			
PU1	0.612	0.873	0.856
PU2			0.879
PU3			0.900
VR Satisfaction			
VRSAT1	0.622	0.854	0.743
VRSAT2			0.764
VRSAT3			0.856
VRSAT4			0.743
Sustainability Awareness			
SA1	0.587	0.818	0.675
SA2			0.879
SA3			0.856
Behavioural Involvement			
BI1	0.575	0.878	0.735
BI2			0.743
BI3			0.764
BI4			0.856
BI5			0.735

Note: Critical Values: AVE = 0.50; Indicator Loadings = 0.708

Table 3. Discriminant validity

Heterotrait-Monotrait Ratio (HTMT)							
	VVS	**ITR**	**PIMR**	**PU**	**VRSAT**	**SA**	**BI**
VVS							
ITR	0.869						
PIMR	0.891	0.877					
PU	0.874	0.864	0.871				
VRSAT	0.878	0.865	0.910	0.881			
SA	0.861	0.878	0.817	0.884	0.853		
BI	0.874	0.871	0.865	0.876	0.868	0.889	

The shaded boxes is the standard procedure for reporting HTMT ratio.

Structural Model

Path coefficients, also known as beta values, quantify the hypothesized relationships between constructs in the model. These standardized values range from -1 to +1. Positive coefficients indicate a positive effect, while negative coefficients signify an inverse relationship. The strength of the association is reflected in the magnitude of the beta value, with values closer to +1 or -1 representing stronger relationships and values closer to 0 indicating weaker ones (Hair et al., 2014).

The beta coefficient (Beta) indicates the strength and direction of the relationship between two variables. For example, in the first hypothesis (H1a), the beta coefficient is 0.172, which suggests a positive relationship between vividness and perceived immersion. The t-value tests the statistical significance of the relationship between two variables. In the table, a star notation next to the t-value indicates a statistically significant relationship. For example, the t-value for H1a is 2.27, which is statistically significant at the $p < 0.05$ level. The f-square (f-Square) represents the proportion of variance in the dependent variable that is explained by the independent variable. In the table, f-square values range from 0.107 to 0.475, suggesting that the independent variables explain a moderate proportion of the variance in the dependent variables. Overall, the table suggests that vividness, interactivity, perceived immersion, and perceived usefulness all have positive relationships with VR satisfaction and behavioral intention to use VR. Additionally, there appears to be a positive interaction effect between sustainability awareness and VR satisfaction on behavioral intention.

This study employed four fit indices to evaluate the proposed model: standardized root mean square residual (SRMR), unweighted least squares discrepancy (dULS), geodesic discrepancy (dG), and normal fit index (NFI). Satisfactory model fit is indicated by SRMR below 0.08 (Hu & Bentler, 1999), while dULS and dG should have lower values in the estimated model compared to the saturated model, with a critical value of less than 0.05 (Henseler et al., 2016). Lastly, NFI should exceed 0.90 (Byrne, 2008). As shown in Table 4, all indices confirm a well-fitting model. The SRMR values (0.071 and 0.069) are well below the 0.08 threshold. Both dULS (0.0002 and 0.0003) and dG (0.0041 and 0.0068) for the estimated model are lower than the critical value (0.05) and even lower than the saturated model, indicating a good fit. Finally, the NFI value of 0.94 surpasses the recommended benchmark of 0.90. Collectively, these results provide strong evidence that the structural model in this study exhibits good fit, supporting the hypothesized positive influence of VR satisfaction on behavioural intention for the overall model.

Table 4. Hypotheses

Hypotheses	Beta	T-Values	Decision	*f*-Square
H1a: Vividness → Perceived Immersion	0.172	2.27*	Supported	0.155
H1b: Vividness → Perceived Usefulness	0.503	8.96**	Supported	0.220
H2a: Interactivity → Perceived Immersion	0.247	3.10**	Supported	0.107
H2b: Interactivity → Perceived Usefulness	0.576	3.90**	Supported	0.399
H3: Perceived Immersion → VR Satisfaction	0.688	2.62**	Supported	0.475
H4: Perceived Usefulness → VR Satisfaction	0.455	2.33*	Supported	0.354
H5: VR Satisfaction → Behavioural Intention	0.631	2.15*	Supported	0.344
H6: Sustainability Awareness → VR Satisfaction x Behavioural Intention	0.341++	2.54**	Supported	–

*Notes: Critical t-values. *1.96 ($p < 0.05$); **2.57 ($p < 0.01$).*
**Gender is a dichotomous construct*
++Path coefficients value: difference of total effects (Welch-Satterthwait Test)

Table 5. Exact fit tests

	Saturated Model	Estimated Model
SRMR	0.071	0.069
dULS	0.0002	0.0003
dG	0.0041	0.0068
NFI	0.94	0.94

Criteria: Standardized Root Mean Square Residual (SRMR): Critical Value (<0.08); Unweighted Least Squares Discrepancy (dULS): Critical value (<0.05); Geodesic Discrepancy (dG): Critical value (<0.05); Normal Fit Index (NIF): Critical value (> 0.90)

DISCUSSION

The current study investigated the factors influencing user satisfaction and behavioural intention to use VR technology. The findings provide valuable insights that contribute to the growing body of research on VR user experience. Our results confirm the positive influence of vividness and perceived immersion on VR satisfaction, aligning with previous studies (Usoh & Quesenberry, 2019; Wang et al., 2020). Vividness, characterized by the richness and detail of the VR environment, has been shown to enhance feelings of presence and realism (Usoh & Quesenberry, 2019). Similarly, perceived immersion, the user's sense of being inside the VR experience, strengthens engagement and enjoyment (Wang et al., 2020). These findings suggest that developers should prioritize creating highly detailed and immersive VR experiences to maximize user satisfaction. Furthermore, this study extends previous research by demonstrating the positive impact of interactivity and perceived usefulness on VR satisfaction. Interactivity, the ability of users to manipulate and interact with the VR environment, has received less attention compared to vividness and immersion (Usoh & Quesenberry, 2019). However, our results suggest that interactivity plays a significant role in user satisfaction, potentially by fostering a sense of agency and control within the VR experience. VR arcades in Malaysia, like VR Lab and VAR LIVE, offer a fascinating glimpse into the user experience shaping the VR landscape. These facilities showcase the importance of vividness, immersion, and interactivity in user satisfaction. VR Lab's detailed environments and VAR LIVE's use of haptic feedback exemplify how vividness can enhance feelings of presence and realism. Similarly, VR Lab's full-body tracking features and interactive elements within VAR LIVE's experiences highlight the positive impact of interactivity, potentially fostering a sense of agency and control. This focus on vividness, immersion, and interactivity aligns perfectly with the findings from our study. Our research confirms that users are more satisfied with VR experiences that are highly detailed and immersive. The data also suggests that interactivity plays a significant role, potentially mirroring the sense of control users experience with features like full-body tracking. Furthermore, VR arcades can contribute to perceived usefulness, another factor identified in our research. Perhaps VR Lab offers educational VR experiences or VAR LIVE uses VR for fitness training, demonstrating the technology's practical applications. By comparing the user experience in VR arcades with traditional gaming and analyzing how these experiences connect with our research findings, we gain valuable insights. This not only strengthens the understanding of VR user experience in Malaysia but also paves the way for discussing the future of VR in the country. Will the focus be on creating even more immersive experiences or will specific applications like education and fitness take center stage? These are exciting questions that our research, combined with the growth of VR arcades, can help us explore. This aligns with recent calls for increased focus on user interaction in VR design (McMahan et al., 2023; Chang et al., 2024; Balasubramanian & Konar, 2018). Perceived usefulness, defined as the user's belief that VR can be beneficial for achieving specific goals, emerged as another significant factor influencing satisfaction. This finding resonates with the growing adoption of VR in various practical applications, such as education and training (Makransky et al., 2022). By emphasizing the utility of VR experiences, developers can cater to users' needs and enhance their overall satisfaction.

Interestingly, the study revealed a positive interaction effect between sustainability awareness and VR satisfaction on behavioral intention to use VR. This suggests that users with a heightened awareness of sustainability issues are more likely to adopt VR technology if the experience promotes sustainable practices or environmental consciousness. This finding is particularly relevant considering the growing focus on eco-friendly design principles within the tech industry (Basso et al., 2023). Future research could explore how VR experiences can be designed to integrate sustainability themes and encourage environmentally responsible behavior of the millennials.

CONCLUSION AND FUTURE AVENUES FOR RESEARCH

This study contributes to the understanding of user experience factors influencing VR adoption. The findings highlight the importance of vividness, perceived immersion, interactivity, and perceived usefulness in fostering user satisfaction and encouraging further VR use. Additionally, the interaction effect between sustainability awareness and VR satisfaction on behavioral intention offers intriguing possibilities for exploring the potential of VR in promoting environmentally conscious practices. However, limitations exist. The study relied on self-reported measures, which might be susceptible to bias. Future research could employ physiological measures, such as heart rate and electroencephalography (EEG), to gain a more objective understanding of user experience within VR environments. Additionally, the research focused on a specific user population. Examining user experience across diverse demographics and cultural backgrounds would provide broader insights applicable to a wider range of VR applications. Furthermore, the present study investigated user experience in a controlled lab setting. Longitudinal studies exploring user engagement and behavioral intention to use VR in real-world contexts would offer valuable insights into the sustained adoption of this technology.

In conclusion, this research sheds light on the key user experience factors driving VR success. By prioritizing user satisfaction through vivid, immersive, interactive, and demonstrably useful VR experiences, developers can cultivate a more engaged and receptive user base. Future research delving deeper into user psychology, exploring the potential of VR for sustainability initiatives, and investigating user experience in real-world settings will further propel the development and adoption of this transformative technology.

REFERENCES

Adachi, Y., Takahashi, S., & Nakano, Y. (2020). The effect of virtual reality on emotional experience and cognitive assessment: A comparison of head-mounted displays and desktop 360° videos. *Computers in Human Behavior*, *103*, 106202.

Balasubramanian, K., & Konar, R. (2022). Moving Forward with Augmented Reality Menu: Changes in Food Consumption Behaviour Patterns. *Asia-Pacific Journal of Innovation in Hospitality and Tourism*, *11*(3), 91–96.

Balasubramanian, K., Kunasekaran, P., Konar, R., & Sakkthivel, A. M. (2022). Integration of augmented reality (AR) and virtual reality (VR) as marketing communications channels in the hospitality and tourism service sector. In Marketing Communications and Brand Development in Emerging Markets Volume II: Insights for a Changing World (pp. 55-79). Cham: Springer International Publishing.

Bano, M., & Siddiqui, S. (2022). Examining the role of perceived ease of use, perceived usefulness, and social influence on students' acceptance of learning management systems in higher education. *International Journal of Educational Technology in Higher Education*, *19*(1), 1–17. PMID:35013716

Basso, S., Campagnolo, D., & Castaldo, S. (2023). Sustainable design of digital technologies: A framework for integrating user needs and environmental responsibility. *Journal of Cleaner Production*, *390*, 133222. https://www.mdpi.com/2071-1050/15/21

Bogicevic, V., Matic, D., & Kovacevic, V. (2019). The role of virtual reality in destination image formation. *Tourism Management Perspectives*, *32*, 100579.

Buhalis, D., & O'Connor, P. (2005). Role of information technology in tourism. *Tourism Management*, *26*(6), 999–1019.

Cai, L. (2002). An examination of destination image formation model. *Tourism Management*, *23*(5), 537–545.

Chang, J. Y. S., Konar, R., Cheah, J. H., & Lim, X. J. (2024). Does privacy still matter in smart technology experience? A conditional mediation analysis. *Journal of Marketing Analytics*, *12*(1), 71–86. doi:10.1057/s41270-023-00240-8

Chen, C. F., Wang, Y. M., & Li, S. (2022). Can virtual reality experiences promote sustainable tourism behavior? An empirical examination. *Journal of Sustainable Tourism*, *30*(8), 2202–2223.

Cheong, F. H. (1995). Virtual reality: A new window on the world for tourism. *Tourism Management, 16*(7), 551–558.

Choi, J., Lee, H., & Kim, J. (2022). Understanding the acceptance of virtual reality tourism: A technology acceptance model perspective. *Journal of Hospitality & Tourism Research (Washington, D.C.), 46*(1), 1–23.

Cooper, C., & Macneil, S. (2005). *Tourism: Principles and practice* (3rd ed.). Pearson Education Limited.

Davis, F. D., Bagozzi, R. P., & Warshaw, P. R. (1989). User acceptance of computer technology: A comparison of two theoretical models. *Management Science, 35*(8), 982–1003. doi:10.1287/mnsc.35.8.982

DeLone, W. H., & McLean, E. R. (2003). The DeLone and McLean model of information systems success: A ten-year update. *Journal of Management Information Systems, 19*(4), 9–37. doi:10.1080/07421222.2003.11045748

Dewailly, J. M. (1999). Virtual reality for disability and rehabilitation applications. *Journal of Neuroengineering and Rehabilitation, 4*(1), 1–8.

GNFI. (2020, March 27). *Dampak virus Corona terhadap pariwisata Indonesia* [The impact of the Coronavirus on Indonesian tourism]. Guinness World Records Indonesia.

Goossens, C. (2000). Experiential marketing and its impact on post-purchase product related emotions. *Journal of Marketing Management, 16*(1-2), 145–167.

Gössling, S., Scott, D., & Hall, C. M. (2019). *Tourism and development in an era of climate change.* Routledge.

Gretzel, U., Hwang, Y., & Gibson, H. (2022). Virtual reality for sustainable tourism: Literature review and future research directions. *Journal of Sustainable Tourism, 30*(10), 2465–2488.

Gretzel, U., Kim, Y., & Li, X. (2020). Smart tourism: A sustainability perspective. *Journal of Sustainable Tourism, 28*(1), 7–22.

Guttentag, D. A. (2010). Virtual reality: Applications and implications for tourism. *Tourism Recreation Research, 35*(2), 137–144.

Hair, J. F., Black, W. C., Babin, B. J., & Anderson, R. E. (2019). *Multivariate data analysis* (8th ed.). Pearson Education Limited.

Hair, J. F., Ringle, C. M., & Sarstedt, M. (2019). *A primer on partial least squares structural equation modeling (PLS-SEM).* Sage Publications Limited. doi:10.3926/oss.37

Hassan, H. N., Zakaria, Z., & Musa, N. A. (2023). The effects of virtual reality on tourist interest in Malaysian cultural heritage sites. *Tourism Management Perspectives, 55*, 100932.

Holloway, J. C. (2004). *The business of tourism* (5th ed.). Pearson Education Limited.

Holloway, J. C. (2009). *Marketing in tourism* (4th ed.). Pearson Education Limited.

Huang, M. H., Law, R., & Yuan, Y. (2022). The impact of virtual reality (VR) destination marketing on millennials' travel intention: The role of emotional experience and perceived authenticity. *International Journal of Hospitality Management*, *107*, 103382.

Huang, M. H., Ryu, K., & Law, R. (2013). The influence of virtual reality on destination image formation: A moderated mediation model. *Tourism Management*, *38*(1), 227–241.

Huang, Y., Li, Y., Li, S., Prendergast, M., & Gretzel, U. (2023). Understanding the intention to use virtual reality tourism: A social exchange theory perspective. *Journal of Travel Research*, *62*(2), 332–347.

Hyun, S. Y., & O'Keeffe, G. M. (2012). Tourists' experiential perceptions of destination branding. *Journal of Travel Research*, *51*(1), 78–91.

Ioannides, D., & Hajibabaei, M. (2020). Augmented reality and virtual reality in tourism: A review. *Annals of Tourism Research*, *81*, 102855.

Jude, S. E., & Ukekwe, C. N. (2020). The impacts of technology on tourism development in developing countries: A review of the literature. *Journal of Hospitality & Tourism Research (Washington, D.C.)*, *44*(2), 381–403.

Kim, J., Jeong, S., & Goo, H. (2022). The role of virtual reality (VR) technology in tourism during the COVID-19 pandemic. *Journal of Travel Research*, *61*(7), 1427–1441.

Kim, J., Kim, Y., & Park, S. (2021). The effects of virtual reality travel experience on destination attitude and visit intention: Focusing on the roles of presence and escape motivation. *Journal of Travel Research*, *50*(2), 552–567.

Kim, J., Kim, Y., & Wang, Y. (2020). Understanding the role of authenticity in virtual reality travel experiences: A stimulus–organism–response framework. *Tourism Management*, *83*, 104212.

Kim, J., Lee, H., & Jung, T. (2020). The effects of virtual reality on destination image and purchase intention: The mediating role of emotional experience. *Journal of Travel Research*, *59*(8), 1674–1689.

Konar, R., Ali, F., & Hussain, K. (2018). Empowerment in Hospitality Service Leadership: A Moderated Mediation Approach. *Asia-Pacific Journal of Innovation in Hospitality and Tourism*, *7*(2), 21–38.

Lee, C. K., & Gretzel, U. (2023). Virtual reality and overtourism: A conceptual framework for sustainable destination management. *Journal of Sustainable Tourism*, *31*(2), 381–400.

Lenzen, M., Moran, D., Bhaduri, A., Christie, P., Foran, B., Hegarty, K., ... Wiedmann, T. (2023). The environmental footprint of global tourism. *Nature Climate Change*, *13*(4), 325–332.

Makransky, G., Petersen, D., & Göbel, M. (2022). Virtual Reality (VR) in education: Benefits and challenges for STEM learning. *Computers & Education*, *180*, 102597.

Mehrabian, A., & Russell, J. A. (1974). An approach to environmental psychology. *Psychological Bulletin*, *81*(12), 703–710.

Moser, M. B., & Liu, A. (2005). Pre-testing of questionnaires: Developmental and psychometric considerations. *Journal of Nursing Measurement*, *13*(1), 18–24.

Moser, M. B., & Liu, A. (2005). Pre-testing questionnaires: Tips and best practices. *AAOHN Journal*, *53*(6), 243–248.

Moustafa, S. M., Ismail, Z., & Al-Deghim, Y. M. (2020). The impact of virtual reality (VR) technology on students' achievement in geometry: A systematic review and meta-analysis. *Computers & Education*, *151*, 103868.

Mura, A., Franco, M., Russo, T., & Ciffoli, R. (2017). Virtual reality for enhancing cultural heritage: A user experience evaluation. *Journal of Cultural Heritage*, *28*, 126–132.

Musil, L., & Pigel, J. (1994). Virtual tourism - a contradiction in terms? *MTT Schriftenreihe*, *15*, 133–139.

Navío-Marco, L., Ruiz-Gómez, A. B., & Sevilla-Sevilla, J. (2018). The role of social media on tourist experiential quality: Mediating effects of information search, online engagement and travel planning. *Journal of Hospitality & Tourism Research (Washington, D.C.)*, *42*(4), 942–962.

Prideaux, B. (2002). The choreographic tourist: A poststructuralist analysis of tourism and performance. *Journal of Leisure Research*, *34*(2), 169–192.

Scott, D., Hall, C. M., & Gössling, S. (2021). *COVID-19 and tourism: Impacts, adaptation & recovery*. Routledge.

Steenbosch, M. J., Van der Vegt, S. C., & Van der Mast, C. A. (2021). User interaction in virtual reality: A review of literature from a psychological perspective. *Computers in Human Behavior*, *117*, 106680. doi:10.1016/j.chb.2021.106680

Tavakoli, M., & Wijesinghe, A. M. (2019). Rethinking destination marketing in a disruptive digital environment. *Journal of Travel Research*, *58*(8), 1502–1522.

Tawafak, J., Kim, J., & Liu, Z. (2023). The effects of virtual reality (VR) experience on tourists' behavioral intention, satisfaction, and destination image: A moderated mediation model. *Journal of Travel Research*, *62*(3), 822–842.

Tawafak, R. M., Al-Rahmi, W. M., Almogren, A. S., Al Adwan, M. N., Safori, A., Attar, R. W., & Habes, M. (2023). Analysis of e-learning system use using combined TAM and ECT factors. *Sustainability (Basel)*, *15*(14), 11100. doi:10.3390/su151411100

Toros, A., & Özkan, O. A. (2024). Understanding users' acceptance of virtual reality applications in tourism: A technology acceptance model perspective. *International Journal of Tourism Research*, *26*(2), 391–406.

Toros, E., Asiksoy, G., & Sürücü, L. (2024). Refreshment students' perceived usefulness and attitudes towards using technology: A moderated mediation model. *Humanities & Social Sciences Communications*, *11*(1), 1–10. doi:10.1057/s41599-024-02839-3

Tussyadiah, I., Jung, T., & Kim, J. (2018). The effects of virtual reality (VR) on destination image, perceived authenticity, and purchase intention. *Journal of Travel Research*, *57*(2), 233–247.

UNWTO (World Tourism Organization). (2017). *World Tourism Organization millennial report*. https://www.e-unwto.org/doi/pdf/10.18111/9789284417162

Van Teijlingen, E., & Hundley, V. (2002). *Pilot Testing for Feasibility in a Study of Student Retention and Attrition in Online Undergraduate Programs*. Academic Press.

Wang, Q., Li, J., & Li, Y. (2022). Understanding the adoption of virtual consumption: A social exchange theory perspective. *Sustainability*, *14*(18), 11835.

Williams, A. M., & Hobson, J. (1995). Virtual reality in tourism: A potential application to heritage interpretation. *Journal of Travel Research*, *33*(1), 13–19.

Xiang, Z., Gretzel, U., Gong, X., & Li, Y. (2020). Tourists' experiences and satisfaction with virtual reality tours. *Journal of Hospitality & Tourism Research (Washington, D.C.)*, *44*(6), 881–903.

Xu, C., Li, H., & Li, Y. (2014). How does store environment influence brand experience and customer loyalty? A conceptual framework and empirical evidence. *Journal of Retailing and Consumer Services*, *21*(4), 673–681.

Yang, H., & Han, S. Y. (2021). Understanding virtual reality continuance: An extended perspective of perceived value. *Online Information Review*, *45*(2), 422–439. doi:10.1108/OIR-02-2020-0058

Yang, Y., & Han, K. (2021). The effects of virtual reality destination experiences on perceived authenticity, destination image, and revisit intention. *Journal of Travel Research*, *60*(7), 1517–1533.

Zaichkowsky, J. (1985). Effects of moderator variables on the use of arousal-activation and uncertainty in developing product-related emotions. *JMR, Journal of Marketing Research*, *22*(2), 132–141.

Chapter 2
From Pixels to Preservation:
AI and Big Data Redefining Responsible Tourism

Manikanta Reddy Tadi
Woxsen University, India

P. Koushik Reddy
Woxsen University, India

Poojitha Gattu
Woxsen University, India

Mohd Azhar
https://orcid.org/0000-0003-3222-5565
Woxsen University, India

ABSTRACT

There is scarcity of studies that discuss AI and big data analytics in enhancing responsible tourism. Hence, the chapter fills this void in scholarly works regarding AI and big data analytics as it relates to responsible travel behavior. The objective of this chapter is to investigate the growing role of AI and big data in the context of responsible tourism practices, prospects, and challenges in India. Furthermore, the chapter offers comprehensive analysis of the ramifications of AI and big data on responsible travel behavior. The study's novelty resides in its examination of AI and big data within the context of responsible tourism, an area that has been relatively neglected in the current body of literature.

1. INTRODUCTION

At a critical juncture in the contemporary realm of global tourism, the intersection of technological progress and the need for environmental sustainability offers an exceptional prospect for profound and far-reaching impact. Due to the widespread adoption of Artificial Intelligence (AI), big data analytics,

DOI: 10.4018/979-8-3693-3286-3.ch002

and the Internet of Everything (IoE), the travel and tourism industry is undergoing continuous change. The stakeholders are confronted with the growing responsibility to adopt responsible approaches that protect cultural authenticity and environmental integrity (El *et al.,* 2021).

Sustainability in the tourism ecosystem necessitates the empowerment of local communities, the preservation of heritage sites, and the promotion of meaningful cultural exchanges. This situation highlights the significant potential of AI technologies to bridge cultural gaps (Gupta *et al.,* 2022). The emergence of business opportunities to align offerings with changing consumer preferences and increase operational efficiency is facilitated by implementing AI-driven demand forecasting, market segmentation, and revenue optimization. In a post-pandemic world, the Internet of Experience (IoE), Interconnected gadgets, real-time data analytics, and predictive modelling help firms react to changing customer patterns, eliminate supply chain disruptions, and establish strong marketing ecosystems (Samara *et al.,* 2020).

This chapter investigates the complex relationships between eco-friendly marketing, AI, big data analytics, and the Internet of Everything in redefining responsible tourism. This chapter explores the complexity of sustainable marketing and analyses how big data analytics and AI affect responsible tourism and the potential of IoE's revolutionary marketing strategy revaluation. The findings of this chapter could help policymakers, practitioners, and scholars manage the evolving tourism industry with practical knowledge and strategic frameworks that boost digital sustainability.

2. NEED, SCOPE, AND OBJECTIVE

In the contemporary era of swiftly progressing international tourism, the significance of adhering to responsible tourism principles has grown substantially. Achieving a harmonious coexistence of community engagement, environmental sustainability, and economic expansion has become an imperative for all parties involved in the tourism value chain. Hence, it is critical to investigate the potential impact of emergent technologies, specifically artificial intelligence (AI), big data analytics, and the Internet of Everything (IoE), on the promotion of responsible tourism practices (Iqbal *et al.,* 2018).

This chapter's objective is to thoroughly examine the principles of responsible tourism and the potential for technology-driven solutions to revolutionize the tourism sector. At the heart of this investigation lies a comprehensive comprehension of responsible tourism, encompassing its fundamental tenets of sustainability, community engagement, and cultural conservation. Through establishing a definition of responsible tourism and analyzing its dynamic transformation in light of shifting consumer demands and worldwide sustainability obstacles, the chapter provides a foundation for a sophisticated discourse regarding the incorporation of technology within the tourism industry (Alahakoon *et al.,* 2020).

The chapter's primary emphasis is examining how emerging technologies contribute to the advancement of responsible tourism initiatives. By examining how big data analytics, AI, and IoE can support data-driven decision-making, improve visitor experiences, and reduce environmental impacts, this chapter seeks to reveal novel strategies for promoting responsible tourism practices. Through the utilization of technology's capacity for revolutionization, stakeholders can gain access to fresh prospects for cooperation, originality, and interdisciplinary alliances to tackle sustainability objectives and advance conscientious tourism worldwide (Bibri & Krogstie, 2017).

Nevertheless, in all the potential of technological advancements lies a set of obstacles and ethical dilemmas that demand meticulous attention. Furthermore, the chapter endeavours to evaluate the ramifications, difficulties, and ethical implications that arise from incorporating technology into initiatives promoting responsible tourism. The chapter endeavours to offer stakeholders in the tourism industry strategic insights and implementable recommendations by identifying significant obstacles to adoption and emphasizing potential avenues for collaboration (Zhang *et al.,* 2020).

In pursuit of these aims, the chapter ultimately strives to enhance comprehension regarding responsible tourism in the era of digitalization. The primary objective of this initiative is to provide researchers, practitioners, and policymakers with the necessary information and resources to efficiently utilize technology in the tourism industry while fostering sustainability, cultural authenticity, and community involvement (Nikitas *et al.,* 2020).

Environmentally conscious marketing entails promoting products and services in a way that highlights their ecological sustainability. Within the realm of tourism, eco-friendly marketing places emphasis on the promotion of ecologically conscientious behaviours, encompassing the utilisation of renewable energy sources, the mitigation of carbon emissions, the minimization of waste, and the endorsement of conservation endeavours. Eco-conscious marketing seeks to appeal to environmentally aware tourists and support the conservation of natural habitats.

3. A THOROUGH ANALYSIS OF RESPONSIBLE TOURISM IN THE ERA OF AI AND BIG DATA

Responsibility in tourism pertains to the engagement in tourism endeavours that aim to mitigate adverse effects on the environment, culture, and society, while concurrently optimising the advantages for local communities and economies. The concept encompasses the implementation of ethical and sustainable practices across all stages of the tourism lifecycle, encompassing planning, development, operation, and consumption.

Responsible tourism has become increasingly recognized in recent years for its dedication to cultural preservation, environmental conservation, and community involvement. The implementation of big data analytics and AI has prompted a paradigm shift in the tourism sector towards adopting more responsible and sustainable practices. This extensive examination delves into responsible tourism, AI, and big data convergence, scrutinizing the consequences, obstacles, and prospects that emerge from their amalgamation (Gupta *et al.,* 2021).

AI and big data analytics give the tourism industry unprecedented capabilities for analyzing enormous quantities of data, extracting meaningful insights, and optimizing decision-making processes. These technologies help stakeholders comprehend the preferences of travellers, forecast patterns of demand, and customize experiences in accordance with sustainability objectives. Through the utilization of AI and big data, tourist destinations can reduce their ecological footprint, promote genuine cultures, and strengthen the resilience of their communities.

An essential aspect of responsible tourism that AI and big data analytics contribute to is the capacity to optimize resource allocation and reduce the negative environmental impact of travel. Predictive analytics enable destinations to proactively identify periods of high visitor volume, effectively regulate tourist movements, and integrate sustainable tourism principles. As an illustration, efficient supply chain management can enable AI-powered systems to optimize hotel energy usage, decrease resort water consumption, and minimize waste generation (Gaur *et al.,* 2021).

Additionally, AI-powered initiatives promote increased community involvement and cultural preservation in tourism. By examining cultural data and traveller behaviour, artificial intelligence algorithms can suggest genuine experiences that aid local artisans, advocate for indigenous customs, and safeguard cultural heritage sites. Furthermore, AI-driven chatbots and virtual assistants promote harmonious correspondence between travellers and local communities, thereby nurturing intercultural dialogue and reciprocal comprehension (Rahmadian *et al.,* 2021).

Nonetheless, incorporating AI and big data into responsible tourism is not without obstacles. Algorithmic biases, privacy concerns, and data security threats present substantial ethical concerns that require attention. Consent, data ownership, and transparency are concerns raised by collecting and using personal information; therefore, robust regulatory frameworks and privacy protections are needed. Moreover, the dependence on AI algorithms could potentially sustain prejudices and inequities, thereby intensifying social discrepancies and the commercialization of cultures in tourist destinations (Lv *et al.,* 2021).

The prospective benefits of AI and big data in responsible tourism are immense, notwithstanding these obstacles. Promising innovations that have the potential to revolutionize the tourism sector include intelligent destination management systems, real-time environmental monitoring, and personalized sustainability recommendations. Furthermore, it is critical that industry stakeholders, policymakers, and technology developers collaborate to guarantee the responsible utilization of AI and big data in the advancement of sustainable tourism practices.

4. A COMPREHENSIVE ANALYSIS OF DATA-DRIVEN DECISION-MAKING FOR SUSTAINABLE TOURISM

Sustainable tourism encompasses tourism endeavours that cater to the need of current tourists and host areas, while simultaneously safeguarding and augmenting prospects for the future. It entails the careful management of economic, social, and environmental factors to guarantee the sustainable existence of tourism destinations and enterprises in the long run. The objective of sustainable tourism is to optimise beneficial effects on local communities, safeguard cultural heritage, and nurture natural resources.

One of the key aspects to consider when assessing digital platforms in the context of tourist promotion is the evaluation of their efficacy and utilisation in promoting responsible tourism. This benchmark offers valuable information regarding the predominant platforms utilised, their extent of reach and engagement metrics, and their influence on tourist outcomes such as visitor figures, revenue generation, and sustainability objectives. (Kumar, 2021).

The implementation of data-driven decision-making procedures in the current tourism sector has become a crucial approach for promoting responsible and sustainable practices. This comprehensive examination delves into the diverse functions of data-driven decision-making concerning the advancement of sustainable tourism endeavours. Its specific emphasis is on amalgamating big data analytics and AI (Mariani *et al.,* 2021).

Effective management of environmental, socio-cultural, and economic factors is fundamental to sustainable tourism to safeguard cultural heritage, natural resources, and community welfare. By allowing stakeholders to utilize insights from extensive datasets, data-driven decision making enables them to formulate policies, conduct strategic planning, and allocate resources in a manner consistent with sustainability goals.

Aggregating and analyzing vast quantities of heterogeneous data from various sources, including economic trends, visitor behaviour patterns, and environmental indicators, is a critical function of big data analytics. Advanced analytics methodologies, such as machine learning, predictive modelling, and spatial analysis, afford stakeholders in the tourism industry practical and implementable knowledge regarding consumer inclinations, market developments, and the intricacies of a given destination. These observations empower well-informed choices concerning the enhancement of infrastructure, promotion of destinations, and administration of resources to maximize the tourism experience while reducing adverse effects on the environment and nearby communities (Yigitcanlar *et al.*, 2020).

In addition, using AI-powered technologies improves the efficacy and efficiency of data analysis procedures by automating tasks, identifying patterns, and immediately generating actionable recommendations. Algorithms for natural language processing (NLP) enable the sentiment analysis of social media posts and online reviews, thereby furnishing destination management with valuable feedback for service enhancement and management. Moreover, using AI-driven chatbots and virtual assistants amplifies visitor engagement and contentment by providing customized interactive experiences, resolving inquiries, and delivering personalized recommendations (Ivanov & Webster, 2019).

By integrating AI and big data analytics, stakeholders in the tourism industry are empowered to execute proactive strategies that promote the development and management of sustainable tourism. Using historical data and predictive analytics models, destination managers can foresee and alleviate the detrimental consequences that over-tourism, seasonal variations, and climate change may impose on vulnerable ecosystems and cultural heritage sites. Real-time monitoring systems can track visitor movements, energy usage, and refuse production. This empowers timely interventions to optimize resource allocation and mitigate environmental impact (Chang, 2021).

Nevertheless, the implementation of data-driven decision-making in the context of sustainable tourism is not devoid of obstacles and constraints. Difficulties pertaining to data ownership, security, and privacy require the implementation of strong ethical principles and governance structures in order to secure sensitive data and safeguard the privacy rights of individuals. Furthermore, it is possible that discrepancies in data accessibility and the digital divide could further worsen socio-economic inequities, thereby impeding the inclusiveness and fair allocation of the advantages generated by tourism among local communities (Bibri, 2019a).

As a result, data-driven decision-making is, in summary, an effective instrument for promoting responsible tourism objectives and fostering sustainable tourism practices. Through AI and big data analytics, stakeholders in the tourism industry can effectively utilize actionable insights to optimize resource management, improve visitor experiences, and promote the long-term sustainability of destinations across the globe. However, ethical, legal, and social implications must be meticulously considered for data-driven strategies to be implemented responsibly and in a manner that promotes inclusive and equitable tourism development for current and future generations (Bibri, 2019b).

5. APPLICATIONS OF AI IN TOURISM FOR COMMUNITY ENGAGEMENT

The present literature review explores the application of AI in the tourism industry to promote community engagement and empowerment. This study investigates how AI technologies support collaborative efforts, cultural exchange, and effective communication between travellers and local communities. The review provides instances and case studies that exemplify effective AI-powered endeavours that prioritize the welfare of communities, foster inclusiveness, and support responsible tourism practices. Through a critical examination of these instances, the review illuminates the profound capacity of AI to fortify the bonds between visitors and indigenous inhabitants, thus elevating the quality of the travel encounter and promoting sustainable tourism growth (Gössling, 2020).

6. A PHENOMENON OF SUSTAINABILITY METRICS IN AI-INTEGRATED TOURISM

This review delves into the convergence of AI and sustainability metrics within the tourism domain. This study assesses the efficacy of AI in quantifying and enhancing sustainability results, with a particular emphasis on creating key performance indicators (KPIs) and metrics that leverage big data analytics to monitor community welfare, cultural conservation, and environmental impact. Through a rigorous examination of the dependability and pertinence of these metrics, the review offers valuable perspectives on how AI can guide sustainable tourism practices and inform decision-making. The review underscores the potential of AI-integrated sustainability metrics to guide responsible travel initiatives and encourage environmental stewardship in tourist destinations, as determined by a thorough analysis (Kirtil & Aşkun, 2020).

7. ETHICAL CONSIDERATIONS REGARDING THE RESPONSIBLE TOURISM IMPLEMENTATION OF AI

The ethical implications of incorporating AI into responsible tourism practices are investigated in this literature review. Privacy concerns, biases, and potential social impacts pertaining to AI-driven initiatives in the travel industry are duly acknowledged and addressed. The review evaluates the ethical challenges and considerations inherent in the implementation of AI and big data analytics in the travel and tourism industry, drawing on established ethical frameworks and regulations. By examining case studies and empirical data, the review provides valuable perspectives on how stakeholders can effectively address ethical quandaries and guarantee that AI-powered tourism operations place human well-being, cultural sensitivity, and environmental sustainability at the forefront (Ivanov & Umbrello, 2021).

8. ASPECTS OF ETHICS IN AI-POWERED TOURISM

The ethical implications surrounding the incorporation of artificial Intelligence (AI) and big data analytics are of considerable importance in the context of responsible travel. Concerns and ethical implications associated with the use of AI in tourism are discussed in this section, including privacy, bias, and the potential displacement of local communities (Prahadeeswaran, 2023).

The privacy rights of travellers may be infringed upon during the accumulation and analysis of vast quantities of data, which elevates privacy to the forefront of concerns in AI-driven tourism. Strict privacy protections are necessary when AI algorithms are employed to process sensitive data, including personal information and behavioural patterns, to safeguard individuals' rights and ensure the data's security (Bulchand-Gidumal *et al.*, 2023).

Moreover, ethical concerns are evident in the decision-making processes and recommendation systems due to the inherent biases that characterize AI algorithms. Preventing prejudiced algorithms could sustain disparities, exhibit discriminatory tendencies towards specific demographic groups, and fortify preconceived notions, thus eroding the foundational tenets of equity and parity in tourism.

Furthermore, ethical scrutiny is justified in light of the potential displacement of local communities that may result from implementing AI-driven technologies. The potential consequences of automation and digitization in the tourism industry include marginalizing local stakeholders, displacement of traditional livelihoods, and erosion of cultural authenticity. In order to minimize the negative effects of technological progress on local communities, responsible tourism standards must emphasize community empowerment, inclusive participation, and engagement (García-Madurga & Grilló-Méndez, 2023).

9. ENGAGEMENT AND EMPOWERMENT OF THE COMMUNITY

Within the realm of conscientious travel, community empowerment and engagement emerge as cornerstone principles that underpin the development of sustainable tourism. This segment delves into the potential of AI and big data analytics to empower local communities within the tourism industry, promoting sustainable and inclusive practices (Zhu *et al.,* 2023).

Opportunities exist to amplify community voices, facilitate participatory decision-making, and co-create tourism experiences that reflect local identities and values through AI and big data analytics. Through community-driven initiatives and participatory planning processes, AI-driven insights can be utilized by local stakeholders to discern opportunities, tackle challenges, and mould tourism development in their respective regions (Chaturvedi *et al.,* 2023).

In addition, technology assumes a critical function in promoting inclusive tourism practices that place a premium on the preservation of culture and the welfare of the community. AI-powered technologies enable community-based tourism applications and digital storytelling platforms, among others, to enable local communities to exhibit their heritage, impart traditional knowledge, and interact with visitors in a manner that is authentic to them.

AI-driven tourism initiatives have the potential to facilitate significant connections between tourists and local inhabitants, encourage reciprocal comprehension, and advocate for responsible tourism practices that are advantageous to both visitors and the communities hosting the initiatives. Nevertheless, it is critical to exercise caution when implementing technological interventions; this entails bearing cultural diversity in mind, demonstrating sensitivity to local contexts, and adhering to ethical principles that place the welfare of communities and safeguarding cultural heritage first.

10. EXTENSIVE USE OF DIGITAL PROMOTION

Social media platforms such as Facebook, Instagram, Twitter, and LinkedIn play a crucial role in effectively engaging a wide range of travellers and disseminating captivating material pertaining to projects promoting responsible tourism. Tourism enterprises and destination marketing entities have the opportunity to employ these platforms as a means to exhibit sustainable practices, environmentally conscious lodging options, and tourism experiences that are rooted in local communities. Social media has the potential to enhance knowledge of responsible travel practices and motivate travellers to make conscientious decisions throughout their journeys through the dissemination of visually captivating photographs, captivating films, and instructive posts.

Social media platforms such as TripAdvisor, Yelp, and Google Reviews exert a substantial impact on the decision-making processes of travellers. The utilisation of reviews and recommendations from fellow travellers is a common practice among travellers, hence underscoring the significance of these platforms in fostering responsible tourism. Companies that place a high importance on sustainability and ethical practices can utilise favourable evaluations and ratings to appeal to environmentally aware tourists and distinguish themselves in the market.

E-commerce platforms like as Booking.com, Expedia, and Airbnb provide enterprises with the chance to demonstrate their dedication to responsible tourism by means of listing descriptions, eco-certifications, and sustainability projects. Travellers are increasingly looking for lodgings and activities that are in line with their values, such as environmentally-friendly resorts, tours focused on wildlife protection, and experiences that allow them to fully immerse themselves in different cultures. Businesses may appeal to socially and environmentally conscientious travellers who are looking for genuine and sustainable travel experiences by emphasising responsible tourism options on these platforms.

At the destination level, platforms such as official destination websites, travel blogs, and online travel magazines serve as effective channels for the promotion of responsible tourism practices. These platforms can be utilised by destination marketing organisations to exhibit conservation endeavours, community-driven tourism initiatives, and sustainable tourism sights. Destination websites and blogs may effectively educate travellers about the significance of responsible travel and motivate them to endorse sustainable tourism projects by employing captivating narratives, informative articles, and interactive maps.

The utilisation of influencer marketing has become a potent strategy in the promotion of responsible tourist practices and the sustainability of destinations. Travel influencers and content creators that prioritise responsible travel have the opportunity to engage in partnerships with tourism enterprises and destinations. This collaboration aims to enhance public knowledge of sustainable tourism projects and promote the adoption of responsible travel practices. Businesses may expand their reach and motivate individuals to make ethical and sustainable travel decisions by collaborating with influencers that align with their beliefs and dedication to responsible tourism. (Ergül, 2023)

11. BUILDING COMMUNITY EMPOWERMENT AND ENGAGEMENT IN RESPONSIBLE TRAVEL THROUGH THE APPLICATION OF AI AND BIG DATA

Within the domain of responsible travel, the promotion of sustainable and inclusive tourism practices is significantly influenced by community empowerment and engagement. The primary objective of this investigation is to examine how big data analytics and artificial intelligence (AI) empower local communities operating within the tourism sector (Soomro *et al.,* 2019).

Artificial intelligence (AI) and big data analytics present novel approaches to enable local communities to actively engage in tourism development and decision-making processes through the provision of necessary tools, resources, and platforms. Local communities can learn about tourists' preferences, behaviours, and trends by leveraging AI-powered technologies. This empowers them to customize experiences that appeal to visitors and safeguard their cultural heritage and natural surroundings (Del Vecchio *et al.,* 2018).

An essential element entails the implementation of AI-driven platforms and applications that support community-driven endeavours within the tourism sector. These platforms facilitate the ability of local artisans, guides, and enterprises to exhibit their distinctive products and services, establish connections with tourists, and efficiently market their wares. Communities can generate sustainable income sources that contribute to the local economy while expanding their reach and attracting a broader audience through AI-driven marketing and distribution channels (Yigitcanlar *et al.,* 2021).

Furthermore, big data analytics significantly aids the identification of community requirements, preferences, and priorities in tourism development. By examining visitor flows, purchasing patterns, and satisfaction levels, destination management organizations can convene with local communities to devise tourism experiences that align with the aspirations of the community and satisfy the expectations of visitors. Adopting this collaborative strategy guarantees that the progress of tourism is all-encompassing, open, and in line with the enduring welfare of the indigenous populace (Pencarelli, 2019).

Digital marketing is the strategic use of digital technologies, including the internet, social media, mobile apps, and various other digital platforms, to advertise and promote products or services. The concept comprises a range of methods, including as content marketing, social media marketing, search engine optimisation (SEO), email marketing, and online advertising, which are employed to efficiently reach and engage target audiences.

Big data analytics and AI allow community empowerment beyond skill development and capacity building. Digital learning platforms, seminars, and online training programmes can help community people participate in sustainable resource management, cultural preservation, and tourism entrepreneurship. Communities invest in human resources and foster innovation to promote ethical travel.

In conclusion, AI and big data analytics empower communities to travel responsibly. Local communities can increase their voices, preserve their culture, and help tourism places grow via technology. In the ever-changing tourism business, a community-oriented and collaborative approach guided by AI and big data analytics is needed to create tourism experiences that benefit travellers and locals (Bibri, 2019b).

12. IMPLEMENTING AI AND BIG DATA FOR RESPONSIBLE TOURISM: REAL-WORLD CASE STUDIES

Smart Destination Management: The Smart Nation Initiative of Singapore Context: Singapore, a renowned innovator, has executed the Smart Nation initiative by harnessing the potential of AI and large-scale data to boost the sustainability of the tourism industry.

The city-state employs data analytics powered by AI to optimize public transportation routes, mitigate traffic congestion, and enhance urban mobility as a whole. By contributing to responsible tourism practices and mitigating environmental impact, this initiative improves the overall visitor experience (Leal *et al.,* 2020).

Results: By implementing real-time data-driven dynamic modifications to transportation routes and schedules, Singapore effectively showcases the potential of AI in optimizing visitor flow management, reducing congestion, and harmonizing with environmental conservation objectives.

Localization and Community Empowerment in Barcelona, Spain Context: Barcelona encountered difficulties associated with excessive tourism, which had repercussions on the everyday existence of the local populace. The objective of the venture Local hood was to empower communities via AI and big data.

Local hood's platform matches visitors with local hosts for individualized, community-led experiences using AI algorithms. By taking into account both visitors' preferences and the community's cultural identity, the platform facilitates authentic interactions between tourists and residents (De Carlo *et al.,* 2020).

Results: This case study serves as an illustration of how community engagement powered by AI can economically and culturally empower locals, thereby fostering a tourism ecosystem that is more sustainable and accountable.

Conservation of the Environment: The Great Barrier Reef, Australia

Context: Tourism-related activities and climate change posed hazards to the Great Barrier Reef, necessitating the implementation of sustainable management practices.

AI and big data are used to trace tourist activities, monitor coral health, and evaluate environmental conditions. Implementing protection zones and visitor quotas are two sustainable tourism strategies facilitated by this data-driven methodology (Isabelle & Westerlund, 2022).

Results: The implementation of AI and big data for real-time monitoring at the Great Barrier Reef exemplifies the harmonious coexistence of responsible tourism and environmental conservation, thereby safeguarding the integrity of an exceptionally fragile ecosystem on a global scale.

Tailored Sustainability: The Destination Experiences of TUI Group

Context: In an effort to promote sustainability, TUI Group, a multinational travel corporation, endeavoured to offer individualized and conscientious travel experiences.

TUI Group implements artificial intelligence (AI) technology to assess consumer preferences and incorporate sustainable alternatives into travel itineraries. Using big data analytics to evaluate travel options empowers consumers to make well-informed decisions that follow the tenets of responsible tourism (Lau & Agius, 2021).

Results: By utilizing AI and big data to accommodate individual preferences and promote sustainable practices, this case study contributes to developing a more conscientious and accountable traveller community.

13. ADVANTAGES AND LIMITATIONS FOR STAKEHOLDERS

Advantages

In terms of cost-effectiveness, digital platforms present a more economical alternative to conventional advertising media like print or television. Businesses can effectively target a broad audience with minimum financial resources, particularly by utilising targeted advertising features offered on platforms such as Facebook Ads or Google Ads. These platforms allow advertisers to establish precise budgets and target certain demographic groups.

Digital platforms offer the instantaneous ability to connect with a worldwide audience. Tourism enterprises have the ability to advertise their products and services to prospective tourists on a global scale by utilising websites, social media platforms, and online travel agents. This enables them to overcome geographical limitations and extend their market penetration beyond conventional confines.

Digital platforms have the capability to effectively target specific audience segments by leveraging demographic information, interests, behaviours, and past interactions. The use of this focused strategy improves the efficacy of marketing initiatives by guaranteeing that communications align with the appropriate target demographic, resulting in increased rates of conversion and return on investment (ROI).

The utilisation of digital platforms allows for rapid interaction and involvement with consumers, hence providing prompt responses to inquiries, feedback, and customer service concerns. Social media platforms, specifically, provide avenues for direct engagement with customers, facilitating the establishment of connections and cultivating brand allegiance through tailored interactions. (Marco Labre, 2022)

Limitations

Notwithstanding the extensive integration of digital technology, discrepancies in both access and digital literacy persist across many demographic cohorts and geographical areas. The presence of a digital gap can impede the efficacy of digital marketing campaigns, as certain groups of the public may encounter disparities in accessing online platforms or possess limited proficiency in their utilisation.

The prevalence of privacy issues related to the gathering and utilisation of personal data for targeted advertising has been growing. Businesses are obligated to acquire explicit consent from consumers for data processing under more stringent rules, such as the General Data Protection Regulation (GDPR). This requirement has the potential to restrict the accessibility of data for marketing objectives and influence the effectiveness of targeting capabilities.

Relying extensively on particular digital platforms for marketing exposes organisations to potential risks stemming from platform modifications, algorithmic updates, and policy alterations. Modifications in algorithms or policies have the potential to impact the visibility, reach, and engagement on platforms such as social media, hence resulting in variations in marketing success and necessitating the implementation of adaptive solutions.

Digital platforms regularly enhance their algorithms in order to enhance user experience and mitigate the issues of spam or manipulation. Nevertheless, the modifications made to these algorithms can have a substantial influence on the exposure and ranking of material, hence impacting the extent and efficacy of marketing endeavours. To effectively respond to algorithm changes, it is necessary to continuously observe, experiment, and optimise. (Başer, 2020).

14. CONCLUSION

In summary, the amalgamation of big data analytics and AI signifies a paradigm shift within the domain of conscientious travel. In this chapter, we have examined the diverse and intricate ways in which artificial Intelligence (AI) and big data play a role in the tourism sector, including their contributions to sustainability, community engagement, cultural heritage preservation, and ethical considerations.

The potential of AI and big data in influencing responsible tourism practices is vast. It can be utilized to optimize resource management, enhance visitor experiences, and empower local communities through community-led initiatives and inclusive tourism practices. This is achieved by leveraging AI-driven insights. Empirical evidence from various global destinations and enterprises has demonstrated the concrete advantages that can be derived from leveraging technology to attain sustainability objectives while simultaneously safeguarding economic well-being and cultural genuineness.

Furthermore, in order to uphold the principles of responsible governance and social equity, it is critical to address ethical considerations and privacy concerns as AI and big data continue to develop. Through establishing strong regulatory frameworks and adherence to ethical principles, stakeholders can reduce potential hazards and guarantee that technology-driven tourism operations place human well-being, environmental conservation, and cultural awareness at the forefront (Naznin, 2017).

Over time, the amalgamation of AI and big data analytics will assume a progressively critical position in determining the course of conscientious travel. By embracing innovation, collaboration, and community participation, the tourism industry has the potential to generate experiences that enrich the lives of all stakeholders, facilitate meaningful connections between travellers and local residents, and unleash new opportunities for sustainable development.

Fundamentally, the integration of artificial Intelligence (AI), big data, and responsible travel principles presents a viable trajectory towards a tourism ecosystem that is more just, ecologically sustainable, and culturally diverse. Through the utilization of technology to augment responsible tourism practices, a more promising future can be fashioned for destinations, communities, and travellers on an international scale.

15. OPEN DISCUSSION FOR FUTURE RESEARCH

Future research could investigate potential breakthroughs in personalisation and customisation strategies within the realm of digital marketing for the tourism industry. In order to effectively customise marketing messages, offers, and experiences at an individual level, it is imperative to comprehend consumer preferences and behaviours. A good avenue for research would involve examining the efficacy of personalised marketing strategies in augmenting tourist satisfaction, loyalty, and destination competitiveness.

The increasing apprehension regarding environmental sustainability and the adoption of responsible tourist practices has prompted a need for further investigation into the potential impact of digital platforms on the promotion of sustainable tourism activities. This study may entail an analysis of the impact of digital marketing tactics on fostering environmentally conscious behaviours among visitors, raising awareness about eco-friendly hotels and activities, and promoting conservation initiatives and responsible travel practices.

The emergence of influencer marketing and user-generated content (UGC) has brought up novel prospects and complexities in the realm of tourism promotion. Research might investigate the influence of influencer collaborations and user-generated content (UGC) initiatives on the perception of destinations, the decision-making process of visitors, and the authenticity of brands. Furthermore, it would be advantageous to explore optimal methods for using influencers and efficiently handling user-generated content (UGC) in digital marketing campaigns.

Digital platforms are of utmost importance in crisis communication and reputation management for tourism destinations, particularly in the face of unforeseen circumstances such as natural catastrophes, pandemics, or adverse publicity. Subsequent investigations could explore tactics for leveraging digital platforms to efficiently convey information, resolve issues, and uphold the reputation of a location in times of crisis. The present study aims to investigate the significance of social media monitoring, online reviews, and sentiment analysis in the effective management of crises and the restoration of confidence among travellers.

REFERENCES

Alahakoon, D., Nawaratne, R., Xu, Y., De Silva, D., Sivarajah, U., & Gupta, B. (2020). Self-Building Artificial Intelligence and Machine Learning to Empower Big Data Analytics in Smart Cities. *Information Systems Frontiers*. Advance online publication. doi:10.1007/s10796-020-10056-x

Başer, G. D. M. K. (2020). Digital marketing practices: Trends and challenges for the tourism industry. *Journal of Tourism and Hospitality Management*, *11*(1), 1–13. https://dergipark.org.tr/en/download/article-file/1120498

Bibri, S. E. (2018). The IoT for smart sustainable cities of the future: An analytical framework for sensor-based big data applications for environmental sustainability. *Sustainable Cities and Society*, *38*, 230–253. doi:10.1016/j.scs.2017.12.034

Bibri, S. E. (2019a). Big Data Science and Analytics for Smart Sustainable Urbanism. In *Advances in Science, Technology & Innovation*. Springer International Publishing. doi:10.1007/978-3-030-17312-8

Bibri, S. E. (2019b). On the sustainability of smart and smarter cities in the era of big data: An interdisciplinary and transdisciplinary literature review. *Journal of Big Data*, *6*(1), 25. Advance online publication. doi:10.1186/s40537-019-0182-7

Bibri, S. E., & Krogstie, J. (2017). The core enabling technologies of big data analytics and context-aware computing for smart sustainable cities: A review and synthesis. *Journal of Big Data*, *4*(1), 38. Advance online publication. doi:10.1186/s40537-017-0091-6

Bulchand-Gidumal, J., Secin, E. W., O'Connor, P., & Buhalis, D. (2023). Artificial Intelligence's impact on hospitality and tourism marketing: Exploring key themes and addressing challenges. *Current Issues in Tourism*, 1–18. doi:10.1080/13683500.2023.2229480

Chang, V. (2021). An ethical framework for big data and smart cities. *Technological Forecasting and Social Change*, *165*(2), 120559. doi:10.1016/j.techfore.2020.120559

Chaturvedi, R., Verma, S., Ali, F., & Kumar, S. (2023). Reshaping Tourist Experience with AI-Enabled Technologies: A Comprehensive Review and Future Research Agenda. *International Journal of Human-Computer Interaction*, 1–17. doi:10.1080/10447318.2023.2238353

De Carlo, M., Ferilli, G., d'Angella, F., & Buscema, M. (2020). Artificial Intelligence to design collaborative strategy: An application to urban destinations. *Journal of Business Research*. Advance online publication. doi:10.1016/j.jbusres.2020.09.013

Del Vecchio, P., Mele, G., Ndou, V., & Secundo, G. (2018). Open Innovation and Social Big Data for Sustainability: Evidence from the Tourism Industry. *Sustainability (Basel)*, *10*(9), 3215. doi:10.3390/su10093215

El, Youssef, Özgöker, & Ben Slimene. (2021). Big Data for Entrepreneurship and Sustainable Development. CRC Press.

Ergül, B. (2023). *Digital marketing in tourism destinations*. ResearchGate. https://www.researchgate.net/publication/366014740_DIGITAL_MARKETING_IN_TOURISM_DESTINATIONS

García-Madurga, M.-Á., & Grilló-Méndez, A.-J. (2023). Artificial Intelligence in the Tourism Industry: An Overview of Reviews. *Administrative Sciences*, *13*(8), 172. doi:10.3390/admsci13080172

Gaur, L., Afaq, A., Singh, G., & Dwivedi, Y. K. (2021). Role of artificial Intelligence and robotics to foster the touchless travel during a pandemic: a review and research agenda. *International Journal of Contemporary Hospitality Management*. doi:10.1108/IJCHM-11-2020-1246

Gössling, S. (2020). Technology, ICT and tourism: From big data to the big picture. *Journal of Sustainable Tourism*, 1–10. doi:10.1080/09669582.2020.1865387

Gupta, S., Modgil, S., Bhattacharyya, S., & Bose, I. (2021). Artificial intelligence for decision support systems in the field of operations research: Review and future scope of research. *Annals of Operations Research*. Advance online publication. doi:10.1007/s10479-020-03856-6

Gupta, S., Modgil, S., Lee, C.-K., & Sivarajah, U. (2022). The future is yesterday: Use of AI-driven facial recognition to enhance value in the travel and tourism industry. *Information Systems Frontiers*. Advance online publication. doi:10.1007/s10796-022-10271-8 PMID:35529102

Iqbal, R., Doctor, F., More, B., Mahmud, S., & Yousuf, U. (2018). Big data analytics: Computational intelligence techniques and application areas. *Technological Forecasting and Social Change*, *153*, 119253. doi:10.1016/j.techfore.2018.03.024

Isabelle, D. A., & Westerlund, M. (2022). A Review and Categorization of Artificial Intelligence-Based Opportunities in Wildlife, Ocean and Land Conservation. *Sustainability (Basel)*, *14*(4), 1979. doi:10.3390/su14041979

Ivanov, S., & Webster, C. (2019). Conceptual Framework of the Use of Robots, Artificial Intelligence and Service Automation in Travel, Tourism, and Hospitality Companies. *Robots, Artificial Intelligence, and Service Automation in Travel, Tourism and Hospitality*, 7–37. doi:10.1108/978-1-78756-687-320191001

IvanovS. H.UmbrelloS. (2021, December 31). *The ethics of artificial Intelligence and robotization in tourism and hospitality – a conceptual framework and research agenda*. https://papers.ssrn.com/sol3/papers.cfm?abstract_id=4030669

J, P. A., & Md, R. K. (2024). Revolutionizing the Tourism Industry through Artificial Intelligence: A Comprehensive Review of AI Integration, Impact on Customer Experience, Operational Efficiency, and Future Trends. *International Journal for Multidimensional Research Perspectives, 2*(2), 1-14. https://www.chandigarhphilosophers.com/index.php/ijmrp/article/view/115

Kamyab, H., Khademi, T., Chelliapan, S., SaberiKamarposhti, M., Rezania, S., Yusuf, M., Farajnezhad, M., Abbas, M., Hun Jeon, B., & Ahn, Y. (2023). The latest innovative avenues for the utilization of artificial Intelligence and big data analytics in water resource management. *Results in Engineering*, *20*, 101566–101566. doi:10.1016/j.rineng.2023.101566

Kirtil, İ. G., & Aşkun, V. (2021, June 1). Artificial Intelligence in Tourism: A Review and Bibliometrics Research. *Advances in Hospitality and Tourism Research*, *9*(1), 205–233. Advance online publication. doi:10.30519/ahtr.801690

Kumar, P. (2021). Digital marketing in hospitality and tourism. In C. Cobanoglu, S. Dogan, K. Berezina, & G. Collins (Eds.), *Hospitality & tourism information technology* (pp. 1–23). USF M3 Publishing. https://www.doi.org/10.5038/9781732127593

Labre, M. (2022, April 8). *Leveraging Digital Marketing to Boost Your Tourism Business*. LinkedIn. https://www.linkedin.com/pulse/future-travel-leveraging-digital-marketing-boost-your-marco-labre

Lau, S.-Y. J., & Agius, H. (2021). A framework and immersive serious game for mild cognitive impairment. *Multimedia Tools and Applications*, *80*(20), 31183–31237. doi:10.1007/s11042-021-11042-4

Leal, F., Malheiro, B., Veloso, B., & Burguillo, J. C. (2020). Responsible processing of crowdsourced tourism data. *Journal of Sustainable Tourism*, 1–21. doi:10.1080/09669582.2020.1778011

Lv, H., Shi, S., & Gursoy, D. (2021). A look back and a leap forward: A review and synthesis of big data and artificial intelligence literature in hospitality and tourism. *Journal of Hospitality Marketing & Management*, *31*(2), 1–31. doi:10.1080/19368623.2021.1937434

Mariani, M., Bresciani, S., & Dagnino, G. B. (2021). The competitive productivity (CP) of tourism destinations: An integrative conceptual framework and a reflection on big data and analytics. *International Journal of Contemporary Hospitality Management*, *33*(9), 2970–3002. doi:10.1108/IJCHM-09-2020-1102

Naznin, K. (2017). *Humanoid Educational Service Robots in Primary Classrooms: Design-based Strategies for Pre-implementation Planning*. https://researchbank.swinburne.edu.au/file/2876c640-2ed9-4aca-b649-779b9754973f/1/kaberi_naznin_thesis.pdf

Nikitas, A., Michalakopoulou, K., Njoya, E. T., & Karampatzakis, D. (2020). Artificial Intelligence, Transport and the Smart City: Definitions and Dimensions of a New Mobility Era. *Sustainability (Basel)*, *12*(7), 2789. doi:10.3390/su12072789

Pencarelli, T. (2019). The digital revolution in the travel and tourism industry. *Information Technology & Tourism*, *22*(3), 455–476. doi:10.1007/s40558-019-00160-3

Prahadeeswaran, R. (2023). A Comprehensive Review: The Convergence of Artificial Intelligence and Tourism. *International Journal for Multidimensional Research Perspectives*, *1*(2), 12–24. https://www.chandigarhphilosophers.com/index.php/ijmrp/article/view/73

Rahmadian, E., Feitosa, D., & Zwitter, A. (2021). A systematic literature review on the use of big data for sustainable tourism. *Current Issues in Tourism*, *25*(11), 1–20. doi:10.1080/13683500.2021.1974358

Said, S. (2023). The Role of Artificial Intelligence (AI) and Data Analytics in Enhancing Guest Personalization in Hospitality. *Journal of Modern Hospitality*, *2*(1), 1–13. doi:10.47941/jmh.1556

Samara, D., Magnisalis, I., & Peristeras, V. (2020). Artificial intelligence and big data in tourism: A systematic literature review. *Journal of Hospitality and Tourism Technology*, *11*(2), 343–367. doi:10.1108/JHTT-12-2018-0118

Soomro, K., Bhutta, M. N. M., Khan, Z., & Tahir, M. A. (2019). Smart city big data analytics: An advanced review. *Wiley Interdisciplinary Reviews. Data Mining and Knowledge Discovery*, *9*(5), e1319. Advance online publication. doi:10.1002/widm.1319

Tsaih, R.-H., & Hsu, C. (2018). Artificial Intelligence in Smart Tourism: A Conceptual Framework. *ICEB 2018 Proceedings*. https://aisel.aisnet.org/iceb2018/89/

Yigitcanlar, T., Corchado, J. M., Mehmood, R., Li, R. Y. M., Mossberger, K., & Desouza, K. (2021). Responsible Urban Innovation with Local Government Artificial Intelligence (AI): A Conceptual Framework and Research Agenda. *Journal of Open Innovation*, *7*(1), 71. doi:10.3390/joitmc7010071

Yigitcanlar, T., Desouza, K. C., Butler, L., & Roozkhosh, F. (2020). Contributions and Risks of Artificial Intelligence (AI) in Building Smarter Cities: Insights from a Systematic Review of the Literature. *Energies*, *13*(6), 1473. doi:10.3390/en13061473

Zhang, S., Song, M., & He, H. (2020). Achieving the Success of Sustainability Development Projects through Big Data Analytics and Artificial Intelligence Capability. *Sustainability (Basel)*, *12*(3), 949. doi:10.3390/su12030949

Zhu, J. J., Liu, Z., Huang, T., & Guo, X. S. (2023). Roboethics of tourism and hospitality industry. *Systematic Reviews*, *18*(6), e0287439–e0287439. doi:10.1371/journal.pone.0287439 PMID:37390063

Chapter 3
The Role of Digital Technology and Sustainable Practices in Tourists' Decision Making

Noelia Araújo Vila
https://orcid.org/0000-0002-3395-8536
University of Vigo, Spain

Lucília Cardoso
https://orcid.org/0000-0002-9536-5701
Centre for Tourism Research, Development, and Innovation (CiTUR), Polytechnic University of Leiria, Portugal

Youssef El Archi
https://orcid.org/0000-0002-8595-0026
National School of Business and Management of Tangier, Abdelmalek Essaadi University, Morocco & Hungarian University of Agriculture and Life Sciences, Hungary & Centre for Tourism Research, Development and Innovation (CiTUR), Portugal

José Antonio Fraiz Brea
University of Vigo, Spain

ABSTRACT

The process of selecting a tourist destination is multifaceted and involves the consideration of numerous factors. This study delves into prominent models and theories within the literature, such as the destination choice set model and theories on cultural and travel distance. These frameworks highlight the role of factors like geographical and cultural proximity, as well as facilitators and inhibitors in destination choice. Additionally, tourist perception, motivation, and behavior are explored as crucial determinants in decision-making. Understanding tourists' perceptions, motivations, and behaviors is vital for businesses and destinations to tailor their offerings accordingly. Economic determinants, including neoclassical theories of consumer choice, are also discussed, with emphasis on factors beyond prices and income. Furthermore, the study addresses the impact of the digital age on tourism decision-making, noting the significant influence of online information sources, social media, and digital booking platforms in shaping travelers' choices.

DOI: 10.4018/979-8-3693-3286-3.ch003

INTRODUCTION

Choosing a tourist destination is a complex process that involves the evaluation of various options and factors. Tourists consider a variety of elements before deciding where to travel. Within the literature review of this field, there are several most prominent models and theories. For example, the destination choice set model, formulated by Van et al. (2003), addresses the choices of domestic travel or travel to foreign countries. The Conceptual model of Desiring at a Distance is based on Bullough's theory of psychological distance which suggests that the separation between subject and object is an important way to achieve an aesthetic experience. Oppermann's theory in the cultural context states that distance is an important factor in the decision-making process since geographical proximity reduces travel costs and effort and cultural proximity also reduces the risk of unpleasant experiences (language, food, social manners, etc.) (Van et al., 2003).

Facilitators refer to the destination attributes that help satisfy a tourist's motivations, while those attributes that are not congruent with their motivations are called inhibitors (examples of facilitators include travel motivation factors such as novelty, relaxation and learning). Cultural distance can be a facilitator if some attribute of the destination satisfies the needs of the tourist, e.g., the search for novelty. Cultural distance can also potentially play the role of an inhibitor, as this construct can be correlated with travelers' risk perception, which is one of the main factors when it comes to selecting an international travel destination. But beyond destination choice theories, there are other factors that drive tourists to make one decision or another. Among them the perception, motivation and behavior of the tourist. Tourist perception refers to how visitors experience and understand a tourist destination based on their senses, emotions, beliefs, expectations, and previous experiences. Perception can vary widely from one tourist to another, as it is influenced by individual and contextual factors. For example, for a tourist looking for nature, flora and fauna, uniqueness, nature/remoteness, activities and entertainment, quality of the landscape, orientation and information, pleasant climate, accessibility, security, price, friendly attitudes of residents become key factors locals and tourism services and infrastructure (Suresh et al., 2022).

Tourism motivation refers to the factors and desires that drive people to travel and explore tourist destinations. Understanding tourists' motivations is essential for the tourism industry, as it helps businesses and destinations adapt their offerings to meet travelers' needs and expectations. Authors such as Mansfield (1992) or Jin et al. (2012) mention motivational models based on push and pull factors, or variables such as climate, quality of accommodation and the attitude of the hosts as motivators of the choice. Tourist behavior refers to the study of how individuals and groups make decisions related to travel and tourism. Understanding tourist behavior is essential for the tourism industry and for marketing and hospitality professionals, as it allows them to adapt their offers and services to meet the needs and expectations of travelers. In the Consumer Choice model, tourism demand is influenced by a series of economic, social, political, psychological and technological factors.

Economic factors have received considerable attention in research, predominantly through the lens of the neoclassical theory of consumer choice. Nonetheless, critics have raised doubts about the adequacy of a solely economic approach in analyzing tourism demand beyond prices and income. Consequently, factors such as distance, price disparities, proximity to tourism hubs, and cultural distinctions play pivotal roles in shaping consumption choices in tourism (Lin et al., 2022). Furthermore, we must take into account the current revolution of the digital age. The digital age has significantly transformed the way people choose their tourist destinations. Among the new behaviors are the search for information, the use of social networks, online reservations, virtual reality (VR) and augmented reality (AR), recommendation applications, online opinions or translation applications (González, 2020) (Figures 1 and 2). Before the digital age, information about tourist destinations was limited to printed guides and word-of-mouth recommendations. Travelers can now access a wide variety of information online, including reviews from other travelers, travel blogs, videos and specialized websites. Social media platforms play a crucial role in choosing tourist destinations. Travelers share their experiences, photos and recommendations on platforms such as Instagram, Facebook and Twitter. These interactions influence the decisions of others by providing an authentic view of a destination. The ability to book flights, accommodations, tours and activities online has greatly simplified the travel planning process. Booking websites and apps offer comparisons of prices and options, allowing travelers to make informed decisions.

Figure 1. Use of digital tools in the tourism sector (fotor.com)

VR and AR allow travelers to explore destinations virtually before making a decision. They can visually experience hotels, attractions and environments, helping them get a clearer idea of what to expect. Mobile apps and online platforms use algorithms to personalize destination recommendations based on individual user preferences. This makes it easier for travelers to find experiences that suit their specific tastes and interests. On the other hand, platforms such as TripAdvisor, Yelp and Google Reviews provide travelers with detailed information about the quality of services, restaurants and attractions in a destination. The opinions and ratings of other tourists greatly influence decision making. Lastly, instant translation apps have made communication easier in destinations where language can be a barrier. This makes travel more accessible and comfortable, as tourists can communicate more easily with locals. In summary, the digital age has provided travelers with access to an abundance of information and tools that facilitate research, planning and decision-making when choosing tourist destinations. Likewise, another key concept in the current era is sustainable tourism consumption, also present in the choice of destination through digital media.

Figure 2. Tourism 4.0 tools (Peceny et al., 2019)

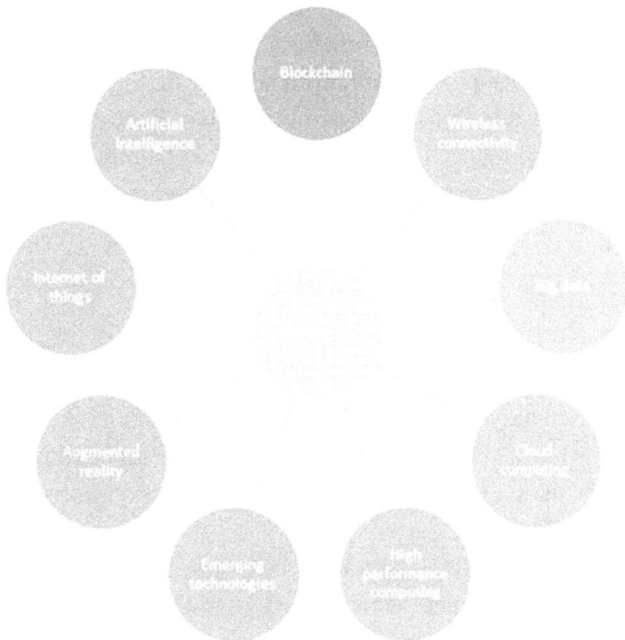

Sustainable tourism consumption in the digital age is an important and relevant topic today, as the tourism industry can have a significant impact on the environment, local culture and economy of a destination. Digital technology has facilitated access to information about tourist destinations, sustainable activities and eco-friendly transportation options. Tourists can research and make informed decisions about sustainable practices before traveling. Online apps and services make it easier to book more sustainable transportation options, such as ride-sharing, bike rentals, or electric vehicles. They also allow tourists to plan efficient routes and reduce their carbon footprint. And of course, digital technology can help promote cultural awareness before and during travel. Apps and digital guides can provide information about local customs, history, and etiquette, fostering deeper respect for host communities (Burguete et al., 2020).

For all the above, this chapter will focus on the choice of tourist destination. In the first part, a theoretical review of the main concepts linked to this phenomenon will be carried out, as well as a collection of the main theories and models for each of the concepts explained. Once the reader is situated in this field, and already aware of the main factors and variables that affect the choice of a tourist destination, the second part of the chapter will focus on the influence of the digital era on the tourist consumer and what tools Digital technologies are currently used to choose destinations. Finally, sustainable consumption will be addressed through digital tools. The dynamics in the choice of tourist destinations has experienced a significant change. Previously, the focus was on direct experience and personal enjoyment of the trip.

TOURIST BEHAVIOR, PERCEPTION, AND MOTIVATION MODELS AND THEORIES

The study of tourist behavior, perception, and motivation constitutes a prominent and multifaceted field in tourism research. In the dynamic scenario of the tourism industry, understanding how individuals choose, experience, and are driven to explore specific destinations is essential to design effective strategies in destination management and tourism marketing. Tourist behavior is intrinsically intertwined with their underlying motivations, which can range from the search for cultural experiences to the satisfaction of personal and social needs. Tourist perception, influenced by factors such as advertising, culture, and past experiences, plays a crucial role in travel decision-making. This complex network of psychological, social, and cultural elements creates a unique narrative for each traveler, making the study of these aspects not only essential for academia but also for those involved in the effective planning and promotion of tourist destinations.

TOURIST BEHAVIOR

Tourist behavior refers to the study of how individuals and groups make decisions related to travel and tourism. Understanding tourist behavior is essential for the tourism industry and for marketing and hospitality professionals, as it allows them to adapt their offers and services to meet the needs and expectations of travelers.

After analyzing the scientific literature on tourist behavior, there are 6 recurrent models used. To simultaneously investigate the importance of individual characteristics and geographical attributes in the participation decision and the main determinants of tourism consumption, the Double-Hurdle Model (HM) has been estimated. In this model, the variables used are age, family, and origin (specific characteristics of the location) (Bernini et al., 2017). The Ideal Self-Images Model (Self-Congruity) or Image Model of Ahn et al. (2013) postulates that tourists evaluate the image of the destination based on their real and ideal self-image (self-congruence). In turn, self-congruence positively influences the choice of destination. Tourists also evaluate the destination's image based on its functional or utilitarian attributes, such as service quality, price, location, and physical attractiveness. The model suggests that functional congruence results from matching destination utility-based performance attributes and tourist-desired performance attributes. A close coincidence positively influences the choice of destination.

In the Consumer Choice Model, tourism demand is influenced by a series of economic, social, political, psychological, and technological factors. Sepecifically, the Li & Chen model contemplates the physical environment, digital environment and social environment in the model (Figure 3). The economic determinants have been widely studied, with the dominant framework being the neoclassical theory of consumer choice. However, critics have questioned the suitability of a purely economic model for examining factors other than prices and income in tourism demand. Thus, the choice of consumption depends on variables such as distance, price differences, distance from tourism development, or cultural distance (Lin & Chen, 2022). The Affection Marketing Model is a model that connects the affect marketing dimension developed by place attachment (and also the tourism marketing perspective) with the consequent satisfaction with ethnic tourism contexts and the behavioral intention to repeat the experience or reinforce the experience with a variation of tourist experiences - loyalty to the destination place (Sousa & Paniza, 2020). In the so-called WOM (Word of mouth) Model, the attitude of residents towards tourists is an important factor in the general attractiveness of a destination and one of the attributes that motivates tourists to make a visit is the "interaction with the local population". This interaction generates a rewarding experience for tourists.

Figure 3. Tourist choice behavior model (Li & Chen, 2022)

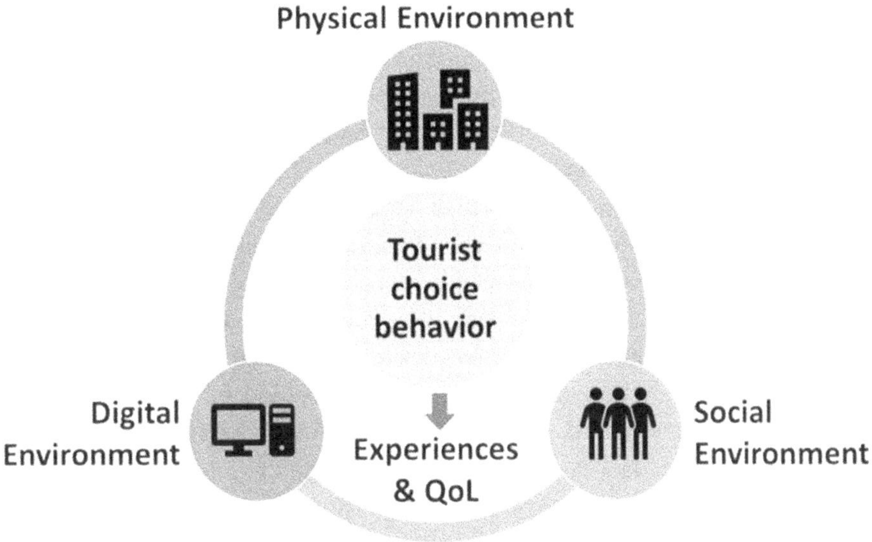

Tourists care about the responses of residents they know and, in cases where they experience a good relationship, they may be more satisfied with the trip and more willing to recommend it by WOM, since one of their most important trip expectations has been accomplished (Lai et al., 2018). Finally, the CATREG Destination Choice Model uses an econometric procedure called CATREG, which is especially useful when handling data sets with a combination of nominal, ordinal and/or interval variables. As for theories, two are the most referenced, Theory of Planned Behavior (TPB) and Demand theory. The Theory of Planned Behavior (TPB) is one of the most used sociopsychological theories in the literature in terms of feasibility, testability, methodological suitability, and validity in the framework of tourism, leisure, and hospitality management. According to the TPB, the decisive factor for behavior is intentions, understood as a motivating factor that influences behavior. The TPB suggests that there are three independent determinants of behavioral intentions: attitude (toward a behavior), subjective norms (understood as an individual's perception of a particular behavior), and perceived behavioral control (which refers to the perceived degree of difficulty in performing a particular behavior) (Jin et al., 2019). In Demand theory, the systemic approach to demand modeling attempts to derive the demand for a particular destination by estimating the demands of a group of destinations considered substitutes or complementary (Divisekera, 2010; Lin & Chen, 2022). These theories provide conceptual frameworks that help understand the complexity of tourist behavior, thus facilitating the formulation of effective strategies in the tourism industry.

We must also mention the effect of COVID-19 on tourist behavior. According to Wachyuni and Kusumaningrum (2020), following the pandemic, travel intention would return to normal after 6 months of the pandemic, in more than 65% of the population. Therefore, the pandemic is now largely forgotten, and travel intentions have returned to normal. Additionally, the use of virtual reality has become increasingly prevalent in developing the perspective of tourist behavior, a rising trend in new models (Onciouiu & Priescu, 2022).

TOURIST PERCEPTION

Tourist perception refers to how visitors experience and understand a tourist destination based on their senses, emotions, beliefs, expectations, and previous experiences. Perception can vary widely from one tourist to another, as it is influenced by individual and contextual factors. The field of study on this topic is broad, with the Destination Choice Model (Suresh et al., 2022) being positioned as prominent models, where the key factors related to the choice of destination for those tourists seeking nature include diversity of flora and fauna, uniqueness, nature/remoteness, activities and entertainment, quality of the landscape, orientation and information, pleasant climate, accessibility, safety, price, friendly attitudes of locals and tourist services and infrastructure. The Plog Personality Model (Bayarsaikhan et al., 2020), in which Plog personality types significantly affect destination choice. Plog personality types have been found to differentiate tourism destination choices when controlling for sociodemographic and travel-related confounding variables. The Risk Perception Model, in which risk perception is fundamental in the decision-making process of tourists, since risk can change conscious decision-making about the choice of destination. Chebli et al. (2020) propose a tripartite theoretical model of tourist attraction to examine the influence of personal factors on tourists' perceptions of attractions that determine a city's appeal. Additionally, the TPB model (Theory of Planned Behavior) is revised by Wang et al. (2002), finding that perceived behavioral control is the main contributor to travel intention for the traditional TPB, while attitude partially mediates the relationship between subjective norm and travel intention for the mediation test.

The Mediation Model, a model that proposes two mediators: the perception of the destination (PC) and the perception of the impacts of the pandemic (PD). Or among others, the Mura's Model (Le & Bui, 2022) or Touristic Destinations' Perceived Risk and Perceived Value as Indicators of E-Wom and Revisit Intentions Conceptual Model (Uslu & Karabulut, 2018). Regarding theories on tourist perception, we will reference the main ones. The Theory of Distance where temporal distance includes a phenomenon or stimulus that has a temporal distance from the individual. Spatial distance is a distance that a phenomenon or stimulus has with the individual in terms of place; and social distance is the distance that the stimulus has with itself or with the individual or is even an event that has occurred in the presence of others. This theory helps explain how an individual perceives a phenomenon or stimulus that is close or far from them (Stephan et al., 2011).

Destination Information comes not only from people returning from their trips. Newspapers and magazines regularly include tourist supplements and travel stories, and travel reports are increasingly being broadcast on television. The phenomenon of electronic social media such as Facebook and its possible tourist influences have given rise to numerous and diverse academic contributions, which are still predominantly exploratory in nature. For example, it is indicated that there has been greater use of electronic word of mouth and other Internet sources in tourism decision-making processes and travel purchase intentions in the first decade of the 21st century (Jacobsen & Munar, 2012). Li et al. (2016) propose a formulation of the user balance model based on static activity. In particular, they adopt the concept of a multi-state supergrid to represent the choice space of activity-travel patterns. A numerical example using hypothetical scenarios is presented to illustrate the proposed model and solution algorithm (Figure 4).

Decision making consists of several temporal sub-decisions: central (choice of destination, trip, time, etc.), secondary (activities and attractions to visit) and route decisions (stops and items to buy, the budget for gifts, etc.). This procedure is influenced by several psychological (internal) and non-psychological (external) variables (Micić et al., 2019). Tourist Perceived Risks, a theory in which seven types of risk are identified in consumer behavior literature, including social, physical, satisfaction, time, psychological, functional or performance, and financial risks (Emami & Ranjbarian, 2019). Karl and Schmude (2017) have generated a framework according to their review of the literature that introduces that risk perceptions operate between the destination and tourist attributes with safety images.

Figure 4. The travel decision making process (Li et al., 2016a)

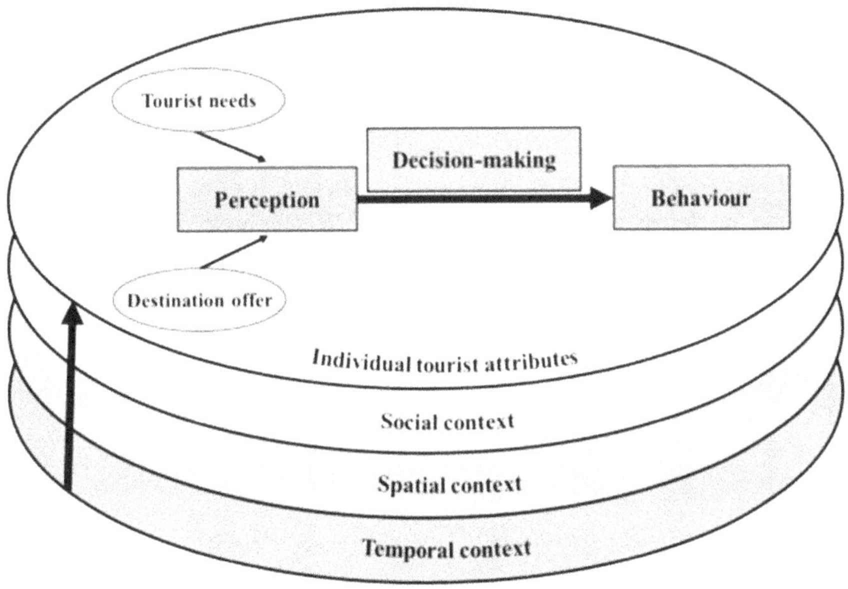

Marketing Strategies and Fear in consumer behavior, according to which fear also plays a fundamental role in marketing strategies. It has been found that consumer behavior can be influenced by a moderate fear stimulus. These practices are based on the recognition that consumer behavior is influenced by the emotional state of an individual. Furthermore, fear is assessed for educational purposes (Mura, 2010). The friction of geographical distance and cognitive distance, where geographical distance is slowing down human activities. The greater the distance, the less frequent the interactions between the two extremes. On the other hand, cognitive proximity facilitates effective communication and improves interactions. And finally mentioning the "unwanted self" as a less desired identity that represents the sum of traits with negative valence, memories of feared experiences, embarrassing situations, frightening events, and unwanted emotions that a person is constantly motivated to avoid, suggested that the dominant implicit standard that people use to evaluate their well-being is their distance from a subjective similarity to their most negative self-image, that is, the unwanted self. Therefore, people tend to keep track of their real, everyday self by implicitly referring to their unwanted self, because "without a tangible unwanted self, the real self would lose its navigation cues" (Ogilvie, 1987, p. 380). In short, tourist perception refers to the interpretation and understanding that individuals have of a tourist destination, its attractions, services, and experiences. This perception is shaped by a combination of subjective and objective factors that influence the way tourists experience and evaluate a specific place. Therefore, understanding tourist perception is essential for tourism destinations and service providers, allowing them to adapt and improve the visitor experience, encourage loyalty and build a positive reputation in the tourism industry.

Tourist Motivation

Tourism motivation refers to the factors and desires that drive people to travel and explore tourist destinations. Understanding tourists' motivations is essential for the tourism industry, as it helps businesses and destinations adapt their offerings to meet travelers' needs and expectations.

There are 4 most common identified models:

i) Push-pull motivation model: Push-pull framework and perceived fit theory. Perceived fit theory is applied to better understand how the interaction between push and pull factors can explain tourism destination choice. A high perceived fit between travel motivations and destination image positively influences the intention to choose the new destination, and this effect is mediated by the perceived attractiveness of the destination (Maggioni et al., 2020) (Figure 5).

Figure 5. Push-pull motivation model (Maggioni et al., 2020)

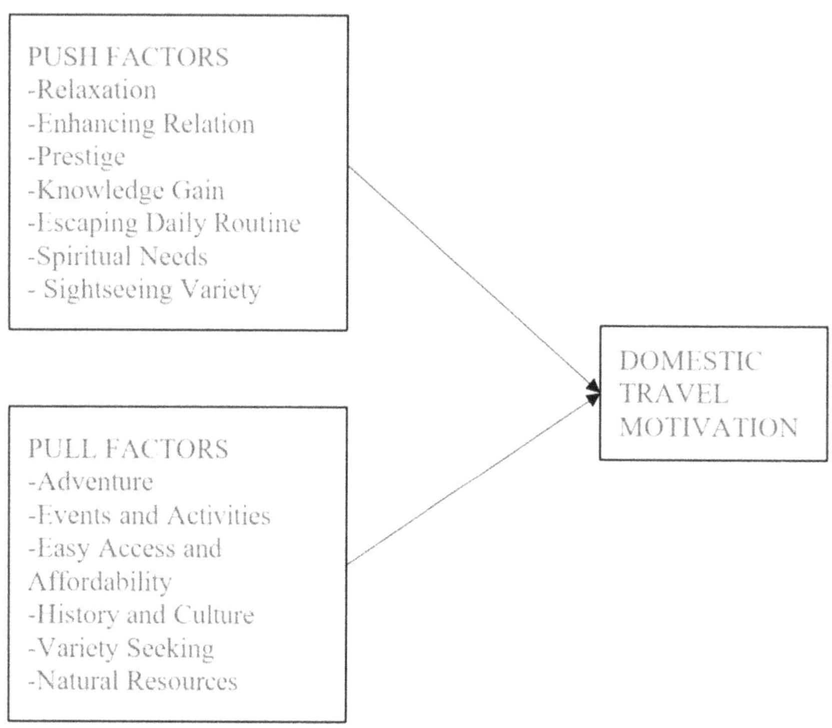

ii) Destination choice model: "Basic tourism conditions, such as weather, actual quality of service, actual quality of accommodation, and attitude of hosts, are sometimes unknown at the time decisions are made. Existing tourism information sources can only present an image of the probability of purchasing a certain service at a specific potential destination based on the image created. In view of the disadvantages of the "normative-rational" approach, therefore, the probabilistic behavioral approach (based on random utility) should be adopted for the study of tourist choice" (Mansfeld, 1992, p.402). The literature on destination choice focuses on the direct impact of destination attributes such as distance and price. Analyze whether motivations act as moderating dimensions of the influence of distance and price in the choice of tourist destinations, defined as intra-country administrative units (Mutinda & Mayaka, 2012).

iii) Model of push and pull motivations of older people: A more complete model of push and pull motivations of older people, which includes the interaction between their emotions and the satisfaction experienced in a particular place. A model that relates these variables and more precisely explains the association between them offers a necessary and useful contribution to the literature on senior tourism; Specifically, the developed model integrates motivations, emotions, satisfaction, and previous experience; Furthermore, it reveals two types of push motivations (maintenance and expansion) and distinguishes push and pull motivations in different people, thus providing information that can improve our understanding of the motivational processes underlying the choice of a tourist destination. This knowledge allows marketers to capture the objective characteristics that a destination offers and its intangible elements, and manage the factors responsible for customer satisfaction, a key requirement for developing the intention to revisit or recommend a destination. in particular (Pestana et al., 2020).

iv) The Analytical Hierarchy Process (AHP) model -. It provides solutions to complex problems and employs hierarchical structures by developing priorities for different alternatives determined by decision-makers (Brushan & Rai, 2004). The AHP approach is used to build an evaluation model for decision making, using weighted criteria (Vujičić et al., 2022).

Furthermore, currently, models are beginning to incorpórate variables such as virtual reality, giving rise to virtual tourist motivation (Yang et al., 2023).

With regard to theories of tourist motivation, the literature recurrently alludes to three theories: i) Decision Theory in which a probability is associated with each diagnosis (the states of nature) and with each treatment (the potential actions). Using any of the standard protocols to construct the customer value function over the set of outcomes we can define a utility function (including uncertainty) that, maximized, identifies the solution that should be adopted (since, by definition, it is the that maximizes the customer's expected utility). The existence of such a function is guaranteed by a certain number of axioms that represent what, following the theory, must be the principles of the rational behavior of a decision maker; ii) Individual Destination Choice, a theory that examines how regional attributes affect individual destination choices for nature tourism within a larger territory. Special attention is devoted to the role of temperature in relation to place of residence and distance (Boto-García et al., 2021) and iii) Motivation to travel, a subjective factor that affects both destination choices and the behaviors of tourists. Specifically, the connection between motivation to travel and destination choice depends on the benefits the traveler seeks and the activities he or she engages in within a destination.

CHOICE OF TOURIST DESTINATION

Choosing a tourist destination is a complex process that involves the evaluation of various options and factors (Douglas & Hoogendoorn, 2024; Utari et al., 2024). Tourists consider a variety of elements before deciding where to travel. Some of the most important factors that influence tourists' choice of destination are formulated in various models (Figure 6). Within the theory of this field, there are 8 main models identified and 8 of the most prominent theories.

Figure 6. Example of destination choice model (Li et al., 2016b)

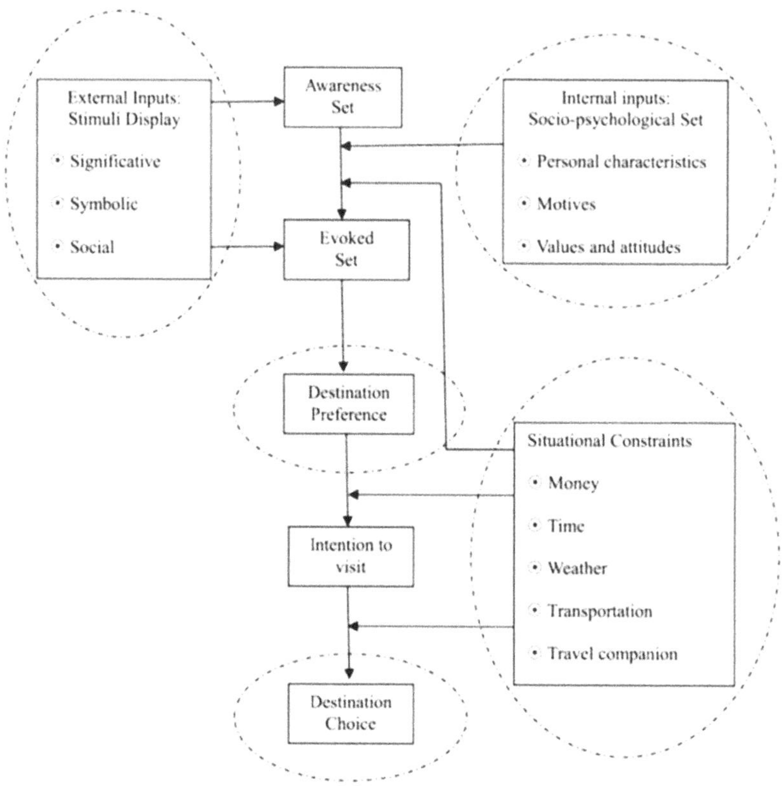

The destination choice set model, formulated by Van, Borgers, and Timmermans (2003) addresses the choices of domestic trips or trips to foreign countries. This model includes four metaregions (water, land north and south, and other) for domestic tourism and two for foreign travel (neighboring and further afield). The Conceptual Model of Desiring at a Distance is based on Bullough's theory of psychological distance which suggests that the separation between subject and object is an important way to achieve an aesthetic experience. The model proposes 3 dimensions that influence the desire for distance and the intention to travel: Attitude, subjective norm, perceived behavioral control (Mao & Cao, 2019). The TPB Model of Jalilvand and Samiei (2012) is based on the Theory of Planned Behavior Model (TPB). The TPB model proposes 3 dimensions that influence intention and behavior: attitude towards a behavior, subjective norms, and behavioral perceptions.

The Destination Choice (DC) Ensemble Model states that according to set theory, DCs are multi-stage processes in which numerous alternative destinations are successively narrowed down in a funnel shape, the concepts of the ensemble model are summarized in a process structure model approach, the total set can be divided into three subcategories: unknown set (i.e., all destinations that a tourist does not know; available knowledge set (i.e., all destinations that a tourist knows and has the ability to visit; unavailable knowledge set (i.e. all known destinations but due to different reasons are not available) (Karl et al., 2020). The set of destination options for families with children in a wheelchair wheels is the model of Nyman et al., (2018) adapted from Decrop 2010 destination set model (which includes thirteen types of destination brand choice sets: universal set, total opportunity set, awareness set, evoked, consideration set, action set, inert set, fog set, inaction set, retention set, rejection set, inept set and conscious unavailable set). According to Cardoso and Dias (2017), the Decrop model takes advantage of classic destination choice models, as it presents a logical sequence of a decision funnel that evolves throughout four cognitive tasks: consideration, evaluation, coping with limitations, and choice. The model by Nyman et al., (2018) proposes the following sequence of sets: universal set, awareness set, consideration set (if you can afford a wheelchair), available set, accessible (if wheelchair friendly) and final choice.

In the tourism realm, facilitators are elements at a destination that fulfill a tourist's desires, while inhibitors are aspects that don't align with their motivations. For instance, facilitators encompass factors like novelty, relaxation, and learning, which cater to travelers' interests. Cultural disparity can act as a facilitator if it fulfills a tourist's need for novelty, but it can also be an inhibitor since it's linked to travelers' risk perception, a critical factor in choosing international destinations. The Travel Risk Theory (Karl et al., 2020; Xie et al., 2023) is based on Karl's 2018 theory and maintains that the characteristics of a destination to satisfy the needs of tourists must be seen in an integrated way, that is, the Risk perception towards an attribute may vary from one destination to another. Furthermore, certain risks are more salient in some tourists' destination choices and consideration should be given to how differences in the perception of risk types are reflected at key stages of the destination choice process. From the point of view of consumer behavior, travel risk perception can be defined as the expected negative impact of the respective risk and the probability of these negative consequences occurring.

The evaluation of both elements is prone to errors and biases that result, for example, in tourists overestimating the probability of terrible events (e.g., terrorism) compared to more mundane events. However, tourists base their decisions on these (biased) perceptions of risk, which may not reflect the real situation of a destination. Karl et al. (2020) proposes three elements to consider: 1) the sociodemographic profile of each tourist influences how travel risk perceptions shape travel decision making; 2) Greater travel experience influences tourists' risk perception; and, 3) Personality and psychographic factors also shape risk perceptions and travel decision making. Cognitive Distance Theory refers to people's beliefs about the distances between places in large-scale spaces, that is, places that are too far away and obscured to not be visible to each other.

On the one hand, tourists often must expend considerable effort (time, money, and physical strength) to overcome the obstacle of distance: the "friction effects of distance." On the other hand, distance can psychologically satisfy tourists' desire to physically distance themselves from home: the "catalytic effects of distance" (Cao et al., 2019). The Theory of Desire at a Distance, where psychological distance refers to the distance between the subject and an aesthetic, emotional or mental object. This distance makes it possible to create a feeling of pleasure by eliminating the subject's practical attitude towards the object, thus giving him a completely new experience. Physical distance and temporal distance are unique forms of the specific psychological distance considered in aesthetic research, as descriptions of personal relationships are often highly emotional in color but peculiar in character. If the subject and object lose distance due to, for example, underdistancing (loss of aesthetic experience due to utilitarian purposes) or overdistancing (weakening of the relationship between subject and object due to a large distance or because the aesthetic object is not easily understood and mastered by the subject), the subject may be unable to obtain a sense of beauty. With the insufficient distance threshold as the lower limit and the excessive distance threshold as the upper limit, the psychological distance within this range can lead to a better aesthetic experience for the subject.

The subject adopts an "action at a distance" (i.e., seeks to remain close or far from the aesthetic object) to seek an appropriate distance from which to enjoy continuous beauty and have the best aesthetic experience). Cao et al. (2019) draws on Bullough's theory to introduce the concept of "distant desire" to represent the willingness of subjects to perform certain actions at a distance to obtain better aesthetic experiences within the interval between excessive distancing and insufficient distancing. Lastly, the theory of past travel perspectives and experiences is based on the work of Crouch et al. (2016), which identified a significant relationship between past travel experience and future destination preferences, and suggested that this experience preference structure may be related to the demographic and psychographic characteristics of tourists (Masiero & Qiu, 2018). Cardoso et al. (2019), alluded to perspective theory and justified that the experience of past memories affects perceptions about the destination and therefore interferes with the choice of the destination. In summary, the choice of tourist destination is a multifaceted process that reflects individual preferences, past experiences, and the interaction with various influencing factors.

INFLUENCE OF DIGITAL ERA ON THE TOURIST CONSUMER

Today, the digital age has revolutionized the way consumers approach and experience tourism. From trip planning to destination immersion, the influence of technology redefines the connection between the traveler and the traveler's experience, shaping new expectations, personalized options, and a more dynamic interaction with the tourist environment. Instant accessibility to information, the influence of social networks, and the incorporation of innovative technologies are transforming the tourism industry and redefining the relationship between technology and the contemporary traveler. The digital revolution is radically changing the world we live in. Smart home sensors can interconnect devices such as thermostats, washing machines, televisions, laptops, tablets, and other objects to Internet of Things platforms.

New digital technologies have introduced important innovations in factories, hospitals, hotels, cities, and territories. Industry 4.0 is signaling the end of well-established patterns and is asking academics, managers, and citizens willing to survive in this ever-changing and increasingly complex environment to look at it through different lenses and new paradigms. The tourism sector, in addition, is closely involved in digital transformations, increasingly qualifying them with expressions such as Tourism 4.0 or Smart Tourism (Pencarelli, 2020). Information technology (IT) has dramatically transformed travel and tourism (Sheldon, 1997). From the development of Computer Reservation Systems (CRS) in the 1960s to Global Distribution Systems (GDS) in the 1980s and the arrival of the Internet in the early 1990s, the tourism industry has always faced to the emergence of new technological developments that posed both opportunities and challenges (Buhalis & Law, 2008).

Consequently, tools such as search engines have become a dominant force influencing travelers' access to tourism products (Xiang et al., 2008). Second, the tremendous growth of social media has changed the dynamics of online communications (Sigala et al., 2012). Third, recent advances in mobile computing, particularly with the adoption of smartphones and their travel applications, create new environments for information seeking and use through which the contextually defined needs of travelers on the move become increasingly prominent in guiding travel decisions (Wang et al., 2012). In the tourism industry, technological progress has gone hand in hand with sectoral changes. Many studies over time have underlined the inevitable impact of ICT development on the growing attractiveness of destinations and cultural sites, increasingly characterized by intensive information exchange and value co-creation (Da Costa Liberato et al., 2018). Hence, the culture of sharing and its participatory implications are increasingly part of the travel experience of experts and academics in the sector. Both recognize that the narration of a place, or a cultural site, can no longer be delegated solely to institutional representatives.

On the contrary, the voice of the tourist who approaches for the first time can become the most honest statement to convey about the entire landscape, along with its cultural and social richness (Frias et al., 2019). When informing themselves about future destinations, visitors look for content capable of involving them by anticipating the travel experience, allowing them to continue during the trip/tour and making them feel part of a community of travelers, more than just tourists. Therefore, the collaborative dimension, which exploits the natural competencies (communicative, territorial, social, etc.) and the reliability of those who have experienced the destination/cultural site first-hand, becomes a broad source of value, encouraging people to resonate with the main characteristics of the place (Bassano et al., 2018). This is why digital became relevant in the dynamics of consumption, through the use of tools such as artificial intelligence and blockchain technology (Porras et al., 2019), as well as other disruptive technologies that allow selecting and offering segmented content, changing the decision-making process of tourists, who are increasingly open to being influenced by these suggestions, instead of following professional advice.

The connection of physical infrastructure and natural and cultural resources with social services/facilities and commercial activities means that tourists – and public/private operators – can take advantage of the collective intelligence of an area to generate advantages (Allam & Newman, 2018). Understanding how IT affects consumer behavior can serve as a critical foundation for companies to identify and develop effective and sustainable marketing communication strategies. Technologies come and go and therefore it is important for companies to identify those technologies with the potential to serve their long-term strategic purposes. Pre-trip travel planning, which is a specific type (and stage) of consumer information search, can be considered a fundamental component of the travel experience in the sense that a traveler often needs to obtain a substantial amount of information. information to develop a travel plan. As such, the information available to individual travelers has a significant impact on several aspects of traveler decision making, especially when choosing a destination to visit (Bieger & Laesser, 2004).

In addition, trip planning helps the traveler make decisions and generate expectations for the next trip (Gretzel, et al., 2006). Therefore, it is argued that travel planning serves as a window through which it is possible to understand how IT has changed travel and tourism. Before the "social media era," consumers traditionally had limited resources to gather information about services and products (Hu & Wei, 2013). Accommodation purchases are mainly based on brochures and travel agencies (Baruca & Civre, 2012). In recent years, traditional sources of information have been replaced by social media and other digital tools, as travelers began to consult user-generated content (UGC) on social media to plan and make decisions about their choice destination, stay in hotels, consumption in restaurants, etc. (Browning &

Sparks, 2013; Stojanovic et al., 2022). In the context of tourism, consumers use digital technologies for a wide range of scenarios, for example, sharing their travel-related experiences, interacting with others, connecting with people from different destinations, and purchasing travel-related products and services (Munar & Jacobsen, 2014). While the proliferation of information and communication technologies (ICT) has always had a significant effect on consumer behavior in the tourism sector, Web 2.0 has transformed the way tourists select and purchase tourism services. online (Gohil, 2015). For example, hotel decision-making has fundamentally changed (Hudson & Thal, 2013), since social networks are used at every stage of the consumer decision-making process.

DIGITAL TOOLS IN DESTINATION CHOICE

The last two decades have been the decades in which the Internet and digital technologies have become fundamentally integrated into our lives. The ubiquity of devices with Internet access and the associated mobile modality have profoundly influenced the tourism industry at an infrastructural and communicative level: we book trips online, we talk to our friends at home, we can imagine distant places in images, texts or virtual realities.

The rise of technology-based solutions that encourage hypermobility of people, places, and spaces has created seismic changes in the way tourists and tourism manifest themselves and, as such, how academic research in this area has progressed. Greater mobility has gone hand in hand with the immediacy of information that is now commonly available through digital media. Through online travel agencies (OTAs) such as booking.com and Trivago, and from peer-to-peer booking sites like Airbnb, we can curate our travel experiences anywhere in the world with a click of a mouse or a tap on a smartphone. We can search for package deals, make travel reservations with airlines, train and ferry operators, search for nearby points of interest, check current weather conditions and read local news. Our itineraries can be planned in a similar way: platforms like Google StreetView allow us to access precise, high-resolution digital images of places around the world, while the artificial intelligence technology present in mapping services uses GPS to provide information about the shortest or most convenient routes to follow (Cinnamon & Gaffney, 2022).

The result is a remarkable profusion of mediated experiences and vicarious travel possibilities that would have seemed miraculous even a few decades ago. The countless previews, evaluations, and debates about travel destinations that exist online imply that the physical trip to a "real" place is only one part of a tourism experience that is increasingly dominated by digital engagement. Contemporary tourism, in short, emerges as a fusion of socio-technological relationships; a ubiquitous co-presence of technologically mediated relationships. Virtuality blends with the physical as tourists and tourism emerge through complex networks of widespread socialization and authentication: maintaining connection with friends and family, influencing other tourists' choices through online reviews, sharing social media content, influenced by experience providers that drive decision-making and purchasing practices through personalized content and supported by peer recommendation (Van Nuenen, 2019).

However, with such a technosocial opportunity, research recognizes that this desire to be more "connected" and "experience" more destinations also impacts the economic and social fabric of the communities and destinations visited (Cheng et al., 2020). If we reflect on the last two decades, it is obvious that we cannot separate our understanding of the relationship between IT and tourism from the global social view of technology. Borrowing the ideas of Negroponte and Friedman, we can consider the development of IT and tourism knowledge to consist of two distinct eras: the first decade can be characterized as "digitalization" (1997-2006) and the second represents the "era of acceleration" (2007-2016). It is important to note that our view of technology in relation to tourism management seems to have changed over the years. In the near future, research related to tourism management will likely be driven by the opportunities and challenges created by the relentless growth of data and information. Obviously, there will be concerns and issues related to the quality, reliability, and ethics of big data, social insights, and artificial intelligence applications. Most importantly, however, information technology is connected to travel and tourism in so many ways that it will give us a new space for imagination and a new space to envision our future.

Thanks to technology, there are now possibilities to access a variety of data, in massive quantities, in different formats and potentially in real time; and, thanks to technology, it is now possible for us to "connect the dots" to map the complex tourism system (Xiang, 2018). In short, today digital tools play a fundamental role in choosing a destination since they provide information, facilitate planning, and allow travelers to make informed decisions. Among the best-known and most used are (Sotomayor, 2019), -whose summary is included in Figure 7-:

1. Travel websites. Platforms such as TripAdvisor, Lonely Planet, Expedia, and Booking.com offer reviews of hotels, activities, and destinations. Users can get opinions from other travelers to make informed decisions.
2. Map applications. Google Maps, MapQuest or similar applications allow you to explore destinations, get directions, know the distance between places of interest and discover nearby restaurants, shops, and attractions.
3. Social networks. Platforms like Instagram, Facebook, and Twitter can provide a visual preview of destinations through photos and experiences shared by other travelers. It is also possible to connect with people who have visited those places and get personalized recommendations.
4. Trip planning platforms. Tools like TripIt or Sygic Travel help you organize itineraries, store important information about flight and hotel reservations, and share travel plans with others.
5. Blogs and travel blogs. Reading travel blogs or watching videos on platforms like YouTube can provide detailed information and personal experiences about specific destinations.
6. Price comparators. Websites and applications such as Skyscanner, Kayak, or Momondo allow you to compare prices for flights, hotels, and car rentals, helping you find the best deals.
7. Language applications. Apps like Duolingo or Babbel can be useful for learning or reviewing basic phrases and vocabulary of the target language.
8. Travel Forums. Platforms like Reddit or Lonely Planet Thorn Tree are great for getting tips and answers to specific questions from other experienced travelers.
9. Climate and weather forecast. Apps like AccuWeather or Weather.com provide up-to-date information about the weather at your chosen destination, which can be crucial for planning activities.
10. Virtual reality. Some companies offer virtual reality experiences that allow users to explore destinations before making final decisions about their trip.

Figure 7. The best-known and most used technology in tourism (wordpress.org, n.d.; sixt.com, n.d.; dreamstime.com, n.d.; lifewire.com, 2024; Booking.com, n.d.; wired.com, n.d.; women on the road, 2021; stock adobe, n.d.; El País, 2023)

CONCLUSION

Tourism destination choice is a multifaceted process influenced by a variety of factors and theories. Models such as destination choice sets and the Conceptual of Distance Desire stand out, which address psychological and cultural aspects in travel decision-making. Cultural distance is presented as an element that can act as both a facilitator and an inhibitor, depending on how it aligns with the individual motivations and perceptions of the tourist. Furthermore, tourist perception, motivation, and behavior are crucial components in this process. Perception varies depending on individual preferences, while motivation drives people to explore tourist destinations. Tourist behavior, influenced by economic, social, political, and technological factors, affects tourism demand and consumption choice.

The digital revolution has significantly transformed this process, providing travelers with access to a wide range of information and tools. Social media, virtual reality, online booking, and recommendation apps have simplified travel planning and influenced decisions through authentic shared experiences. Technology has also boosted sustainable tourism consumption by providing information on sustainable practices and eco-friendly transportation options. The instant accessibility of information through the Internet, the ubiquity of search engines, the exponential growth of social media, and the adoption of mobile technology have changed the way travelers access and use travel-related information.

The culture of sharing, driven by the voice of tourists themselves, has altered the traditional narrative of destinations and cultural sites. The tourism industry is immersed in the era of Tourism 4.0 or Smart Tourism, where artificial intelligence, blockchain technology, and other disruptive innovations are transforming the way segmented content is selected and offered. The connection between physical infrastructure and natural and cultural resources with commercial activities and social services takes advantage of collective intelligence to generate competitive advantages.

Trip planning, a critical stage in the travel experience, has become increasingly reliant on technology. The information available through search engines, social networks, and mobile planning tools significantly impacts traveler decision-making and expectations. Before the era of social media, traditional sources of information such as brochures and travel agencies were limited.

The use of new technologies in tourism has brought with it a number of significant benefits, but also presents certain threats that must be considered, such as over-reliance on technology, concerns about data privacy, depersonalization of customer services or the existence of a digital divide, leaving certain groups excluded from the benefits of tourism. To mitigate these threats, it is important to implement appropriate cybersecurity measures, promote digital literacy, ensure the protection of personal data, encourage ethical business practices and seek a balance between the adoption of technology and the preservation of authenticity and sustainability in the tourism.

Consumers are now turning to user-generated content on social media for authentic information and shared experiences about destinations, accommodations, and restaurants. In conclusion, the choice of tourist destination in the current era is a complex process that involves psychological, cultural, economic, and technological factors, with sustainability emerging as a key factor. Digital technology plays a critical role in providing tourists with tools to research, plan, and make informed decisions while encouraging more sustainable and conscious practices while traveling.

ACKNOWLEDGMENT

This research is financed by national funds through FCT—Foundation for Science and Technol-ogy, I.P., within the scope of the project https://doi.org/10.54499/UIDB/04470/2020.

REFERENCES

Ahn, T., Ekinci, Y., & Li, G. (2013). Self-congruence, functional congruence, and destination choice. *Journal of Business Research*, *66*(6), 719–723. doi:10.1016/j.jbusres.2011.09.009

Allam, Z., & Newman, P. (2018). Redefining the smart city: Culture, metabolism and governance. *Smart Cities*, *1*(1), 4–25. doi:10.3390/smartcities1010002

Baruca, P. Z., & Civre, Z. (2012). How do guests choose a hotel. *Academica Turistica*, *5*(1), 75–84.

Bassano, C., Piciocchi, P., & Pietronudo, M. C. (2018). Managing value co-creation in consumer service systems within smart retail settings. *Journal of Retailing and Consumer Services*, *45*, 190–197. doi:10.1016/j.jretconser.2018.09.008

Bayarsaikhan, T., Kim, S. T., & Gim, T. H. T. (2020). International tourists' destination choice differences according to Plog's personality types: Analyzing the case of Mongolia based on the recreation opportunity spectrum. *International Journal of Urban Sciences*, *24*(4), 485–515. doi:10.1080/12265934.2020.1771195

Bernini, C., Cracolici, M. F., & Nijkamp, P. (2017). Place-based attributes and spatial expenditure behavior in tourism. *Journal of Regional Science*, *57*(2), 218–244. doi:10.1111/jors.12308

Bhushan, N., & Rai, K. (2004). Strategic decision-making. In *Strategic Decision Making: Applying the Analytic Hierarchy Process* (pp. 3–10). Springer London. doi:10.1007/978-1-85233-864-0_1

Bieger, T., & Laesser, C. (2004). Information sources for travel decisions: Toward a source process model. *Journal of Travel Research*, *42*(4), 357–371. doi:10.1177/0047287504263030

Booking (n.d.). https://www.booking.com/

Boto-Garcia, D., Mayor, M., & De la Vega, P. (2021). Spatial price mimicking on Airbnb: Multi-host vs single-host. *Tourism Management*, *87*, 104365. doi:10.1016/j.tourman.2021.104365

Browning, V., So, K. K. F., & Sparks, B. (2013). The influence of online reviews on consumers' attributions of service quality and control for service standards in hotels. *Journal of Travel & Tourism Marketing*, *30*(1-2), 23–40. doi:10.1080/10548408.2013.750971

Buhalis, D., & Law, R. (2008). Progress in information technology and tourism management: 20 years on and 10 years after the Internet—The state of eTourism research. *Tourism Management*, *29*(4), 609–623. doi:10.1016/j.tourman.2008.01.005

Burguete, J. L., Hidalgo-Alcazár, C., & Lanero, A. (2020). Turismo sostenible y emprendimiento digital: Una aproximación a los nuevos perfiles de turistas. *Economía Industrial*, *417*, 59–68.

Cao, K., Li, D., & Xie, L. (2019). Bearing-ratio-of-distance rigidity theory with application to directly similar formation control. *Automatica*, *109*, 108540. doi:10.1016/j.automatica.2019.108540

Cardoso, L., Dias, F., de Araújo, A. F., & Marques, M. I. A. (2019). A destination imagery processing model: Structural differences between dream and favourite destinations. *Annals of Tourism Research*, *74*, 81–94. doi:10.1016/j.annals.2018.11.001

Chebli, A., Othmani, M. C., & Said, F. B. (2020). Market segmentation in urban tourism: Exploring the influence of personal factors on tourists' perception. *Journal of Tourism and Services*, *11*(20), 74–108. doi:10.29036/jots.v11i20.144

Cheng, M., Mackenzie, S. H., & Degarege, G. A. (2022). Airbnb impacts on host communities in a tourism destination: An exploratory study of stakeholder perspectives in Queenstown, New Zealand. In *Platform-mediated tourism* (pp. 194–212). Routledge. doi:10.4324/9781003230618-11

Cinnamon, J., & Gaffney, A. (2022). Do-it-yourself Street Views and the urban imaginary of Google Street View. *Journal of Urban Technology*, *29*(3), 95–116. doi:10.1080/10630732.2021.1910467

Crouch, G. I., Huybers, T., & Oppewal, H. (2016). *On the Choice of Tourism Destination versus Tourism Experience: Insights from an Analysis of Past Choice and Future Interest*. Academic Press.

da Costa Liberato, P. M., Alén-González, E., & de Azevedo Liberato, D. F. V. (2018). Digital technology in a smart tourist destination: The case of Porto. *Journal of Urban Technology*, *25*(1), 75–97. doi:10.1080/10630732.2017.1413228

Decrop, A. (2010). Destination choice sets: An inductive longitudinal approach. *Annals of Tourism Research*, *37*(1), 93–115. doi:10.1016/j.annals.2009.08.002

Divisekera, S. (2010). Economics of tourist's consumption behaviour: Some evidence from Australia. *Tourism Management*, *31*(5), 629–636. doi:10.1016/j.tourman.2009.07.001

Douglas, A., Hoogendoorn, G., & Richards, G. (2024). Activities as the critical link between motivation and destination choice in cultural tourism. *Journal of Hospitality and Tourism Insights*, *7*(1), 249–271. doi:10.1108/JHTI-09-2022-0442

Dreamstime. (n.d.). https://www.dreamstime.com/social-media-background-network-people-connecting-modern-technology-devices-image108829611

El Archi, Y., Benbba, B., Zhu, K., El Andaloussi, Z., Pataki, L., & Dávid, L. D. (2023). Mapping the Nexus between Sustainability and Digitalization in Tourist Destinations: A Bibliometric Analysis. *Sustainability (Basel)*, *15*(12), 9717. doi:10.3390/su15129717

El País. (2023). *Bono Cultural Joven: qué es, cómo solicitarlo y en qué tipo de eventos se puede utilizar.* Disponible online: https://elpais.com/escaparate/2023-08-02/todo-sobre-el-bono-cultural-joven.html

Emami, A., & Ranjbarian, B. (2019). The perceived risk of iran as a tourism destination (A mixed method approach). *Iranian Journal of Management Studies*, *12*(1), 45–67.

Erol, I., Neuhofer, I. O., Dogru, T., Oztel, A., Searcy, C., & Yorulmaz, A. C. (2022). Improving sustainability in the tourism industry through blockchain technology: Challenges and opportunities. *Tourism Management*, *93*, 104628. doi:10.1016/j.tourman.2022.104628

Frias, C., Silva, C., Amante, M. J., & Seabra, C. (2019). Storytelling tourist experiences in the websites of world heritage historical centers. *Experiential Consumption and Marketing in Tourism within a Cross-Cultural Context*, 174.

Gohil, N. (2015). Role and impact of social media in tourism: A case study on the initiatives of Madhya Pradesh State Tourism. *International Journal of Research in Economics and Social Sciences*, *5*(4), 8–15.

González, A. (2020). La influencia de los medios de comunicación en el turismo. *Proceedings Book*, *1*(4), 6.

Gretzel, U., Fesenmaier, D. R., & O'Leary, J. T. (2006). The transformation of consumer behaviour. *Tourism business frontiers: Consumers, products and industry, 9*, 18.

Halkiopoulos, C., Antonopoulou, H., Papadopoulos, D., Giannoukou, I., & Gkintoni, E. (2020). Online reservation systems in e-Business: Analyzing decision making in e-Tourism. *Journal of Tourism Heritage & Services Marketing*, *6*(1), 9–16.

Hu, F. H., & Wei, G. (2013). *The impact of the knowledge sharing in social media on consumer behaviour.* Academic Press.

Hudson, S., & Thal, K. (2013). The impact of social media on the consumer decision process: Implications for tourism marketing. *Journal of Travel & Tourism Marketing*, *30*(1-2), 156–160. doi:10.1080/10548408.2013.751276

Jacobsen, J. K. S., & Munar, A. M. (2012). Tourist information search and destination choice in a digital age. *Tourism Management Perspectives*, *1*, 39–47. doi:10.1016/j.tmp.2011.12.005

Jin, X., Xiang, Y., Weber, K., & Liu, Y. (2019). Motivation and involvement in adventure tourism activities: A Chinese tourists' perspective. *Asia Pacific Journal of Tourism Research*, *24*(11), 1066–1078. doi:10.1080/10941665.2019.1666152

Karl, M., Muskat, B., & Ritchie, B. W. (2020). Which travel risks are more salient for destination choice? An examination of the tourist's decision-making process. *Journal of Destination Marketing & Management*, *18*, 100487. doi:10.1016/j.jdmm.2020.100487

Lai, I. K. W., Hitchcock, M., Lu, D., & Liu, Y. (2018). The influence of word of mouth on tourism destination choice: Tourist–resident relationship and safety perception among Mainland Chinese tourists visiting Macau. *Sustainability (Basel)*, *10*(7), 2114. doi:10.3390/su10072114

Li, H., Meng, F., & Zhang, Z. (2016b). Non-participation of Domestic Tourism: Analyzing the Influence of Discouraging Factors. *International Journal of Tourism Research*, *18*(6), 567–578. doi:10.1002/jtr.2074

Li, Q., Liao, F., Timmermans, H. J., & Zhou, J. (2016a). A user equilibrium model for combined activity–travel choice under prospect theoretical mechanisms of decision-making under uncertainty. *Transportmetrica A: Transport Science*, *12*(7), 629–649. doi:10.1080/23249935.2016.1189718

Lifewire. (2024). https://www.lifewire.com/best-travel-planner-apps-4175910

Lin, Y. C., & Chen, T. C. T. (2022). Type-II fuzzy approach with explainable artificial intelligence for nature-based leisure travel destination selection amid the COVID-19 pandemic. *Digital Health*, *8*, 20552076221106322. doi:10.1177/20552076221106322 PMID:35707268

Lomas, K. R., & Trujillo, C. A. (2018). Environmental educational model for community tourism of fakcha llakta community-Ecuador. *International Journal of Professional Business Review: Int. J. Prof. Bus. Rev.*, *3*(1), 95–110.

Loureiro, S. M. C., Guerreiro, J., & Ali, F. (2020). 20 years of research on virtual reality and augmented reality in tourism context: A text-mining approach. *Tourism Management*, *77*, 104028. doi:10.1016/j.tourman.2019.104028

Ma, L., & Cao, J. (2019). How perceptions mediate the effects of the built environment on travel behavior? *Transportation*, *46*(1), 175–197. doi:10.1007/s11116-017-9800-4

Maggioni, I., Sands, S. J., Ferraro, C. R., Pallant, J. I., Pallant, J. L., Shedd, L., & Tojib, D. (2020). Consumer cross-channel behaviour: Is it always planned? *International Journal of Retail & Distribution Management*, *48*(12), 1357–1375. doi:10.1108/IJRDM-03-2020-0103

Mansfeld, Y. (1992). Group-differentiated perceptions of social impacts related to tourism development. *The Professional Geographer*, *44*(4), 377–392. doi:10.1111/j.0033-0124.1992.00377.x

Mansfield, E. (1998). Academic research and industrial innovation: An update of empirical findings. *Research Policy*, *26*(7-8), 773–776. doi:10.1016/S0048-7333(97)00043-7

Masiero, L., & Qiu, R. T. (2018). Modeling reference experience in destination choice. *Annals of Tourism Research*, *72*, 58–74. doi:10.1016/j.annals.2018.06.004

Micić, J., Denda, S., & Popescu, M. (2019). The significance of the risk-related challenges in tourist destination choice. *Journal of the Geographical Institute "Jovan Cvijić". SASA*, *69*(1), 39–52.

Mousazadeh, H., Ghorbani, A., Azadi, H., Almani, F. A., Zangiabadi, A., Zhu, K., & Dávid, L. D. (2023). Developing sustainable behaviors for underground heritage tourism management: The case of Persian Qanats, a UNESCO world heritage property. *Land (Basel)*, *12*(4), 808. doi:10.3390/land12040808

Munar, A. M., & Jacobsen, J. K. S. (2014). Motivations for sharing tourism experiences through social media. *Tourism Management*, *43*, 46–54. doi:10.1016/j.tourman.2014.01.012

Mura, P. (2010). 'Scary… but I like it!' Young tourists' perceptions of fear on holiday. *Journal of Tourism and Cultural Change*, *8*(1-2), 30–49. doi:10.1080/14766825.2010.482209

Mutinda, R., & Mayaka, M. (2012). Application of destination choice model: Factors influencing domestic tourists destination choice among residents of Nairobi, Kenya. *Tourism Management*, *33*(6), 1593–1597. doi:10.1016/j.tourman.2011.12.008

Novas, N., Gázquez, J. A., MacLennan, J., García, R. M., Fernández-Ros, M., & Manzano-Agugliaro, F. (2017). A real-time underground environment monitoring system for sustainable tourism of caves. *Journal of Cleaner Production*, *142*, 2707–2721. doi:10.1016/j.jclepro.2016.11.005

Nyman, E., Westin, K., & Carson, D. (2018). Tourism destination choice sets for families with wheelchair-bound children. *Tourism Recreation Research*, *43*(1), 26–38. doi:10.1080/02508281.2017.1362172

Ogilvie, D. M. (1987). The undesired self: A neglected variable in personality research. *Journal of Personality and Social Psychology*, *52*(2), 379–385. doi:10.1037/0022-3514.52.2.379

Oncioiu, I., & Priescu, I. (2022). The use of virtual reality in tourism destinations as a tool to develop tourist behavior perspective. *Sustainability (Basel)*, *14*(7), 4191. doi:10.3390/su14074191

Peceny, U. S., Urbančič, J., Mokorel, S., Kuralt, V., & Ilijaš, T. (2019). Tourism 4.0: Challenges in marketing a paradigm shift. In Consumer behavior and marketing. IntechOpen.

Pencarelli, T. (2020). The digital revolution in the travel and tourism industry. *Information Technology & Tourism*, *22*(3), 455–476. doi:10.1007/s40558-019-00160-3

Pestana, M. H., Parreira, A., & Moutinho, L. (2020). Motivations, emotions and satisfaction: The keys to a tourism destination choice. *Journal of Destination Marketing & Management*, *16*, 100332. doi:10.1016/j.jdmm.2018.12.006

Porras-Bueno, N., Plaza-Mejía, M. D. L. Á., & Vargas-Sánchez, A. (2019). Quality of life and perception of the effects of tourism: A contingent approach. *Best Practices in Hospitality and Tourism Marketing and Management: A Quality of Life Perspective*, 109-132.

Sheldon, P. J. (1997). 18 Tourism information technology. *International handbook on the economics of tourism, 399*.

Sigala, M., Christou, E., & Gretzel, U. (Eds.). (2012). *Social media in travel, tourism and hospitality: Theory, practice and cases*. Ashgate Publishing, Ltd.

Sixt. (n.d.). https://www.sixt.com/magazine/tips/top-free-navigation-apps/

Sotomayor Granda, M. F. (2019). Propuesta de difusión turística mediante herramientas web y estrategias de marketing digital. Caso de estudio: Cantón Loja, Ecuador. *Siembra*, *6*(1), 68–84. doi:10.29166/siembra.v6i1.1717

Sousa, A., & Paniza Prados, J. L. (2020). Visitor management in world heritage destinations before and after Covid-19, Angkor. *Sustainability (Basel)*, *12*(23), 9929. doi:10.3390/su12239929

Stephan, E., Liberman, N., & Trope, Y. (2011). The effects of time perspective and level of construal on social distance. *Journal of Experimental Social Psychology*, *47*(2), 397–402. doi:10.1016/j.jesp.2010.11.001 PMID:21836728

Stock adobe (n.d.). https://stock.adobe.com/

Stojanovic, I., Andreu, L., & Curras-Perez, R. (2022). Social media communication and destination brand equity. *Journal of Hospitality and Tourism Technology*, *13*(4), 650–666. doi:10.1108/JHTT-11-2020-0302

Suresh, K., Wilson, C., Quayle, A., Managi, S., & Khanal, U. (2022). Can a tourist levy protect national park resources and compensate for wildlife crop damage? An empirical investigation. *Environmental Development*, *42*, 100697. doi:10.1016/j.envdev.2021.100697

Sutarman, A., Rahardja, U., Oganda, F. P., Millah, S., & Azizah, N. N. (2023). The Role of Information Technology in Empowering the Creative Economy for Sustainable Tourism. [ATT]. *Aptisi Transactions on Technopreneurship*, *5*(2sp), 175–185. doi:10.34306/att.v5i2sp.352

Um, S., & Crompton, J. L. (1992). The roles of perceived inhibitors and facilitators in pleasure travel destination decisions. *Journal of Travel Research*, *30*(3), 18–25. doi:10.1177/004728759203000303

Uslu, A., & Karabulut, A. N. (2018). *Touristic destinations' perceived risk and perceived value as indicators of E-WOM and revisit intentions*. Academic Press.

Utari, B. A., Arsallia, S., Ramdani, M. A., Rahmafitria, F., Belgiawan, P. F., Dirgahayani, P., & Nasution, R. A. (2024). Tourist destination choice on five priority destinations of Indonesia during health crisis. *Journal of Destination Marketing & Management*, *32*, 100880. doi:10.1016/j.jdmm.2024.100880

Valeri, M., & Baggio, R. (2021). Social network analysis: Organizational implications in tourism management. *The International Journal of Organizational Analysis*, *29*(2), 342–353. doi:10.1108/IJOA-12-2019-1971

Van Middelkoop, M., Borgers, A., & Timmermans, H. (2003). Inducing heuristic principles of tourist choice of travel mode: A rule-based approach. *Journal of Travel Research*, *42*(1), 75–83. doi:10.1177/0047287503254116

Van Nuenen, T. (2019). Algorithmic authenticity: Sociotechnical authentication processes on online travel platforms. *Tourist Studies*, *19*(3), 378–403. doi:10.1177/1468797619832314

Vujičić, M. D., Stankov, U., & Vasiljević, Đ. A. (2022). Tourism at a crossroads–ignoring, adopting, or embracing alternative pathways for more sustainable post-pandemic tourism development. In *Crisis management, destination recovery and sustainability* (pp. 11–21). Routledge. doi:10.4324/9781003295839-3

Wachyuni, S. S., & Kusumaningrum, D. A. (2020). The effect of COVID-19 pandemic: How are the future tourist behavior? *Journal of Education. Society and Behavioural Science*, *33*(4), 67–76.

Wang, D., Park, S., & Fesenmaier, D. R. (2012). The role of smartphones in mediating the touristic experience. *Journal of Travel Research*, *51*(4), 371–387. doi:10.1177/0047287511426341

Wang, L. H., Yeh, S. S., Chen, K. Y., & Huan, T. C. (2022). Tourists' travel intention: Revisiting the TPB model with age and perceived risk as moderator and attitude as mediator. *Tourism Review*, *77*(3), 877–896. doi:10.1108/TR-07-2021-0334

Wired. (n.d.). https://www.wired.com/gallery/best-language-learning-apps/

Women on the road. (2021). https://www.women-on-the-road.com/best-travel-forums.html

WordPress. (n.d.). https://es.wordpress.org/themes/the-travel-booking/

Xiang, Z. (2018). From digitization to the age of acceleration: On information technology and tourism. *Tourism Management Perspectives*, *25*, 147–150. doi:10.1016/j.tmp.2017.11.023

Xiang, Z., Wöber, K., & Fesenmaier, D. R. (2008). Representation of the online tourism domain in search engines. *Journal of Travel Research*, *47*(2), 137–150. doi:10.1177/0047287508321193

Xie, C., Zhang, J., & Huang, S. (2023). Effect of risk message framing on tourists' travel intention: Roles of resilience and impulsivity. *Journal of Travel Research*, *62*(4), 802–819. doi:10.1177/00472875221095212

Yang, T. T., Ruan, W. Q., Li, Y. Q., & Zhang, S. N. (2023). Virtual tourist motivation: The differences between virtual tourism and on-site tourism. *Tourism Review*, *78*(5), 1280–1297. doi:10.1108/TR-07-2022-0331

Chapter 4
Digital Technologies for Sustainable Tourism:
The Case of Ecobnb

Mario Ossorio
University of Campania "Luigi Vanvitelli", Italy

ABSTRACT

This chapter explores the role of digital technologies in favouring the match of the demand and the supply of sustainable tourism activities. The evolution of the concept of tourism and the main traits of sustainable tourism are described. The chapter also sheds light on how digital technologies afford smart travellers to satisfy their needs of tourism activities more respectful of the environment and the local communities. Lastly, the chapter analyses the case of Ecobnb, an Italian digital platform that gathers all the sustainable accommodations in Italy and proposes travel itineraries that respect the environment.

INTRODUCTION

Conciliating economic, social and environment performance is the main challenging goal of all industries around the world. Given their impact, tourism and hospitality entrepreneurs have been facing for last decades new way of arranging their activities in order to reduce their ecological footprint and conversely promote the preservation of the planet and the engagement of local stakeholders.

This chapter has the aim to explore the role of the digital platforms to stimulate a way of arranging tourism activities that, while afford travellers to enhance their experiences during the travel, are respectful of the environment and local community and consequently are consistent with the concept of sustainable tourism.

The chapter is structured in 4 sections. The first section explores how throughout the years the concept of tourism has been turning into sustainable tourism, the resaons behind the evolution, the difficulties in implementing and the comparison with responsible tourism. More specifically, sustainable tourism is characterized by two main goals which are: 1) the need to enhance the quality of people life rather

DOI: 10.4018/979-8-3693-3286-3.ch004

than exclusively the economic develpoment; 2) to plan actions consistently with the fact that natural resources are finite (Ciegis et al., 2009; D'Arco et al., 2021).

The second section illustrates the segments of tourism, such as ecotourism and cultural tourism, that show some similarities with sustainable tourism. In addition, the difficulties of measuring the performance of sustainable tourism and identifying its efficient indicators are discussed.

The third section points out the role of digital technologies in matching the demand and the supply of sustainable tourism activities and the concept of the smart tourism that in the opinion of some scholars conciliate the needs connected to the transition both digital and social and environmental.

The last section describes the digital platform Ecobnb that represents the meeting place between responsible travellers and accommodation facilities that are investing in a better future. Nowadays the platform constitutes the community dedicated to sustainable tourism in Italy and Europe. The success of the platform is linked to the idea of the founders to respond to a fundamental question to find online eco-friendly structures. When they realized that such a search was not easy at all, they also identify the platform that could satisfy the needs of those who want to travel in a green way.

1. SUSTAINABLE TOURISM: ROOTS AND EVOLUTION OF THE INDUSTRY

Over the last eighty years, the tourism industry has increased worldwide, and, particularly, since the 1970s, it has become a relevant economic sector, growing at an exponential pace (Mak, 2008). Because of huge revenues generated, tourism represents a strategic cornerstone industry for an economy (Yan, Shah, Sharma, Chopra, Fareed, & Shahzad, 2022). United Nations World Tourism Organization (UNWTO) estimates that global tourism industry constitutes more than 10% of global gross domestic product (GDP) and generates for one in ten jobs in the world (2018).

However, if, on one hand, tourism industry represents a relevant stimulus for economic development, on the other, it also constitutes a threat in term of water pollution, land use, energy waste, degradation of natural resources and significant change in local customs and traditions (Mao et al., 2014; Jones and Wynn, 2019).

Concern about the relationship between tourism and environment received attention in the early 1970s due to the studies of George Young that pointed out the advantages and the threats of this industry (Young, 1973) and Claude Kaspar who invited a reflection on "*a new dimension of tourism debate*" (Kaspar, 1973, p. 139). Two decades later, Butler called for "*… more responsibility for the effects of travel and behavior on host environments, both physical and human*" (Butler, 1995, p.5). This emphasis on responsibility stimulated several tourism forms as alternative to the mainstream mass tourism (Mihalic, 2006).

At the end of 1980s the literature began to shed light on the relationship between tourism industry and sustainability that represented the roots of "alternative tourism" (Aall, 2014). The concept of sustainability is often defined as "*whatever is done now does not harm future generation*" (Bell and Morse, 2004, p.5) or also as "*long term economic, environmental and community health*" (Vehbi, 2012, p.103).

At the middle of 1990s, UNWTO, the World Travel & Tourism Council (WTTC), and the Earth Council (EC) prepared and published Agenda 21 for tourism. Drawing on this document, Davos Declaration on "Climate Change and Tourism responding to Global Challenges" was proclaimed in 2007. Finally, at Rio + 20 Conference in 2012 the "Future We Want" document incentivized the role of tourism in a context of sustainable development and the fight against poverty (Pan et al., 2018).

Butler (1999) defines sustainable tourism as: "*tourism which is developed and maintained in an area (community, environment) in such a manner and at such a scale that it remains viable over an infinite period and does not degrade or alter the environment (human and physical) in which it exists to such a degree that it prohibits the successful development and well-being of other activities and processes*" (1999, p.29).

Putting together the two concepts, WTO defines sustainable tourism development as a "*development that meets the needs of present tourists and host regions while protecting and enhancing opportunity for the future*" (WTO 2004: 19).

Scholars state that tourism has been a fast adopter of sustainable practices and strategies, probably given its effects, positive and negative, on tourist destinations, by taking into account the resources-consuming consequences (Boulhila et al., 2022) as well as economic leakages and cultural erosion (Gemar et al., 2022).

The concept of sustainable tourism has generated a relevant debate among researchers. More specifically, if, on one hand, it has highlighted the relevance in considering the trade-off between economic and environmental consequences of tourism activities (Baloch et al. 2022), on the other, it shows scarce practical implementation (Tien et al., 2021). In fact, the association of the concept of sustainability to the tourism has stimulated an inspiring debate on the environment and social respectful tourism development models (Mondino & Beery, 2019) with consequent many good practices, such as energy and emissions savings, recycling, reduction of waste and improvement of quality life of local community but with simultaneous difficulties to turn into practical applications (Chettiparamb and Kokkranikal, 2012). The outcome is that the adoption of good practices occurs at a slow pace and a relevant part of tourism industry is "*alarmingly unsustainable*" (Higgins-Desbiolles, 2010, p. 117).

Under these circumstances, academics and policy makers are paying attention to the concept of responsible tourism which "*is about everyone involved taking responsibility for making tourism more sustainable*" (Goodwin, 2011; p. 31), pointing out the capacity to act in favor of making tourism more sustainable (Mihalich, 2016).

Some authors point out that responsible tourism represents the outcome of the negative influence of mass tourism on the host environment and communities (Mondino & Beery, 2019). Put differently, mass tourism in an unaware way has generated some of the problems connected to tourism sustainable development models and forms (Camilleri, 2014). Some authors highlight that customers are the main actors of responsible tourism, when they prefer, for example, small-scale tourism destinations than mass tourism (Legrand et al., 2022). This kind of tourism strategy can more easily be applied at a local community level rather than at national or international level (Yu et al., 2017) and anyway it is not appealing for emerging countries, which consider the relevant profit opportunities connected to mass tourism, while it could be more easily implemented by developed economies.

The growing threat of mass tourism on social and cultural structures as well as on the natural environment of a country led to an increase of attention on responsible tourism as a reaction to the mass tourism (Tribe, 2020).

2. SIMILARITIES AND DIFFERENCES BETWEEN SUSTAINABLE TOURISM AND OTHER SEGMENT OF TOURISM INDUSTRY

However, there are different definitions of tourism activities consistent with sustainable development. Sustainable tourism concerns tourism activities which reconcile environmental, economic and socio-cultural dimensions in order to support its long-term sustainability (UNEP and UNWTO, 2005).

Griggs et al. (2013), incorporating the UN's three pillars of SD paradigms (economic, social and environmental), find that material use, clean air, water cycles are all global must-haves and state the sustainable development can be considered as "*development that meets the needs of the present while safeguarding the Earth's life support system, on which the welfare of current and future generations depends*" (2013, p.306).

The term Ecotourism poses the emphasis on environmental and ecological sustainability and directly involves local communities; in doing this, it promotes employment and economic development in remote locations (Fennell, 2020).

With some analogies with ecotourism, rural tourism sheds light on healthy activities immersed in a context of culture, fresh water and clean air, healthy food (D'Alessandro, 2016). The current form of rural tourism occurs in existing farms and consequently increases farm revenues.

Instead, cultural heritage tourism refers to experience cultures, such as festivals, legends and dances, fixed historical sites (parks and archeological places), and historical monuments (paintings, sculpture and handcraft) that are witnesses of the heritage of a specific geographical area (Light, 2017).

Lastly, community tourism is a term used for describing tourism activities finalized to get better life quality of the local community and simultaneously to offer an impressive experience for tourists (Lai & Hitchcock, 2017).

In an effort to disentangle different definitions of sustainable tourism, Swarbrooke (1999) identifies some of its characteristics that are not possessed by non-sustainable tourism. Low and controlled development models, adequate scale, long-term-horizon, an emphasis on quality and local control are all features of sustainable tourism, in opposite to fast and uncontrolled development, inadequate scale, short term horizon, the emphasis on the quantity and a remote control. With regard to the development strategies, plan before development (vs development without planning), the engagement of local developers and employers (vs outside developers and imported labor) and vernacular architecture (vs non-vernacular one) differentiate sustainable from non-sustainable tourism. Lastly, tourists inclined to experience sustainable tourism have some mental preparation, have propensity to learn local language and are tactful and sensitive. Conversely, tourists not willing to a sustainable tourism have little or no mental preparation, are insensitive and loud and have no interest to learn local language.

Consistently to Swarbrooke (1999) definitions, Fennel and Cooper (2023) identify value systems at the base both Corporatized Tourism and Alternative Tourism: competition, individualism, materialism, self-interest and consumer-choice represent the values characterizing Corporatized Tourism; cooperation, interdependence, meaningfulness/spirituality, share existence, responsibilities, rules and duties constitute values concerning with Alternative Tourism.

Scholars point out that calculating social, competitive and economic performance is fundamental to spur the organizations engaged (e.g.: tourism firms and cooperatives, tour operators, local public authorities, cultural associations, etc.) to rise up their attention in order to achieve sustainability performance in a local community and its tourist destinations (Bandoi et al., 2020). Franzoni (2015) proposes a framework which supports the identification and selection of indicators rooted on three levels (the community, the tourist destination, and the autonomous organizations) and different dimensions (social, competitive, economic). Agyeiwaah et al. (2017), after noticing the great number of indicators that overwhelm tourism industry and lead to inaction, find out the commonalities among them that may represent the base to undertake a more sustainable path (Baggio & Valeri, 2022). The authors propose seven key indicators: with regard to the economic dimension, the indicators are job creation, business viability; quality of life represents the social dimension; water quality and waste management and energy conservation constitute the environmental dimension and lastly, with regard to cultural dimension, the authors propose the maintenance of community integrity.

3. DIGITAL TECHNOLOGIES, SUSTAINABLE AND SMART TOURISM, AND MAIN IMPLICATIONS FOR TOURISM INDUSTRY

Over the last decades, hi-tech industries have been generating a greater computer capacity and innovative technologies. The hospitality industry represents an early adopter of technologies by adopting computer reservation systems in the 70s, distribution systems in 80s and the Internet in 90s (Buhalis and Law, 2008). Information and communication technologies (ICTs) have generated advantages both for travelers and companies. The former may produce and share knowledge, by increasing the opportunity to access information and reducing the relative costs (Mihajlovic, 2012). Companies have benefited due to lower costs, higher revenue, easier marketing research and database development (Morrison et al., 1999). Internet of things (IoT) (e.g.: Nadkarni et al., 2019), virtual reality (VR) (Kim et al., 2020), augmented reality (AR) (Nayyar et al., 2018), big data (Nadkarni et al., 2019), artificial intelligence (AI) and robotics (Kuo et al., 2017) represent the pillars of Industry 4.0 that have changed the dynamics characterizing the hospitality industry. After the pandemic, the increased adoption of industry 4.0 technologies has generated a deep structural change. In a recent literature review, El Archi et al. (2023a e b) emphasize that digital technologies represent a relevant driver of sustainable tourism development, given that they offer new opportunities for enhancing destination management and travelers experiences and stimulating tourists' sustainable behaviors. More specifically, digital technologies have the potential to restructure the tourist industry with regard to communication and engagement with tourists and stakeholders (Hays et al., 2013; Pencarelli, 2020).

In the last decade, many things have been given the label "smart" by indicating things working in the latest technologies: smart phones, card, watches, TV, lenses, economy city, etc. Smartness favors the building products, actions, processes and services in real-time, by involving several stakeholders at the same time to maximize the common performance and produce solutions and value for all (Jasrotia & Gangotia, 2018).

The association of term smart to the tourism entails that destinations ease information sharing amongst the stakeholders, which could improve their decision-making process (Ye et al., 2020). This will in turn facilitate the tourism service providers to offer new improved services to tourists (Shafiee et al., 2021).

Smart tourism destinations require a platform that affords the interconnection between stakeholders which in turn co-create value, pleasure, and experiences for the tourist (Mehraliyev et al., 2019). This entails, first, to engage local community in order to obtain the local stakeholders engagement. In addition, digital centralized platform favors the exchange of information between tourists and tourism organizations that allow the latter to better satisfy customers' needs and expectations. Consequently, smart tourism enhances on one hand tourists experience on the other hand the quality of life of the destination (Buhalis et al., 2023).

Put differently, smart tourism indicates al the technologies derived by the fourth industrial revolution afford to achieve sustainability principles in order to enhance both quality life of people and customer's tourism experience (Pencarelli, 2020). Sustainable mobility, social inclusion, more efficient waste management in tourist destinations are some of the concepts at the base of smart tourism, that pose a particular emphasis on quality of life of people in a tourism destination.

In this scenario, it is crucial for tourism to offer enriching experiences that contribute to well-being and sustainability and tourism providers to carefully consider the impact of implementing new technologies, rather than rushing into it for economic reasons. They should seek evidence on the value of digital technologies and conduct thorough assessments of how technologies may affect tourists. Moreover, the widespread use of digital technologies necessitates involving a variety of players in the tourism industry, not just focusing on tourists themselves. The role of digital technologies extends beyond just using systems during particular service or product delivery times, but it also involves enhancing intricate technology-driven tourist experiences within the broader tourism system.

4. ENHANCING TRAVELERS' EXPERIENCE IN SUSTAINABLE TOURISM DESTINATION: THE ECOBNB PLATFORM

Everything, including travel, impacts the environment in more or less profound ways, and becoming aware of this is the best way to enjoy the wonders of the planet while also taking care of it. Sustainable tourism is a new way of experiencing the world, in a time increasingly attentive to eco-sustainability and eco-compatibility of life.

Ecobnb is an online platform created by four young people from Emilia with a passion for nature. It gathers all the sustainable accommodations in Italy and proposes travel itineraries that respect the environment. Its main aim is to change the way we travel and highlight the accommodation facilities that are invested in a better future. The Ecobnb platform promotes responsible and low-impact tourism, stays in eco-sustainable structures, organic and local food, environmentally friendly itineraries, non-polluting transportation, as well as the rediscovery of nearby and authentic places. Some examples are the treehouse, the organic farm stay, the hotel spread across ancient Italian villages, the zero-emission mountain refuge, the farm, the igloo among the ice.

Through the platform, tourists, both Italian and foreigners, can get to know all those environmentally friendly realities. In many cases, these are small local structures that offer high-quality services. The rule for joining and offering tourists the opportunity to stay in their structure is to adhere to at least 5 out of 10 environmental requirements recognized internationally by those involved in ecotourism: organic food, bioarchitecture, 100% electricity from renewable sources, solar panels for hot water, eco-friendly cleaning products, recycling of over 80%, car-free accessibility, water-saving flow reducers, low-energy light bulbs, and reuse of rainwater. The quality of the proposals is certified by users and Ecobnb's creators who verify that the structures meet the declared characteristics.

Simone recalls that some years ago while he was looking online for a sustainable accommodation to spend a short stay in Potenza for work, he found out that it was difficult to find one on the internet. Some months later, when he projected the website for the Bed & Breakfast Parma association, he indicated including a search criterion connected to the eco-sustainability of the structures, which worked - and is still working - really well, for its efficiency. With this in mind, Simone shared with some friends the idea to put online on a single portal all the eco-sustainable travel structures.

Before founding Ecobnb, there was a project called "ViaggiVerdi," in which public and private funding have believed in it to the point of turning it into what it is today, a partner of large companies working in the green economy sector, as well as solidarity purchasing and responsible tourism associations.

Visiting the Ecobnb website, users can find slow travel itineraries, evaluate alternative ways of accommodation, and discover how to reduce the environmental impact of the holidays. Through Ecobnb, the founders intend to change the way of traveling and to bring out, by putting them online, the possibilities of environmentally friendly tourism, respectful of the economy and local communities. Nowadays, Ecobnb has widened the spectrum of its activities, by becoming a partner of companies operating in the green economy field, associations committed to promoting responsible tourism, Solidarity Purchasing Groups, and the international association dedicated to ecotourism TIES, The International Ecotourism Society. It keeps on pursuing its main goal to stimulate responsible tourism with reduced environmental impact, stays in eco-friendly accommodation, organic food, environmentally friendly itineraries, non-polluting transportation, and the rediscovery of nearby and authentic locations. The same Ecobnb is 0% CO2 emissions and 100% Clean Energy. Its servers are powered 100% by renewable energy sources produced on site (Hetzner online). A sustainable and zero km choice.

CONCLUSION

The tourism industry has seen significant growth over the past eighty years, becoming a major economic sector contributing to global GDP and job creation. However, this growth has also led to negative impacts on the environment, such as water pollution, energy waste, and cultural erosion. Sustainable tourism entails tourism activities which reconcile environmental, economic, and socio-cultural dimensions in order to support its long-term sustainability. Scholars have emphasized other kinds of tourism, such as ecotourism, rural tourism, cultural heritage tourism, and community tourism, which prioritize different aspects of sustainability. Overall, sustainable tourism is characterized by low and controlled development, a long-term focus, an emphasis on quality and local control, integrated planning, local involvement, vernacular architecture, and tourists with a willingness to learn and be respectful. In contrast, non-sustainable tourism is characterized by fast and uncontrolled development, short-term focus, quantity over quality, lack of planning, external developers and labor, non-vernacular architecture, and tourists who are unprepared, insensitive, and uninterested in local culture.

The hospitality industry has been an early adopter of technology, with advancements such as computer reservation systems, distribution systems, and the internet. Technologies like IoT, VR, AR, big data, AI, and robotics have revolutionized the industry, especially after the pandemic. The concept of smart tourism destinations involves leveraging technology to enhance communication, decision-making, and overall tourist experiences. It aims to create value, pleasure, and sustainability for both tourists and local stakeholders. However, it is important for tourism providers to carefully consider the impact of implementing new technologies on people's quality of life and sustainability, rather than solely focusing on economic reasons. Involving various players in the tourism industry and leveraging digital technologies to enhance tourist experiences are key aspects of successful smart tourism initiatives.

An interesting example of a digital platform which stimulates responsible and low-impact tourism is represented by Ecobnb, an online platform founded by four young people from Emilia with a great passion for nature. It assembles information on all the sustainable accommodations in Italy and offers travel itineraries that preserves the environment. Its main scope is to change the travel habits and point out the accommodation facilities that are working for a better future. The Ecobnb platform sponsors stays in eco-sustainable constructions, typical food, zero-emission mountain refuge, environmentally friendly journeys, non-polluting transportation, as well as the rediscovery of local places.

REFERENCES

Aall, C. (2014). Sustainable tourism in practice: Promoting or perverting the quest for a sustainable development? *Sustainability (Basel)*, *6*(5), 2562–2583. doi:10.3390/su6052562

Agyeiwaah, E., McKercher, B., & Suntikul, W. (2017). Identifying core indicators of sustainable tourism: A path forward? *Tourism Management Perspectives*, *24*, 26–33. doi:10.1016/j.tmp.2017.07.005

Baggio, R., & Valeri, M. (2022). Network science and sustainable performance of family businesses in tourism. *Journal of Family Business Management*, *12*(2), 200–213. doi:10.1108/JFBM-06-2020-0048

Baloch, Q. B., Shah, S. N., Iqbal, N., Sheeraz, M., Asadullah, M., Mahar, S., & Khan, A. U. (2023). Impact of tourism development upon environmental sustainability: A suggested framework for sustainable ecotourism. *Environmental Science and Pollution Research International*, *30*(3), 5917–5930. doi:10.1007/s11356-022-22496-w PMID:35984561

Băndoi, A., Jianu, E., Enescu, M., Axinte, G., Tudor, S., & Firoiu, D. (2020). The relationship between development of tourism, quality of life and sustainable performance in EU countries. *Sustainability (Basel)*, *12*(4), 1628. doi:10.3390/su12041628

Bell, S., & Morse, S. (2008). *Sustainability indicators Measuring the immeasurable?* (2nd ed.). Earthscan Publication for Sustainable Future.

Bonink, C. (1992). *Cultural tourism development and government policy* (Unpublished ma dissertation). Rijksuniversiteit, Utrecht, The Netherlands.

Boulhila, S., Alouat, M., Rezzaz, M. A., & Schmitz, S. (2022). Towards a development model of local cultural tourism through the involvement of local actors (Province of Constantine, Algeria). *Geo Journal of Tourism and Geosites*, *40*(1), 9–19. doi:10.30892/gtg.40101-797

Budeanu, A., Miller, G., Moscardo, G., & Ooi, C. S. (2016). Sustainable tourism, progress, challenges and opportunities: An introduction. *Journal of Cleaner Production*, *111*, 285–294. doi:10.1016/j.jclepro.2015.10.027

Buhalis, D., & Amaranggana, A. (2014). Smart tourism destinations. In Z. Xiang & I. Tussyadiah (Eds.), *Information and communication technologies in tourism 2014* (pp. 553–564). Springer.

Buhalis, D., & Law, R. (2008). Progress in information technology and tourism management: 20 years on and 10 years after the Internet—The state of eTourism research. *Tourism Management*, *29*(4), 609–623. doi:10.1016/j.tourman.2008.01.005

Buhalis, D., O'Connor, P., & Leung, R. (2023). Smart hospitality: From smart cities and smart tourism towards agile business ecosystems in networked destinations. *International Journal of Contemporary Hospitality Management*, *35*(1), 369–393. doi:10.1108/IJCHM-04-2022-0497

Butler, R. W. (1993). Tourism – An Evolutionary Perspective. In Tourism and Sustainable Development: Monitoring, Planing, Managing. University of Waterloo Press.

Butler, R. W. (1999). Sustainable tourism: A state-of-the-art review. *Tourism Geographies*, *1*(1), 7–25. doi:10.1080/14616689908721291

Camilleri, M. (2014). Advancing the sustainable tourism agenda through strategic CSR perspectives. *Tourism Planning & Development*, *11*(1), 42–56. doi:10.1080/21568316.2013.839470

Cater, E., & Goodall, B. (1992). Must Tourism Destroy Its Resource Base? In A. M. Mannion & S. R. Bowlby (Eds.), *Environmental Issues in the 1990s* (pp. 309–324). Wiley.

Chettiparamb, A., & Kokkranikal, J. (2012). Responsible tourism and sustainability: The case of Kumarakom in Kerala, India. *Journal of Policy Research in Tourism, Leisure & Events*, *4*(3), 302–326. doi:10.1080/19407963.2012.711088

Chettiparamb, A., & Kokkranikal, J. (2012). Responsible tourism and sustainability: The case of Kumarakom in Kerala, India. *Journal of Policy Research in Tourism, Leisure & Events*, *4*(3), 302–326. doi:10.1080/19407963.2012.711088

Ciegis, R., Ramanauskiene, J., & Martinkus, B. (2009). The concept of sustainable development and its use for sustainability scenarios. *The Engineering Economist*, *62*(2), 28–37.

Cooper, C., Fletcher, J., Gilbert, D., & Wanhill, R. S. (2008). *Tourism principles and practice* (4th ed.). Longman Publishing.

D'Alessandro, F. (2016). Green Building for a Green Tourism. A new model of eco-friendly agritourism. *Agriculture and Agricultural Science Procedia*, *8*, 201–210. doi:10.1016/j.aaspro.2016.02.094

D'Arco, M., Presti, L. L., Marino, V., & Maggiore, G. (2021). Is sustainable tourism a goal that came true? The Italian experience of the Cilento and Vallo di Diano National Park. *Land Use Policy*, *101*, 105198. doi:10.1016/j.landusepol.2020.105198

El Archi, Y., Benbba, B., Kabil, M., & Dávid, L. D. (2023a). Digital Technologies for Sustainable Tourism Destinations: State of the Art and Research Agenda. *Administrative Sciences*, *13*(8), 184. doi:10.3390/admsci13080184

El Archi, Y., Benbba, B., Nizamatdinova, Z., Issakov, Y., Vargáné, G. I., & Dávid, L. D. (2023b). Systematic Literature Review Analysing Smart Tourism Destinations in Context of Sustainable Development: Current Applications and Future Directions. *Sustainability (Basel)*, *15*(6), 5086. doi:10.3390/su15065086

Fennell, D. A. (2020). *Ecotourism*. Routledge. doi:10.4324/9780429346293

Franzoni, S. (2015). Measuring the sustainability performance of the tourism sector. *Tourism Management Perspectives*, *16*, 22–27. doi:10.1016/j.tmp.2015.05.007

Frey, N., & George, R. (2010). Responsible tourism management: The missing link between business owners' attitudes and behaviour in the Cape Town tourism industry. *Tourism Management*, *31*(5), 621–628. doi:10.1016/j.tourman.2009.06.017

Gemar, G., Sánchez-Teba, E. M., & Soler, I. P. (2022). Factors determining cultural city tourists' length of stay. *Cities (London, England)*, *130*, 103938. doi:10.1016/j.cities.2022.103938

Goodwin, H. (2011). Taking responsibility for tourism. Goodfellow Publishers Limited.

Guo, Y., Liu, H., & Chai, Y. (2014). The embedding convergence of smart cities and tourism internet of things in China: An advance perspective. *Advances in Hospitality and Tourism Research*, *2*(1), 54–69.

Hays, S., Page, S. J., & Buhalis, D. (2013). Social Media as a Destination Marketing Tool: Its Use by National Tourism Organisations. *Current Issues in Tourism*, *16*(3), 211–239. doi:10.1080/13683500.2012.662215

Higgins-Desbiolles, F. (2010). The elusiveness of sustainability in tourism: The culture-ideology of consumerism and its implications. *Tourism and Hospitality Research*, *10*(2), 116–129. doi:10.1057/thr.2009.31

Jasrotia, A., & Gangotia, A. (2018). Smart cities to smart tourism destinations: A review paper. *Journal of Tourism Intelligence and Smartness*, *1*(1), 47–56.

Jones, P., & Wynn, M. (2019). The circular economy, natural capital and resilience in tourism and hospitality. *International Journal of Contemporary Hospitality Management*, *31*(6), 2544–2563. doi:10.1108/IJCHM-05-2018-0370

Jones, P., & Wynn, M. G. (2019). The circular economy, natural capital and resilience in tourism and hospitality. *International Journal of Contemporary Hospitality Management*, *31*(6), 2544–2563. doi:10.1108/IJCHM-05-2018-0370

Kaspar, C. (1973). Fremderverkehrsoekologie—eine neue Dimension der Fremdenverkehslehre [Tourism Ecology—A New Dimension of Tourism Science]. *Festscrift zur Vollendung des, 65.*

Kim, M. J., Lee, C. K., & Jung, T. (2020). Exploring consumer behavior in virtual reality tourism using an extended stimulus-organism-response model. *Journal of Travel Research*, *59*(1), 69–89. doi:10.1177/0047287518818915

Kuo, C. M., Chen, L. C., & Tseng, C. Y. (2017). Investigating an innovative service with hospitality robots. *International Journal of Contemporary Hospitality Management*, *29*(5), 1305–1321. doi:10.1108/IJCHM-08-2015-0414

Lai, I. K. W., & Hitchcock, M. (2017). Local reactions to mass tourism and community tourism development in Macau. *Journal of Sustainable Tourism*, *25*(4), 451–470. doi:10.1080/09669582.2016.1221413

Legrand, W., Chen, J. S., & Laeis, G. C. (2022). *Sustainability in the hospitality industry: Principles of sustainable operations*. Routledge.

Light, D. (2017). Progress in dark tourism and thanatourism research: An uneasy relationship with heritage tourism. *Tourism Management*, *61*, 275–301. doi:10.1016/j.tourman.2017.01.011

Mak, J. (2008). *Developing a dream destination: Tourism and tourism policy planning in Hawaii*. University of Hawaii Press. doi:10.21313/hawaii/9780824832438.001.0001

Mao, X., Meng, J., & Wang, Q. (2014). Modeling the effects of tourism and land regulation on land-use change in tourist regions: A case study of the Lijiang River Basin in Guilin, China. *Land Use Policy*, *41*, 368–377. doi:10.1016/j.landusepol.2014.06.018

Mehraliyev, F., Choi, Y., & Köseoglu, M. A. (2019). Progress on smart tourism research. *Journal of Hospitality and Tourism Technology*, *10*(4), 522–538. doi:10.1108/JHTT-08-2018-0076

Mihajlovic, I. (2012). The impact of information and communication technology (ICT) as a key factor of tourism development on the role of Croatian travel agencies. *International Journal of Business and Social Science*, *3*(24).

Mihalič, T. (2006). *Tourism and its environments: ecological, economic and political sustainability issues*. Ekonomska fakulteta.

Mondino, E., & Beery, T. (2019). Ecotourism as a learning tool for sustainable development. The case of Monviso Transboundary Biosphere Reserve, Italy. *Journal of Ecotourism*, *18*(2), 107–121. doi:10.1080/14724049.2018.1462371

Morrison, A. M., Taylor, S., Morrison, A. J., & Morrison, A. D. (1999). Marketing small hotels on the World Wide Web. *Information Technology & Tourism*, *2*(2), 97–113.

Mousavi, S. S., Doratli, N., Mousavi, S. N., & Moradiahari, F. (2016, December). Defining cultural tourism. In *International Conference on Civil, Architecture and Sustainable Development* (Vol. 1, No. 2, pp. 70-75). Academic Press.

Nadkarni, S., & Teare, R. (2019). Reflections on the theme issue outcomes: Expo 2020: what will be the impact on Dubai? *Worldwide Hospitality and Tourism Themes*, *11*(3), 346–350. doi:10.1108/WHATT-03-2019-0013

Nayyar, A., Mahapatra, B., Le, D., & Suseendran, G. (2018). Virtual Reality (VR) & Augmented Reality (AR) technologies for tourism and hospitality industry. *International Journal of Engineering & Technology, 7*(2.21), 156-160.

Niemczyk, A. (2013). Cultural tourists: "An attempt to classify them". *Tourism Management Perspectives*, *5*, 24–30. doi:10.1016/j.tmp.2012.09.006

Pan, S. Y., Gao, M., Kim, H., Shah, K. J., Pei, S. L., & Chiang, P. C. (2018). Advances and challenges in sustainable tourism toward a green economy. *The Science of the Total Environment*, *635*, 452–469. doi:10.1016/j.scitotenv.2018.04.134 PMID:29677671

Pencarelli, T. (2020). The Digital Revolution in the Travel and Tourism Industry. *Information Technology & Tourism*, *22*(3), 455–476. doi:10.1007/s40558-019-00160-3

Shafiee, S., Rajabzadeh Ghatari, A., Hasanzadeh, A., & Jahanyan, S. (2021). Smart tourism destinations: A systematic review. *Tourism Review*, *76*(3), 505–528. doi:10.1108/TR-06-2019-0235

Swarbrooke, J. (1999). *Sustainable Tourism Management*. Cabi. doi:10.1079/9780851993140.0000

Tao, T. C. H., & Wall, G. (2009). Tourism as a sustainable livelihood strategy. *Tourism Management*, *30*(1), 90–98. doi:10.1016/j.tourman.2008.03.009

Tien, N. H., Viet, P. Q., Duc, N. M., & Tam, V. T. (2021). Sustainability of tourism development in Vietnam's coastal provinces. *World Review of Entrepreneurship, Management and Sustainable Development*, *17*(5), 579–598. doi:10.1504/WREMSD.2021.117443

Tribe, J. (2020). *The economics of recreation, leisure and tourism*. Routledge. doi:10.4324/9780429278259

UNWTO. (2017). Discussion Paper on the Occasion of the International Year of Sustainable Tourism for Development 2017. *United Nations World Tourism Organization*.

Vehbi, B. O. (2012). A model for assessing the level of tourism impacts and sustainability of coastal cities. *Strategies for tourism industry-Micro and Macro Perspectives*, 99-114.

World Tourism Organization and Organization of American States. (2018). *Tourism and the Sustainable Development Goals – Good Practices in the Americas*. UNWTO.

Ye, B. H., Ye, H., & Law, R. (2020). Systematic review of smart tourism research. *Sustainability (Basel)*, *12*(8), 3401. doi:10.3390/su12083401

Young, G. (1973). *Tourism: Blessing or blight?* Penguin.

Yu, Y., Li, X., & Jai, T. M. (2017). The impact of green experience on customer satisfaction: Evidence from TripAdvisor. *International Journal of Contemporary Hospitality Management*, *29*(5), 1340–1361. doi:10.1108/IJCHM-07-2015-0371

Chapter 5
Digital Journey Through Fez's Historic Medina:
Embracing the Core of Responsible Tourism

Mohamed Zaifri
https://orcid.org/0000-0003-1780-2803
Laboratory of Informatics, Signals, Automatics, and Cognitivism (LISAC), Faculty of Sciences Dhar El Mahraz, Sidi Mohamed Ben Abdellah University, Morocco

Boukil Mehdi
https://orcid.org/0009-0007-2439-3039
Laboratory of History, Heritage and Territory, Faculty of Letters and Human Sciences Dhar el Mahraz, Sidi Mohamed Ben Abdellah University, Morocco

ABSTRACT

This study aims to explore the role of digital technologies in enhancing sustainable tourism within the historic Medina of Fez, with a focus on preserving its cultural heritage while fostering responsible tourism practices. Through a mixed-methods approach that combines a comprehensive literature review with qualitative analysis—comprising interviews with key tourism stakeholders and analysis of online reviews—this research investigates the adoption and impact of online platforms, interactive digital guides, augmented reality (AR) applications, and artificial intelligence (AI) systems on the tourism experience in Fez. The main findings reveal that digital interventions significantly improve visitor engagement, facilitate appreciation of cultural heritage, and support sustainable tourism practices. Specifically, tools like AR and AI-powered guides offer personalized and immersive experiences that connect visitors more closely with the Medina's rich history, contributing to the conservation of the site.

DOI: 10.4018/979-8-3693-3286-3.ch005

1. INTRODUCTION

Over the last twenty years, a wide range of information and communication technologies (ICT) have been implemented to add value, improve services, and boost the tourist experience at all stages of their travel – before, during, and after their visit. ICT integration plays a crucial role in each phase of a tourist's journey (Knani et al., 2022). The digital transformation of tourism services, framed by the concept of Tourism 4.0, has revolutionized the industry over the past two decades, significantly improving the tourist experience and sustainability, while also having the potential to boost innovation, tourism expansion, and employment opportunities (Kindzule-Millere & Zeverte-Rivza, 2022).

On the other hand, tourism, a key component of Morocco's economic and societal growth, accounted for 7% of its GDP and supported 565,000 jobs in 2019 (MTAESS, n.d.). Founded in the 9th century and flourishing under the Marinids in the 13th and 14th centuries, Fez is considered the cultural and spiritual capital of Morocco. The city's ancient medina, recognized as a UNESCO World Heritage Site in 1981 by decree no. 2-81-25 of October 22, 1981 (UNESCO, n.d.), captivates international tourists with its dense collection of religious, civil and military monuments, including madrasas, fondouks, palaces, mosques and fountains, all set in its narrow streets.

From accommodation booking to destination marketing on social networks, to improving visitor engagement through mobile guiding apps and intelligent tourism recommendation systems, Fez is experiencing a remarkable transition from traditional to digital tools (El-Khadrawy et al., 2020). This digital transformation follows initiatives launched by official authorities to use digitization of different aspects of the tourism experience to enhance its cultural heritage, sustainable practices and ensure long-term growth and preservation.

Nevertheless, tourism in Fez is mainly based on passive, organized trips and focuses on a single product: guided tours, aimed primarily at French and national visitors (Alami et al., 2017). In addition, tourists often rely on printed brochures and maps, or on navigation systems that do not offer tourism-specific guidance. The ancient medina of Fez faces challenges such as overcrowding, deteriorating buildings, harassment by unofficial guides, restricted access to religious monuments, and the socio-economic impact of natural disasters such as COVID-19 and earthquakes.

Studies that explore and analyze the effectiveness of digital interventions in reducing the challenges facing tourism in Morocco and ensuring responsible tourism practices are rare. This highlights a potential knowledge gap in research between tourists and local stakeholders regarding the availability and use of digital tools.

In this article, we aim to study the adoption of tourism 4.0 technologies, including artificial intelligence, extended realities, social media and mobile technologies, to drive the digital transformation of tourism in Fez. Moreover, to better understand how different stakeholders and tourists perceive this digital transformation, we aim to perform a qualitative study by directly interviewing different tourism stakeholders in Fez, followed by an analysis of online reviews from tourists who have already visited the medina.

This article proceeds as follows: Section 2 presents our dual methodology, which includes a qualitative study comprising stakeholder interviews and analysis of online tourist feedback. Section 3 presents the results obtained from the applications studied and the feedback from tourists and stakeholders. In Section 4, we discuss the contributions of these digital platforms to cultural preservation, sustainable development and visitor management, as well as broader implications for future tourism practices and policy development. Finally, section 5concludes with a summary of our key findings.

2. METHODOLOGY

2.1. Interviews With Stakeholders

In our pursuit of qualitative research, we embarked on a comprehensive exploration of digital tourism initiatives within the Fez region. Our methodological approach involved structured interviews and engagements with stakeholders to uncover the landscape of digital innovation in tourism.

Interviews served as a pivotal means of data collection, conducted through dialogues between the researcher and interviewees, aimed at obtaining insights relevant to the analysis and development of the research path and problem. Firstly, we engaged with five artificial intelligence engineers from the Euro-Mediterranean University, key figures behind the development of Historia, an innovative artificial intelligence software enhancing the tourist experience within the historic Medina of Fez.

Subsequently, we expanded our inquiry to include three esteemed researchers specializing in augmented reality and virtual reality technologies. Their contributions provided valuable perspectives on the evolving intersection between digital technologies and tourism experiences. Complementing these academic dialogues, we convened meetings with key stakeholders representing various facets of the tourism ecosystem in Fez, including proprietors of the Ziyarates Fez homestay project, prominent travel agencies, the director of the Agency of Development and Rehabilitation of the Medina and the president of association of tour guides in Fez.

Furthermore, our investigation delved into the digital tourism strategy advocated by the Regional Council of Tourism through a dialogue with the Digital Project Manager. This phase aimed to illuminate the organization's efforts in promoting digital tourism through innovative mobile applications.

Overall, our multifaceted investigation aims to conduct a thematic analysis of the interview content, developing an analytical framework for various ideas and opinions expressed. This framework offers a nuanced understanding, highlighting the digital transformations within the tourism sector in Fez, and emphasizing the symbiotic relationship between technological innovation and experiential enrichment for visitors to the region. Subsequently, we successfully integrated the outputs into the research text to assist in analysis.

It is worth mentioning that this methodology is not new in tourism geography, as previous researchers have also adopted it, such as Asmaa Bouainate, in her doctoral thesis (Bouaouinate, 2008). The author conducted field interviews with stakeholders in the tourism sector in the southern desert oases, emphasizing the scientific importance of such qualitative studies in her academic qualification moreover (Bouaouinate, 2016).

2.2. Analysis of Reviews of International Tourists

We conducted an analysis of comments from international tourists during their visits to Fez, focusing on their feedback shared on platforms such as TripAdvisor or Booking. This sample comprised 230 international tourists of various nationalities, with opinions recorded between March and May 2023, considered the peak tourist season in Fez.

Given the growing importance and widespread use of electronic media among tourists, their remote feedback has become invaluable for studying and evaluating tourist destinations. These comments provide insights into the attractions enjoyed and challenges encountered, aiding in destination evaluation and facilitating numerous academic studies. We conducted a qualitative study by individually examining each comment, categorizing them into themes based on intersecting keywords, and presenting select results using tools like Word and Excel for analysis and visualization.

3. RESULTS

3.1. Overview of Digital Interventions in Fez

This section focuses on the collection of digital interventions that are transforming the tourism experience in Fez, including mobile AR guides, VR tours, AI-powered personal tourism assistance and accommodation booking systems, highlighting their principal purpose and characteristics and how they are being implemented to enhance cultural tourism in Fez.

3.1.1. FEZAR: Mobile Augmented Reality Pocket Guide

Augmented reality (AR) combines various software and hardware technologies to enhance our view of the real world. Combining this technology with tourism has produced impressive results (Zaifri et al., 2018). AR is a system with three fundamental characteristics: the combination of real and virtual aspects, real-time interactivity and support for 3D representation; it is a variation of the virtual environment that aims to enrich real environments rather than replace them entirely, as in virtual reality (VR) (Azuma, 1997).

FEZAR is a location-based mobile augmented reality prototype developed by Zaifri et al. (Zaifri et al., 2023) to support autonomous guidance for tourists visiting the medina of Fez. The prototype leverages AR technology to guide, inform, and amuse tourists while exploring the open museum of the medina of Fez. Major features of FEZAR include:

- Scene Recognition and Tracking: Utilizes the device's camera along with the Wikitude API for real-time detection of surrounding monuments, combining GPS and sensor data for accurate positioning.
- POI Display: Shows distinct virtual markers for each monument type, aligning with the user's location and view direction (refer to Figure 1.a).
- User Interface Customization: Offers a control panel for users to adjust the display radius and select monument types for display (see Figure 1.b).
- Monument Information: Delivers detailed historical narratives and imagery for each monument when a user selects its virtual marker (see Figure 1.c).
- Navigation Radar: Provides a gamified overview of nearby POIs within the user-defined radius.
- Directional Guidance: Assists users in navigating to chosen POIs with a visual directional arrow.
- Site Adaptability: Enables application customization for various historical sites by updating the POI database with new GPS locations and information.

3.1.2. Virtual Reality Tour Systems

The standard definition of Virtual Reality (VR) involves a computer-generated 3D environment that allows users to navigate and interact, simulating one or more of the five senses in real-time. The three primary characteristics of VR include visualization, immersion, and interactivity (Yung & Khoo-Lattimore, 2019).

On the other side, several of Fez's historic and cultural landmarks are off-limits to foreign tourists due to multiple factors: religious restrictions limit access to mosques, historical educational institutions are exclusive to students, and some monuments are currently under rehabilitation.

In this situation, VR offers tourists an excellent opportunity to explore inaccessible monuments from a distance. It can be used for a variety of cultural heritage activities, such as virtual museums, explorations and reconstructions. Tourists can immerse themselves in a virtual

Figure 1. Captured images of the FEZAR application: (a) virtual markers of monuments, (b) control panel to adjust POI distance and type, (c) description of Moulay Abdellah Mosque monument

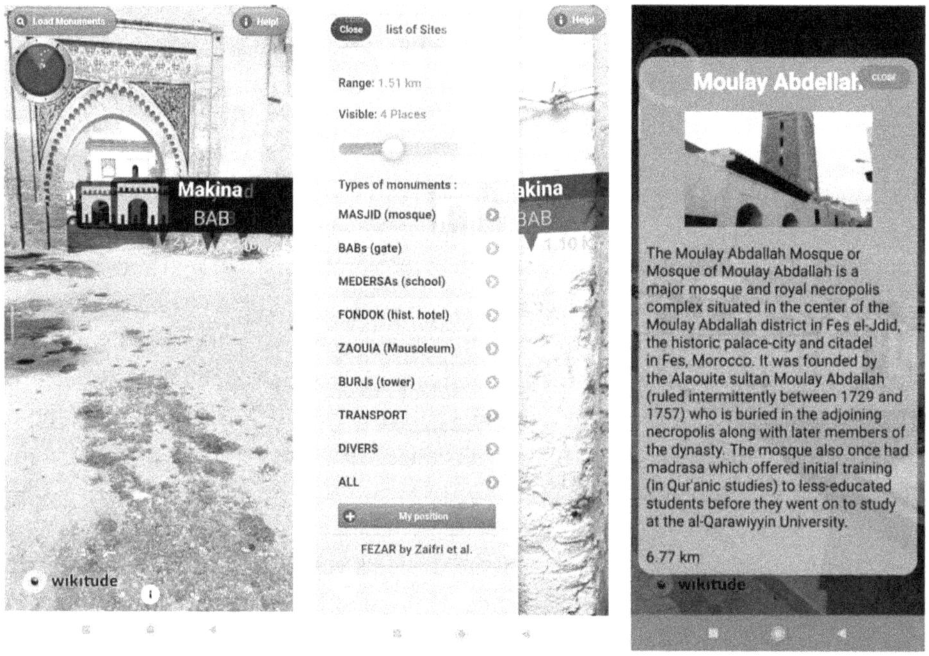

(a) (b) (c)

Maach et al. explored in (Maach et al., 2018) the creation of a 3D VR tour centered on the "Mausoleum of Moulay Idriss", the founder of the city. This tour is accessible as a cross-platform web application compatible with desktops, mobile device, and VR headsets such as Google Cardboard as depicted in Figure 2. The proposed solution enables tourists to:

- Take a virtual reality tour of the "Moulay Idriss mausoleum", which is closed to non-Muslim tourists.
- Navigate inside the monument using realistic images.
- Interact with the mausoleum's various objects in each scene.

Figure 2. Stereoscopic vision of the virtual visit inside the mausoleum of Moulay Idriss
Maach et al. (2018)

Kadri et al. introduced a project called V-Museum (Kadri et al., 2020), a virtual museum application accessible on mobile devices that utilizes AR, VR, and 3D modeling to enhance the cultural tourism experience in Fez. V-Museum introduces users to a 'virtual door' in the real world (Figure 3.a), an AR generated gateway that leads to a comprehensive virtual museum. This space showcases valuable 3D artifacts and historical elements unique to Fez (Figure 3.b). V-Museum allows tourists to:

- Navigate smoothly in a virtual museum and interact with historic artifacts of Fez.
- Explore Moroccan culture in the museum's design…

3.1.3. Histouria: AI-Powered Personal Tour Guide

Given the growing importance of AI for practitioners and academics, research into its applications in the tourism and hospitality sector has grown considerably, to help tourists by providing more relevant information, improving their decision-making process and offering personalized travel experiences (Knani et al., 2022).

Histouria (*Histouria*, n.d.) is a travel planning platform designed to enhance tourism experiences through personalized travel plans. It leverages cutting-edge technologies such as AI and recommendation algorithms to deliver a customized exploration experience. By combining the user's location, preferences and a thorough historical and cultural database, Histouria creates personalized travel itineraries that cater to individual interests.

This project was led by a team of students in the field of artificial intelligence engineer at Euro Med University in Fez. They were among the winners of a Hackathon held during the closing session of the 10th edition of Mediterranean Tourism Forum (https://ccisfm.ma/meditour2023/), on October 13th and 14th, 2023. The goal was to foster innovation in Mediterranean tourism.

The Histouria platform functions by first identifying a user's location through manual input or GPS tracking. It then gathers data on the user's travel preferences and interests, such as activity types, travel style, cultural interests, and dietary needs. Finally, Histouria utilizes an extensive database containing information on historical sites, cultural events, dining, lodging, transportation, and other tourist attractions, adapting suggestions to the user's specific location and interests. Key features include:

Figure 3. V-Museum application. (a) AR port entrance, (b) 3D and 2D virtual artifacts inside the virtual museum
Source: Kadri et al. (2020)

- AI StoryTeller: an AI-powered chatbot that suggests places to visit and activities to enjoy, adapting to the user's preferences. Users can also express themselves vocally or in writing, and explore the past through rich stories and images about historical sites. AI StoryTeller acts as a virtual tour guide, enabling users to interactively explore historical sites, whether they are on-site or remotely. Consequently, this story teller serves as a valuable educational tool, providing users with an interactive method to learn about history. The platform facilitates deep exploration and understanding of historical topics.
- Personalized trips: as shown in Figure 4.a, based on the analysis of user's preferences and location using machine learning techniques, Histouria generates a personalized travel itinerary tailored to the user's specific location, preferences, and constraints. The itinerary may include recommendations for activities, attractions, dining options, accommodation, transportation routes, and other relevant information.
- Real-time guidance and information: Histouria offer real-time recommendations and updates to users during their travels, including notifications about changes in operating hours, special events, weather, traffic, and other pertinent information to enhance their journey and ensure they don't miss important sites and opportunities, as illustrated in Figure 4.b.

3.1.4. Visit Fez-Meknes Mobile Application

This application serves as a key promotional instrument for the Regional Tourism Council (CRT) responsible for promoting tourism in Fez-Meknes Region (see Figure 5). The application is accessible in four languages: French, English, Spanish, and German. It offers travelers a tailored experience of the region, aiming to simplify visitor planning and navigation with innovative features such us:

- Offline Mode: One remarkable aspect of this application is its offline capability, enabling users to search for attractions and map out routes without needing an internet connection. This functionality caters to the common challenges travelers encounter regarding online accessibility, ensuring exploration even in areas with limited connectivity.
- Places to Explore: visit Fez Meknes presents a comprehensive directory of landmarks, historical sites, and dining establishments within the area. Featuring a user-friendly interface, individuals can designate their preferred region and access detailed maps of attractions, thereby streamlining their exploration.
- Itinerary Organization: the itinerary planner serves as a valuable resource for travelers seeking to maximize their visit. By specifying their preferences and travel
- dates, users can devise personalized routes, simplifying the logistical aspects of their journey.
- Orientation: the application introduces an innovative orientation functionality, enabling users to pinpoint nearby attractions through their device's camera.

Figure 4. Historia application. (a) Personalized itinerary recommendation. (b) Real-time information powered by IA.
Source: https://historia-landing.vercel.app/

(a) (b)

Additionally, it is worth noting that the Regional Tourism Council also has a website (https://www.visitfesmeknes.ma/). This site offers also the possibility to plan trips in the region based on the number of days to spend by presenting the various regional attractions to discover.

3.1.5. Empowering Fez's Online Reputation: Utilizing Electronic Reviews and Collaborative Promotion Strategies

The web has become an enormous deposit of data, to which online users add new information every day. A part of that information is represented by online reviews (Dharel, 2020). People now read these reviews and are influenced by them in the process of acquiring a product or service (Bucur, 2015). It is then important to improve the online reputation of the destination of Fez by closely monitoring content and reviews related to the destination on booking sites, social media and influencer blogs, while collaborating with them to promote a positive image of the destination by creating visually appealing content, such as photos and videos, which they will then share on their social media platforms and blogs. They can also write detailed blog posts, travel guides, and other content that provides useful information and tips to travelers interested in Fes.

Figure 5. Screen capture of the user interface of the "Visit Fez-Meknes" mobile application

We notice also that accommodations, restaurants, tour guides and the different stake holders in tourism of Fez should take care the online platform that allows users to share their exotic experiences and exchange opinions.

For example, in Figure 6, Booking.com uses six criteria to evaluate accommodation establishments in Fez: cleanliness, comfort, location, staff, facilities, and value for money. Booking.com uses these criteria to calculate an overall rating for the establishment, displayed on the booking page to help travelers choose the best option for their stay in Fez.

Figure 6. Criteria adopted by Booking.com to evaluate accommodation services
Source: Booking.com

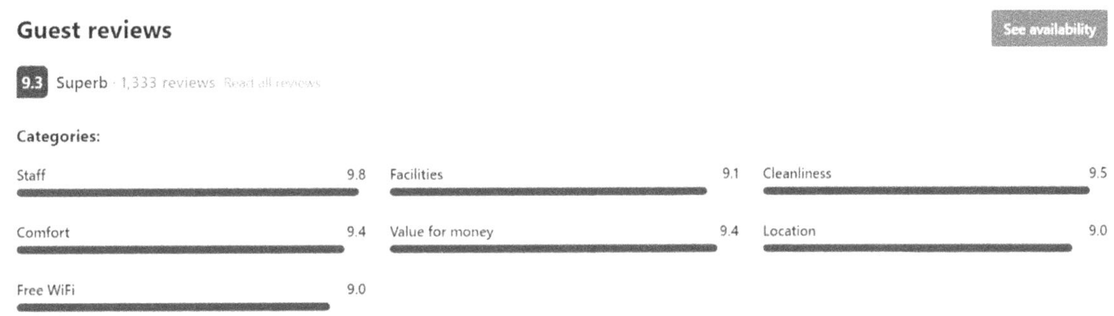

Figure 7. Architectural refinement of a Riad guesthouse in the Medina of Fez (a) Room (b) W.C (c) central patio adorned with mosaics, fountains, and plants

One particular type of accommodation is Riads, which benefit from a significant campaign of restoration to become beautiful guesthouses nestled in the heart of the Medina. With their ancient charm and enchanting atmosphere, they captivate visitors who express their satisfaction through enthusiastic reviews. Thanks to their peaceful ambiance and attentive service, these havens of tranquility allure travelers in search of authenticity and comfort (Figure 7).

We undertook the process of analyzing a sample of 73 comments to evaluate the impressions of tourists who have stayed in Riads in Fez Medina. It is under these circumstances that we calculated the frequency of repetition of keywords that have been synthesized (Boukil, 2023).

3.1.6. Ziyarates Fez Homestay Accommodation Project

Ziyarates Fez exemplifies responsible tourism practices by offering a unique lodging experience within the heart of the Medina of Fez, where guests are welcomed into a family home. These families provide one or multiple chambers within their dwellings, ensuring personalized comfort for guests to fully immerse themselves in the local community and engage with residents, fostering a culture of cross-cultural interaction.

This initiative serves a dual purpose: firstly, it supports underprivileged populations by encouraging them to remain in the Medina, thus preserving their family homes; secondly, it offers an innovative tourism program focused on culture, spirituality, and solidarity, thereby encouraging visitors to prolong their stay (Alami et al., 2017).

The project has been strengthened through training provided to families on tourist reception techniques and with the establishment of an association to efficiently manage reservations across various projects. Reservations can now be made through a dedicated platform (www.ziyaratesfes.com), which showcases the different guest houses included in the program, complete with photos and prices, enabling clients to make informed decisions. Clients also have the opportunity to leave reviews as shown in Figure 8, enhancing transparency and trust.

This form of accommodation provides tourists with the chance to partake in the daily lives of local residents, offering a unique experience where they can explore the tangible and intangible cultural heritage of the old Medina of Fez. By immersing themselves in the culture and lifestyle of the locals, including cuisine, shopping, and traditional customs, tourists gain a deeper understanding of the destination (Fauveaud et al., 2018).

Various themes and keywords, such as spiritual tourism, tangible and intangible heritage, culture, human development, intercultural dialogue, and solidarity tourism, are utilized to promote, describe, and market the Ziyarates Fez network (Carboni & Idrissi Janati, 2016). According to the official website, the current project comprises 34 riads, 95 rooms, 11 restaurants, 71 staff members, 15 suites, and 9 terraces.

3.1.7. Tourist Information System Enriched With Audio Guide

Furthermore, the different roles played by The Agency of Development and Rehabilitation (ADER) on preserving the cultural heritage of the Medina in its various aspects. It has also contributed to develop an electronic tourist information system, utilizing an audio guide for 412 historical monuments in four languages: French, English, Spanish, and Arabic. This system -illustrated in Figure 9- is designed to operate automatically when tourists approach within 10 meters of the monument. Alternatively, tourists can manually access the information by scanning a QR code assigned to each monument.

Figure 8. Example of a Ziyarates Fez Homestay
Source: https://www.ziyaratesfes.com/riad-a-fes-34.html

Figure 9. Simulation of a guided tour with audio assistance (ADER Fes)

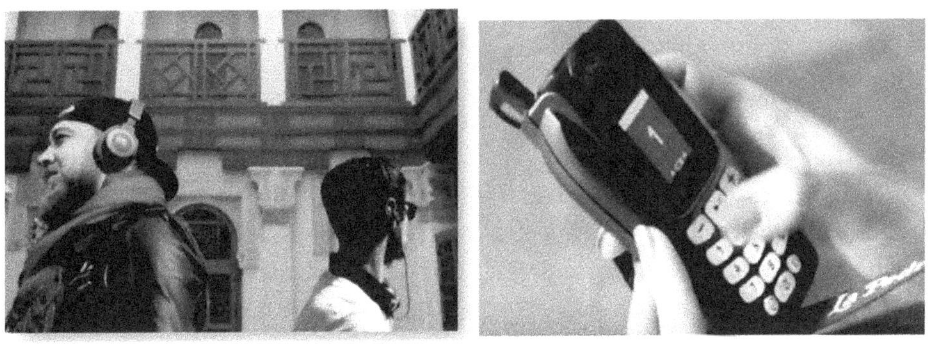

Depending on an interview we conducted with the director of the agency (S. Fouad, personal communication, September 21, 2020) the management of this tool will be entrusted to an operating company following a call for bids with the collaboration of the Ministry of Tourism. The intention is to distribute this system to tourists at the tourist reception and information centers, which are set to be established in the 8 parking areas undergoing renovation in the medina.

3.1.8. Role of Tour Guides in Enhancing the Digital Experience

With the increasing prevalence of digital tools and audio guides, however, we cannot overlook the services provided by tour guides, who play a crucial role in creating an authentic and enriching tourist experience. Their local expertise, profound knowledge of the history, cultural heritage, and traditions of the Medina enable them to contextualize the information provided on online websites and digital platforms, tailoring it to the needs and interests of visitors.

Referring to a study we managed on a sample of 150 reviews from international tourists visiting the Medina of Fez and leaving their comments on the Trip Advisor website, we revealed that 39% reported a high risk of getting lost in the labyrinthine streets and alleys of the old city. It is for this reason that they preferred to use the services of an official tour guide. The Medina is a kind of maze that can derail those who discover it for the first time, in the absence of a knowledgeable guide.

Furthermore, when we studied the reviews upon tour guides in particular, we reveal a majority of satisfied tourists who reported that their respective guides generally met or exceeded their expectations. For example, one tourist expressed, "I couldn't have asked for a better day exploring the Medina of Fez with Rachid." Tour guides were considered essential for understanding the intricate history, secrets, and hidden gems of Fez's ancient Medina, described as a labyrinth that could confuse first-time visitors without knowledgeable guidance. Notably, various guides were praised for their friendliness, professionalism, punctuality, and multilingual abilities (including French, English, and German), which further enriched tourists' experiences and appreciation for the cultural and historical insights provided. However, some comments also commented on potential overpricing and deception, with concerns raised about being led to expensive markets.

In an interview with an experienced tour guide (B. Samir, personal communication, March 29, 2024), it was emphasized that tour guides also serve as cultural mediators, facilitating interactions between tourists and local residents. Through their ability to establish intercultural dialogue, they contribute to creating authentic and meaningful experiences for visitors, allowing them to understand and appreciate the Medina beyond its tourist surface.

Moreover, tour guides are capable of adjusting itineraries and activities based on local conditions and tourists' preferences, providing a customized experience that meets their expectations and enhances their stay in the Medina.

Another function assigned to guides is to reduce the problem of unwanted unauthorized guides. A significant proportion of electronic reviews (34%) on TripAdvisor highlight the problem of fake guides who disrupt tourists in the streets and paths of the old city. Tourists complain about their attempts at forced orientation and their sometimes-aggressive behavior in case of refusal. Tourists also report unpleasant interactions with street vendors and beggars, who can be intrusive and insist heavily on selling their products as well as beggars asking for money. Therefore, potential tourists, when they read these negative comments, may raise concerns and an excess of caution that can create a psychological barrier, thereby reducing the positive impacts of responsible tourism, especially regarding cultural dialogue with the local population as well as economic repercussions on local commerce and tourism in general. Our study of comments from international tourists also indicates that, 22% of them complain about high prices, saying that many vendors pressure customers to buy expensive products, sensing that some may try to deceive or defraud them, advising other tourists to bargain and negotiate for everything they wish to purchase (driven by their growing cultural awareness and their sense of solidarity with other electronic tourists).

4. DISCUSSION

Table I demonstrates how the digital platforms discussed in this paper, from AR and VR systems to AI-powered guide and online reputation management tool, address the challenges facing the tourism sector in Morocco, focusing on cultural preservation, sustainable development, and managing visitors. Table I highlights also a significant transformation in Fez's tourism offer, pointing to a new digital era characterized by sustainability, immersive experiences and inclusivity in tourism.

The proposed digital interventions can be expected to have an impact on future tourism practices, stakeholder engagement and academic research. Technologies such as FEZAR and Historia personalize the tourism experience with direct information and itinerary recommendations, which can increase visitor satisfaction and encourage them to stay longer, as tourists seek authentic experiences using the latest technologies. On the other hand, virtual reality tours democratize access to inaccessible heritage sites for people with mobility problems or those unable to travel, so they can experience virtual tourism from the comfort of their home (Khalloufi et al., 2024).

The success of platforms such as Ziyarates Fez Homestay and the audio guide system indicates a rising need for active involvement from multiple stakeholders, including local communities, tourism authorities, technology companies and policymakers (El Archi et al., 2023). Local communities, in particular, could benefit from increased engagement and economic opportunities offered by official digital platforms such as the "visit Fez-Meknes" app.

Table 1. The role of digital platforms in improving cultural preservation, sustainable development, and visitor management in Moroccan tourism

Technology	Cultural Preservation	Sustainable Development	Visitor Management
FEZAR: Mobile augmented reality pocket guide	• Enhances site appreciation with virtual elements. • Provides detailed stories and images for cultural education.	• Reduces site alterations. • Reduces the need for brochures by digitizing information delivery.	• Offers personalized visitor experiences, improving traffic distribution.
Virtual reality tour systems	• Enables virtual reconstructions. • Virtual explorations of inaccessible sites.	• Facilitates cultural engagement without physical presence, minimizing environmental impact of travel.	• Provides off-site tourism option • Prevents site degradation.
Historia: AI powered personal tour guide	• Provides interactive learning and exploration of historical sites through AI StoryTeller.	• Uses AI for travel planning, minimizing unnecessary travel.	• Redirects tourists from overcrowded areas, controlling visitor flow in real time.
Visit Fez-Meknes App	• Offers a comprehensive guide to landmarks and historical sites with detailed maps and information.	• Enables exploration in offline mode, reducing the need for physical guides or maps.	• Uses innovative wayfinding functionality to help tourists discover nearby attractions.
Electronic reviews and collaborative promotion strategies	• Increase the destination's reputation with relevant content and positive reviews.	• Highlights responsible stakeholder efforts. • Drives awareness of restoration campaigns.	• Orients tourists to highly-rated services and accommodation.
Ziyarates fez	• Preserves local traditions by integrating tourists into the heart of Medina life.	• Boosts local economy. • Supports underprivileged families.	• Offers a dedicated local booking platform.
Audio guides	• Improves access to historical knowledge in several languages	• Minimizes physical presence on site through digital engagement. • Reduces dependence on paper brochures	• Promotes guided tours through automatic information delivery

The digital interventions listed in Table 1 represent a rich field for scientific research, in particular to assess how digital tools influence tourist behavior, heritage preservation and local community economies. There is a strong need for empirical research to assess the effectiveness of these digital solutions in enhancing tourism experiences and advancing sustainable tourism initiatives. A relevant case study in the context of this stage is the study by (Zaifri et al., 2023), which explored the effects of the AR interface "FEZAR" on visitor experiences in the medina of Fez, comparing it with traditional guidance methods. The results indicate that FEZAR improves user engagement and makes the visit both more educative and more enjoyable.

During the interviews with the President of the Association of Guesthouses and the President of the Ziyarates Fes project, both expressed enthusiasm for the concept of responsible tourism, acknowledging its various facets and complexities. They highlighted the importance of cultural dialogue within the unique setting of the Medina, emphasizing the need to preserve and promote Fez's rich heritage while accommodating modern travel trends. Additionally, they emphasized the role of digital empowerment in enhancing the visibility and accessibility of guesthouses, enabling them to reach a wider audience and compete in the global market. Both presidents emphasized the importance of responsible tourism practices in sustaining the cultural and economic vitality of Fez, while also recognizing the transformative potential of digital technologies in amplifying the impact of these initiatives.

From an environmental perspective, the rehabilitation of Riads in the Medina and homestays involve the use of local and sustainable materials. By valuing existing built heritage and avoiding the construction of new infrastructures, these establishments also contribute to preserving urban landscapes and alleviating pressure on natural resources. Socially, the rehabilitation of Riads and guesthouses provides economic opportunities for local communities by creating jobs in the tourism sector. Additionally, these establishments often have a positive impact on surrounding neighborhoods by revitalizing historical areas and stimulating local economic activities such as handicrafts and the sale of local products. Furthermore, rehabilitated Riads and guesthouses play a crucial role in preserving local culture and tradition. By offering visitors the opportunity to stay in authentic accommodations steeped in history, these establishments raise awareness among travelers about Fez's cultural heritage and promote respect for local traditions. Thus, by highlighting rehabilitated Riads and guesthouses on digital platforms and encouraging positive tourist reviews, the tourism industry can strengthen its commitment to responsible tourism and significantly contribute to the sustainable development of Fez and its region.

In the context of responsible tourism practices in the Medina of Fez, also the issue of price gouging in bazaars presents a significant challenge. Despite efforts to empower tourists digitally and promote ethical consumption, the motivation for some vendors to maximize profits through inflated prices persists. For instance, a scenario where a shop owner offers a product at a vastly inflated price exemplifies this dilemma. Such practices not only deter tourists from making repeat purchases but also hinder the city's overall benefit from tourism activity. By addressing these challenges, stakeholders can better align digital empowerment initiatives with responsible tourism practices, ensuring a more equitable and sustainable tourism experience in Fez's Medina.

It becomes therefore imperative for local authorities and tourism stakeholders to take measures to mitigate these problems. For example, strengthening police controls in tourist areas to deter harassers and protect visitors, especially during periods of high tourist influx. It is also important to encourage tourists to report cases of harassment or aggression. Emergency numbers and reception points can be set up to facilitate reporting and assistance for victims, while using video surveillance and new technologies to detect cases of harassment or aggression.

5. CONCLUSION

In this study, we have examined the transformative impact of digital empowerment on the promotion of responsible tourism in the historic medina of Fez, drawing on a holistic methodology that includes a comprehensive literature review and qualitative analysis through stakeholder interviews and online tourist reviews. Highlighting the integration of advanced digital technologies ranging from AI and AR to social media and mobile apps, our research identifies a strategic shift to enhance economic growth and the cultural tourism experience in Fez, while promoting sustainable tourism.

Our dual methodological approach enabled us to study perceptions of digital transformation among stakeholders and tourists. Interviews with key stakeholders, including AI engineers, AR/VR specialists and representatives of the tourism sector in Fez, as well as analysis of feedback from 230 international tourists, revealed widespread approval of digital tools to improve visitor engagement, simplify navigation and offer immersive cultural experiences. These results not only validate the effectiveness of digital interventions in enhancing the tourism experience, but also point to the collective aspiration for sustainable and responsible tourism practices.

Feedback from stakeholders and tourists highlights the crucial role of digital technologies in redefining the tourism landscape of the Medina of Fez. This suggests a growing need for continued investment in digital tools that meet the changing expectations of modern travelers while preserving the authenticity and integrity of the cultural heritage. The lessons learned from our research are invaluable for other historic destinations wishing to leverage digital technologies for responsible tourism.

However, the transition to a digital tourism model in Fez has not been without its difficulties. Our study recognizes the obstacles posed by infrastructure weaknesses, gaps in digital literacy and resistance to change on the part of some tourism stakeholders. Overcoming these challenges requires concerted efforts in developing digital infrastructure, training stakeholders and promoting an ecosystem of collaboration and innovation.

This study, while highlighting the transformative potential of digital interventions in the Medina of Fez, encounters certain limitations, including the focus on specific digital tools, the use of qualitative perspectives rather than quantitative data, and the concentration on a singular historic site. Future research could include exploring a wide range of emerging technologies, incorporating quantitative methodologies, involving a wider range of stakeholders, and broadening the geographic scope to include other historic sites for comparative information. In addition, longitudinal studies could provide valuable insights into the evolving impact of digital technologies over time, enriching the discourse on sustainable tourism and cultural preservation.

REFERENCES

Alami, Y. H., Khazzan, B. E., & Souab, M. (2017). Heritage and cultural tourism in Fes (Morocco). *International Journal of Scientific Management and Tourism*, *3*(2). https://ojs.scientificmanagementjournal.com/ojs/index.php/smj/article/view/293

Azuma, R. T. (1997). A Survey of Augmented Reality. *Presence (Cambridge, Mass.)*, *6*(4), 355–385. doi:10.1162/pres.1997.6.4.355

Bekele, M. K., Pierdicca, R., Frontoni, E., Malinverni, E. S., & Gain, J. (2018). A Survey of Augmented, Virtual, and Mixed Reality for Cultural Heritage. *Journal on Computing and Cultural Heritage*, *11*(2), 1–36. doi:10.1145/3145534

Bouaouinate, A. (2008). *Les acteurs locaux du tourisme de désert au Maroc: Cas de l'erg Chebbi et de Zagora-M'hamid.* https://www.academia.edu/89480928/Les_acteurs_locaux_du_tourisme_de_d%C3%A9sert_au_Maroc_Cas_de_lerg_Chebbi_et_de_Zagora_Mhamid

Bouaouinate, A. (2016). *La mise en tourisme des espaces oasiens du Maroc d'un tourisme de masse à un tourisme alternatif Rapport de Synthèse des Travaux scientifiques et pédagogiques* [Thesis, Université Hassan II Casablanca; Faculté des Lettres et des Sciences Humaines Mohammedia]. https://theses.hal.science/tel-01353766

Boukil, M. (2023). La mise en tourisme de la médina de Fès: le cas de la réhabilitation et transformation des Riads en maisons d'hôtes. *Journal of urban and territorial planning* (N°18, p149-164), Democratic Arabic Center http://democraticac.de/?page_id=61723

Boukil, M., Aarab.H, Serghini. B., Chaouki, M. (2021) Le tourisme international à Fès est-il vers un nouvel essor ? Revue territoire et développement numéro 9-10, Faculté des Lettres et des sciences Humaines Dhar el Mahraz Fès.

Bucur, C. (2015). Using Opinion Mining Techniques in Tourism. *Procedia Economics and Finance*, *23*, 1666–1673. doi:10.1016/S2212-5671(15)00471-2

Dharel, Y. (2020). Online reviews and their impact in tourism business. Centria University of Applied Sciences.

El Archi, Y., Benbba, B., Kabil, M., & Dávid, L. D. (2023). Digital Technologies for Sustainable Tourism Destinations: State of the Art and Research Agenda. *Administrative Sciences*, *13*(8), 184. doi:10.3390/admsci13080184

El-Khadrawy, R. K., Attia, A. A., Khalifa, M., & Rashed, R. (2020). Sustainable Tourism and Culture: A Symbiotic Relationship. *International Journal of Environmental Science & Sustainable Development*, *5*(1), 54–67. doi:10.21625/essd.v5i1.717

Histouria. (n.d.). *Unlocking History, One Chat at a Time.* Retrieved March 15, 2024, from https://historia-landing.vercel.app/

Kadri, M., Khalloufi, H., & Azough, A. (2020). V-Museum: A Virtual Museum Based on Augmented and Virtual Realities for Cultural Heritage Mediation. *2020 International Conference on Intelligent Systems and Computer Vision (ISCV)*, 1–5. 10.1109/ISCV49265.2020.9204253

Khalloufi, H., Zaifri, M., Kadri, M., Benlahbib, A., Kaghat, F. Z., & Azough, A. (2024). El-FnaVR: An Immersive Virtual Reality Representation of Jemaa El-Fna in Marrakech for Intangible Cultural Heritage Experiences. *IEEE Access : Practical Innovations, Open Solutions*, *12*, 9331–9349. doi:10.1109/ACCESS.2023.3347195

Kindzule-Millere, I., & Zeverte-Rivza, S. (2022). Digital transformation in tourism: Opportunities and challenges. *Economic Science for Rural Development Conference Proceedings, 56*, 476–486. https://doi.org/10.22616/ESRD.2022.56.047

Knani, M., Echchakoui, S., & Ladhari, R. (2022). Artificial intelligence in tourism and hospitality: Bibliometric analysis and research agenda. *International Journal of Hospitality Management*, *107*, 103317. doi:10.1016/j.ijhm.2022.103317

Maach, I., Azough, A., & Meknassi, M. (2018). Development of a use case for virtual reality to visit a historical monument. *2018 International Conference on Intelligent Systems and Computer Vision (ISCV)*, 1–4. 10.1109/ISACV.2018.8354052

MTAESS. (n.d.). *Tourisme – Ministère du Tourisme, de l'Artisanat et de l'Economie Sociale et Solidaire*. Retrieved March 31, 2024, from https://mtaess.gov.ma/fr/tourisme/

UNESCO. (n.d.). *Medina of Fez*. UNESCO World Heritage Centre. Retrieved March 31, 2024, from https://whc.unesco.org/en/list/170/

Yung, R., & Khoo-Lattimore, C. (2019). New realities: A systematic literature review on virtual reality and augmented reality in tourism research. *Current Issues in Tourism*, *22*(17), 2056–2081. doi:10.1080/13683500.2017.1417359

Zaifri, M., Azough, A., & El Alaoui, S. O. (2018). Experimentation of visual augmented reality for visiting the historical monuments of the medina of Fez. *2018 International Conference on Intelligent Systems and Computer Vision (ISCV)*, 1–4. 10.1109/ISACV.2018.8354051

Zaifri, M., Khalloufi, H., Kaghat, F. Z., Benlahbib, A., Azough, A., & Zidani, K. A. (2023). Enhancing Tourist Experiences in Crowded Destinations through Mobile Augmented Reality: A Comparative Field Study. *International Journal of Interactive Mobile Technologies*, *17*(20), 20. Advance online publication. doi:10.3991/ijim.v17i20.42273

Chapter 6
Breaking Chains, Building Futures:
Blockchain's Impact on Sustainable Tourism

Bassam Samir Al-Romeedy
Faculty of Tourism and Hotels, University of Sadat City, Egypt

ABSTRACT

This chapter discusses the potential impact of blockchain technology on sustainable tourism, highlighting the increasing importance of sustainability in the tourism industry. Blockchain's decentralized and transparent nature offers transformative solutions to address environmental, social, and economic challenges. The chapter explores how blockchain can contribute to sustainable practices, including resource management, community empowerment, and responsible supply chains. It emphasizes the role of blockchain in enhancing sustainability through transparency, traceability, and accountability. Furthermore, blockchain-based systems streamline the measurement and verification of sustainability indicators, promoting responsible practices and providing reliable information to travelers. Overall, the chapter provides valuable insights on utilizing blockchain to advance sustainability in tourism destinations, attracting environmentally conscious travelers, and fostering sustained socio-economic development.

INTRODUCTION

Blockchain, characterized by its decentralized and unalterable nature, facilitates secure and transparent transactions as well as the exchange of information among multiple parties. In the tourism industry, blockchain has attracted considerable attention due to its potential to transform various aspects of the sector. One key contribution of blockchain is its ability to elevate transparency in tourism. With its decentralized and tamper-proof ledger, blockchain allows the recording of all transactions and data related to the tourism ecosystem. This heightened transparency builds trust among stakeholders, including tourists, travel agencies, accommodations, and local communities. Consequently, tourists can have

DOI: 10.4018/979-8-3693-3286-3.ch006

increased confidence in the authenticity of services and products, while businesses can ensure equitable and transparent transactions (Abdellah, 2023).

The immutable nature of blockchain ensures that recorded information remains unaltered or erased without consensus from network participants. This quality plays a crucial role in combating fraud and enhancing security within the tourism industry. It effectively addresses issues like fraudulent bookings, identity theft, and financial fraud, thereby protecting the interests of both tourists and businesses. Furthermore, blockchain supports the efficient management of digital identities in the tourism sector. Through blockchain-based identity systems, travelers can securely store and handle their personal information and travel credentials. Additionally, blockchain facilitates the automation and execution of smart contracts within the tourism industry. Smart contracts are self-executing agreements that automatically enforce predetermined terms and conditions (Abdel Maksoud, 2023).

The role of blockchain in the tourism industry extends to promoting sustainable practices (Sarfraz et al., 2023). Blockchain technology allows for the creation of transparent supply chains, enabling tourists to trace the origin and sustainability of products and services. This advancement encourages responsible tourism by empowering travelers to make well-informed choices in line with their values, such as supporting local communities and endorsing eco-friendly initiatives (Baralla et al., 2021; Dutta et al., 2020). The significance of blockchain in the tourism sector lies in its ability to enhance transparency, security, efficiency, and sustainability. By leveraging blockchain technology, the tourism industry can overcome traditional challenges, build trust, and establish a travel experience that is both seamless and responsible for both tourists and industry stakeholders (Sarfraz et al., 2023; Erol et al., 2022).

Blockchain technology has the potential to greatly enhance sustainable tourism by introducing enhanced transparency, accountability, and efficiency across various aspects of the industry. Through securely storing transaction records on a decentralized ledger, blockchain ensures that sustainability claims, such as eco-certifications or carbon offsets, can be verified and remain tamper-proof, thereby fostering trust among consumers. It facilitates fair trade and equitable payment practices for local communities by utilizing smart contracts that guarantee direct and immediate compensation for provided services. Moreover, blockchain streamlines sustainable supply chain management, enabling businesses to transparently track the origin and environmental impact of goods and services, thereby ensuring compliance with environmental standards and promoting ethical consumerism (Abdel Maksoud, 2023).

This chapter aims to examine the ways in which blockchain technology supports the tourism industry and contributes to enhancing transparency, accountability, and sustainability in the sector. While blockchain technology holds great potential, its implementation in sustainable tourism practices is not without challenges. Therefore, this chapter will shed light on the key obstacles associated with the utilization of blockchain technology in the tourism industry. The recommendations provided in this chapter aim to expand current understanding and assist decision-makers in making informed choices regarding the adoption of blockchain technology in the tourism sector.

Blockchain: Definition and Components

Blockchain technology is characterized by its decentralized and distributed ledger system, enabling multiple parties to maintain a shared and immutable record of transactions or data. Unlike traditional systems, blockchain operates without a central authority or intermediary, relying instead on a network of participants known as nodes. These nodes collaborate to validate and record transactions, ensuring the integrity and security of the network (Abdel Maksoud, 2023).

Blockchain operates on several fundamental principles that contribute to its functionality. One such principle is immutability, which ensures that once a transaction or data entry is recorded on the blockchain, it becomes permanent and cannot be modified or tampered with. This is achieved through the use of cryptographic hashing, where each block in the chain contains a unique identifier, or hash, generated from the data it contains. Any alteration to the data within a block would result in a different hash value, alerting the network to the presence of tampering (Abdellah, 2023).

Transparency is another key principle of blockchain technology. The complete transaction history or data record is openly accessible to all participants in the network. This transparency fosters trust and accountability, as every participant has the ability to verify the accuracy and integrity of the data or transactions stored on the blockchain (Centobelli et al., 2022; Bhushan et al., 2021). Additionally, the consensus mechanism plays a vital role in maintaining the integrity of the blockchain (Lashkari & Musilek, 2021). Consensus algorithms, such as Proof of Work or Proof of Stake, ensure that all participants reach an agreement on the validity of transactions and the order in which they are added to the blockchain (Nguyen et al., 2019). This consensus is achieved through a process where nodes either compete or collaborate to validate and append new blocks to the chain, thus ensuring consistency across the network (Bouraga, 2021).

Decentralization is another fundamental principle of blockchain, designed to prevent any single entity from having control over the network (Zarrin et al., 2021). Decentralization lies at the core of blockchain technology, setting it apart from conventional centralized systems. In a decentralized blockchain, authority and decision-making are distributed across a network of nodes instead of being concentrated within a single entity or location. This decentralized structure enhances security and resilience by eliminating a single point of failure or attack vulnerability. Furthermore, decentralization promotes transparency and trust as data on the blockchain is openly verified and agreed upon through consensus among all participants. This consensus mechanism makes it exceedingly difficult to alter data without the agreement of the majority, thus minimizing the risks associated with data tampering, fraud, and misuse by centralized authorities. By adopting a democratic and distributed approach to data management, blockchain empowers users and cultivates a more equitable digital ecosystem (Abdel Maksoud, 2023). The distributed nature of the blockchain, with multiple nodes located in different locations, significantly reduces the risk of single points of failure and censorship. This decentralized structure enhances the security and reliability of the blockchain (Politou et al., 2020; He et al., 2019).

The functionality of blockchain technology relies on several interconnected components. These components include blocks, nodes, consensus mechanisms, and smart contracts (Banafa, 2022). Blocks serve as the foundational units of a blockchain, containing a collection of transactions or data that are sequentially added to the chain. Each block is typically assigned a unique identifier, known as a hash, which is derived from the data it contains. This hash plays a crucial role in ensuring the immutability of the block and preserving the integrity of the entire blockchain (Gayvoronskaya & Meinel, 2021).

Nodes, also referred to as network participants, are individual computers or devices that collectively form the blockchain network. Nodes, also known as network participants, are individual computers or devices that collectively form a blockchain network. Each node has a vital role in preserving the integrity and functionality of the network by storing, validating, and transmitting transaction data. Operating in a decentralized manner, nodes maintain either a complete or partial copy of the entire blockchain, ensuring transparency and redundancy. This redundancy helps safeguard the network against failures and malicious attacks. Through active participation in the consensus process, nodes reach agreement on the validity of transactions and the overall state of the blockchain. This agreement enables the establishment of a secure, accurate, and immutable ledger, which serves as the foundation for the functioning of the blockchain system (Abdel Maksoud, 2023). These nodes have a vital role in maintaining the blockchain by verifying and validating transactions. They store a copy of the entire blockchain and actively participate in the consensus mechanism to establish agreement on the validity of transactions and the chronological order in which they are appended to the chain. Nodes can be categorized as full nodes, which store a complete replica of the blockchain, or lightweight nodes, which depend on full nodes for transaction verification (Li et al., 2019; Bashir, 2017).

Consensus mechanisms refer to the protocols or algorithms that facilitate nodes in achieving a consensus on the state of the blockchain. These mechanisms ensure that all participants in the network maintain a unified and consistent perspective of the blockchain. Notable consensus mechanisms include Proof of Work (PoW), where nodes engage in a competitive process of solving intricate mathematical puzzles to append new blocks, and Proof of Stake (PoS), where nodes are selected to validate blocks based on their cryptocurrency holdings (Judmayer et al., 2022; Bashir, 2017).

Smart contracts are contractual agreements that are self-executing and have their terms and conditions directly encoded into computer code. These contracts are stored and executed on the blockchain. Through the use of smart contracts, automation is enabled, and predefined actions can be executed automatically when certain predetermined conditions are met. Smart contracts eliminate the need for intermediaries, providing transparency and security in the execution of contracts. They are often associated with blockchain platforms like Ethereum, which facilitate the development and deployment of decentralized applications (DApps) (Corrales et al., 2019; Modi, 2018; Vigliotti & Jones, 2020).

The impact of Blockchain on Sustainable Tourism

Blockchain technology has a significant impact on sustainable tourism, driven by various factors. One of the primary drivers is the rising demand for transparency and authenticity within the tourism industry. Travelers are increasingly aware of the environmental and social consequences of their decisions and seek reliable information on sustainable practices. Blockchain's decentralized and tamper-proof nature addresses this demand by ensuring the accuracy and verifiability of information related to sustainable certifications, supply chain traceability, and impact measurements. Consequently, blockchain fosters trust among travelers, businesses, and other stakeholders, ultimately resulting in a positive impact on sustainable tourism (Gomes & Lopes, 2023; Vazquez Melendez et al., 2024; Dionysis, 2023).

Improved supply chain management is another key factor contributing to the impact of blockchain on sustainable tourism. In the tourism industry, a wide range of suppliers, including accommodations, transportation providers, and local tour operators, play a crucial role. However, ensuring the authenticity and sustainability of these offerings can be a complex task. Blockchain technology offers a transparent and traceable platform for recording and verifying the origin and credentials of products and services throughout the supply chain. This enables businesses to validate sustainable claims and reduce the risk of misleading environmental claims (greenwashing). As a result, blockchain enhances the integrity of the supply chain, promoting a more sustainable and accountable tourism sector (Di Vaio & Varriale, 2020; Tyan et al., 2021).

The rising emphasis on community engagement and empowerment in sustainable tourism is a significant driver of blockchain's impact. Travelers now have a growing desire for genuine and meaningful experiences that directly contribute to the well-being of local communities. Blockchain technology plays a facilitative role in this regard by enabling tokenization and the establishment of community-based projects. Through blockchain-based platforms, travelers have the opportunity to directly engage in these projects by purchasing tokens or participating in activities that generate tokens. This direct involvement serves to empower local communities, stimulate economic growth, and ensure a more equitable distribution of tourism benefits. By harnessing the power of blockchain, sustainable tourism initiatives can make a tangible and positive difference in the livelihoods of local communities (Tham & Sigala, 2020; Pranita et al., 2023).

The imperative to tackle climate change and decrease carbon emissions is a key factor influencing the impact of blockchain technology on sustainable tourism. The tourism industry is a substantial source of greenhouse gas emissions, and there is an increasing demand to offset these emissions through carbon offset initiatives. Blockchain offers a transparent and efficient platform for tracking, trading, and verifying carbon credits. By tokenizing carbon credits on a blockchain, businesses and travelers can more easily trace and trade carbon offsets, establishing a decentralized and trusted marketplace. This fosters wider participation in carbon offsetting endeavors, ultimately resulting in a reduction of the tourism industry's carbon footprint (Abdel Maksoud, 2023).

Collaboration and knowledge sharing among industry stakeholders play a vital role in driving the impact of blockchain on sustainable tourism. Addressing the intricate challenges of sustainability necessitates collective endeavors and the exchange of best practices, research outcomes, and success stories. Blockchain-based platforms offer a secure and transparent environment for stakeholders to share information and collaborate on sustainable tourism solutions. By securely sharing data related to environmental impact, tourist behavior, and destination management, stakeholders can collectively analyze challenges and devise targeted strategies. This collaborative approach and knowledge sharing accelerate the adoption of sustainable practices and foster innovation within the industry, leading to a positive impact on sustainable tourism (Farahat, 2023).

Blockchain technology has the capacity to transform sustainable tourism by improving transparency, traceability, and accountability across the industry. The decentralized and unchangeable nature of blockchain enables secure and tamper-proof recording of transactions and data, creating a reliable platform for collaboration and the promotion of sustainable practices among various stakeholders, including travelers, businesses, and governments (Pranita et al., 2023).

One of the notable impacts of blockchain on sustainable tourism lies in supply chain management. Through blockchain, the entire supply chain of tourism products and services, such as food, accommodation, and transportation, can be traced from its origin to its destination. This capability empowers businesses to verify the authenticity of sustainable claims, such as organic or fair-trade certifications. By ensuring accurate information, blockchain builds trust and confidence among travelers, enabling them to make informed and sustainable choices (Tyan et al., 2021; Filimonau & Naumova, 2020).

Blockchain's smart contract functionality has the potential to streamline and automate sustainable practices within the tourism industry. Smart contracts are self-executing agreements with predefined conditions, and they can be leveraged to enforce sustainability criteria and certifications. For instance, a smart contract can ensure that accommodations meet specific energy efficiency standards or that tour operators adhere to responsible wildlife tourism guidelines. By automating compliance, blockchain reduces the risk of greenwashing and promotes genuine sustainable practices (Li & Kassem, 2021; Ante, 2021).

Additionally, blockchain technology can empower local communities and incentivize their involvement in sustainable tourism initiatives. Through tokenization, blockchain enables the creation of digital currencies or tokens that can be utilized for community-based projects. Travelers have the opportunity to directly contribute to these projects by purchasing tokens or participating in activities that generate tokens, such as volunteering or engaging in cultural experiences. This not only enhances community engagement but also ensures a more equitable distribution of the economic benefits derived from tourism (Peng & Huang, 2022; Mohamed et al., 2022).

Furthermore, blockchain technology has the potential to facilitate carbon offsetting and carbon credit trading within the tourism industry. Through tokenization, carbon credits can be recorded on a blockchain, allowing businesses and travelers to transparently track and trade carbon offsets. This establishes a decentralized marketplace for carbon credits, enabling businesses to offset their emissions and travelers to compensate for their carbon footprint. Blockchain-based carbon markets provide enhanced visibility and trust, simplifying stakeholder participation in global efforts to address climate change (Stuit et al., 2022; Chen, 2018).

By providing a decentralized and tamper-proof system, blockchain technology enhances trust and security within the realm of sustainable tourism. It establishes a secure storage mechanism for information related to sustainable practices, certifications, and impact measurements, preventing unauthorized alteration or manipulation. This transparency engenders trust among stakeholders such as travelers and businesses, fostering a more accountable and sustainable tourism ecosystem (Farahat, 2023). Also, by leveraging blockchain-powered platforms, community-based tourism initiatives can be facilitated through establishing a direct and transparent connection between travelers and local communities. These decentralized platforms enable communities to showcase their distinctive cultural heritage, provide genuine experiences, and receive equitable compensation. This empowerment of local communities not only contributes to the preservation of traditional practices but also stimulates economic growth and mitigates the adverse effects of mass tourism (Kotler et al., 2023; Holotescu, 2018).

Decentralized platforms powered by blockchain technology can play a crucial role in promoting responsible destination marketing by highlighting lesser-known or off-the-beaten-path locations. These platforms enable smaller destinations to increase their visibility, attract sustainable tourism, and diversify their visitor base. This approach helps alleviate the strain on overcrowded tourist hotspots and encourages a more sustainable distribution of tourism benefits. Blockchain technology has the potential to transform the tracking and management of carbon footprints within the tourism industry. Through the integration of blockchain with Internet of Things (IoT) devices, data concerning energy consumption, transportation emissions, and other carbon-related factors can be securely recorded and stored. This integration allows for precise measurement of carbon footprints at different stages of a traveler's journey, enabling the implementation of focused emission reduction strategies and carbon offsetting initiatives (Farahat et al., 2023, 2022).

Blockchain technology has the potential to facilitate the adoption of circular economy practices within the tourism industry. Through the creation of tokenized systems, blockchain enables transparent and traceable transactions within circular economy networks, supporting resource sharing, waste management, and recycling initiatives. By leveraging blockchain, businesses can promote resource efficiency, minimize waste, and foster collaborations for more sustainable resource management (Rejeb et al., 2022; Rehman Khan et al., 2022).

Blockchain Applications in Sustainable Tourism

Eco-tourism, responsible tourism, or sustainable tourism (Chan et al., 2021) is a travel approach that aims to reduce adverse effects on the environment, conserve cultural heritage, and generate social and economic advantages for local communities. Grounded in sustainability principles, it strives to fulfill the requirements of current tourists and host communities while safeguarding resources for the well-being of future generations (Edgell, 2019).

Transparent Supply Chains

Blockchain technology holds the potential to revolutionize transparency and traceability in supply chains, including the tourism industry, by providing verifiable evidence for sustainability claims. Through the use of a decentralized and tamper-proof ledger, blockchain ensures the secure recording of transactions and information. Each entry is digitally signed and connected to the preceding one, forming an immutable chain of blocks. This transparency enables all participants in the supply chain to access and validate information, making manipulation or falsification difficult. Smart contracts, which are self-executing agreements on the blockchain, can incorporate predefined rules for automated verification processes. In the context of sustainability claims, smart contracts can automatically assess adherence to specific standards or certifications. For example, if a hotel claims to be environmentally friendly, the smart contract can verify its compliance with energy efficiency or waste management criteria (Cao et al., 2023a, b; Ante, 2021).

Blockchain enables comprehensive traceability by documenting each transaction and the movement of goods or services on the chain, allowing stakeholders to monitor the origin, production processes, and distribution. Within the tourism supply chain, this includes tracking the sourcing of local products, the carbon footprint of transportation, and the responsible practices of tour operators and accommodations. Blockchain serves as a platform for storing and authenticating certifications and credentials related to sustainability. Certifying bodies can issue digital certificates on the blockchain, accessible and verifiable by supply chain participants or consumers. This ensures that assertions of sustainability or compliance with specific standards are backed by transparent and verified evidence. Blockchain technology enables consumers to make well-informed decisions by offering transparent information regarding the sustainability of products or services. Through blockchain, consumers can utilize QR codes or unique identifiers to retrieve comprehensive details about sustainability attributes, encompassing certifications, environmental impact, and social responsibility (Abdel Maksoud, 2023; Jabbar et al., 2021).

Blockchain fosters collaboration and trust within the supply chain by enabling stakeholders to share information on a distributed ledger. This transparency cultivates trust and encourages joint efforts toward achieving sustainability goals. It facilitates cooperation, enabling partners to pinpoint areas for improvement and collaborate on initiatives to advance sustainability. The utilization of blockchain in the tourism industry enhances transparency and traceability across the supply chain, ensuring the verifiability of sustainability claims. The technology ensures the availability of reliable, tamper-proof information accessible to all stakeholders, instilling consumer trust, encouraging responsible decision-making, and promoting the adoption of sustainable practices throughout the tourism supply chain (Wan et al., 2020; Cao et al., 2023c).

Fraud Prevention

Blockchain technology has the potential to substantially reduce fraud in bookings, reviews, and claims related to sustainable tourism practices (Bulchand-Gidumal & Melián-González, 2023). The secure and transparent nature of blockchain is particularly impactful in recording bookings, as information stored on the blockchain becomes highly resistant to tampering or manipulation. This significantly mitigates fraudulent activities such as double bookings, unauthorized modifications, or falsification of reservation details. The decentralized structure of blockchain ensures accessibility to booking information by all relevant parties, concurrently preserving the integrity of the data (Bodkhe et al., 2020; Rahman et al., 2022).

Blockchain technology holds the potential to tackle the challenge of fake or biased reviews in online systems by introducing a decentralized and transparent review platform. Through blockchain, reviews can be associated with specific transactions and undergo verification by involved parties, ensuring authenticity and trustworthiness. This methodology strengthens the reliability of reviews for potential travelers. Furthermore, blockchain facilitates the verification and authentication of claims related to sustainable tourism practices. Establishments such as hotels or tour operators can document their sustainability practices, certifications, and audits on a blockchain, offering convenient access for verification by consumers and regulatory bodies. This transparency reduces the risk of false claims and greenwashing, as the information stored on the blockchain is subject to independent verification (Karode & Werapun, 2022; Li et al., 2021).

The smart contract capability of blockchain technology can play a significant role in promoting accountability within sustainable tourism practices. By utilizing smart contracts, specific conditions related to sustainability commitments can be programmed and enforced. For instance, penalties can be automatically triggered if businesses fail to meet predetermined environmental or social targets. This mechanism holds businesses accountable for their sustainability claims and serves as a motivating factor for them to uphold their commitments. Furthermore, blockchain technology strengthens data security and privacy by enabling the storage of personal information, such as customer details, on blockchain networks with encrypted access controls. This mitigates the risk of data breaches and unauthorized access, instilling trust in the booking process and safeguarding traveler privacy. Moreover, blockchain has the potential to facilitate collaborative governance and the establishment of industry-wide sustainability standards. Through the utilization of blockchain networks, stakeholders from different sectors can collaborate to define and enforce standards for sustainable tourism practices. This collective approach reduces fragmentation, fosters transparency, and enhances the credibility of sustainability claims (Abdel Maksoud, 2023).

Through the utilization of blockchain technology, the tourism industry can effectively address fraud issues in bookings, reviews, and claims associated with sustainable tourism practices. Blockchain enhances transparency, accountability, and trust by offering secure and verifiable records, verifying reviews, validating sustainability claims, and enabling collaborative governance. Ultimately, blockchain has the potential to foster a more dependable and responsible tourism industry that benefits both travelers and sustainable destinations (Abdellah, 2023).

Smart Contracts for Fair Compensation

Smart contracts play a crucial role in automating equitable compensation for sustainable tourism practices by incorporating transparent and pre-defined terms within their code. These terms, specifically related to sustainable tourism practices, are automatically executed based on specific conditions outlined in the smart contract. For example, the contract can establish criteria for evaluating sustainability performance, such as reducing carbon emissions or supporting local communities (Karamitsos et al., 2023; Abdellah, 2023).

Smart contracts improve sustainability assessment by integrating with IoT devices, sensors, and external APIs to collect real-time data on metrics such as energy consumption and waste management. By enabling continuous monitoring, smart contracts accurately evaluate the effectiveness of sustainable practices. Utilizing predefined terms and real-time data, these contracts automatically calculate fair compensation, taking into account factors such as achieved sustainability goals, environmental or community impact, and cost savings. This automated process ensures efficiency and reduces the likelihood of errors or disputes in the compensation process for sustainable tourism practices (Abdel Maksoud, 2023).

As well, smart contracts optimize the compensation process in sustainable tourism by leveraging tokenized payments. These tokens, which represent value or rewards, provide transparency, traceability, and seamless transferability within the ecosystem. Operating on blockchain networks, smart contracts guarantee trust and transparency through decentralized and unchangeable ledgers. The inclusion of incentive mechanisms encourages ongoing enhancements in sustainability practices. This transparent and auditable compensation process cultivates trust among stakeholders, reducing disputes, and fostering a collaborative environment among tourism businesses, local communities, and travelers (Treiblmaier, 2022; Onder, 2023).

Smart contracts transform compensation procedures in sustainable tourism by automating payment processing without the need for intermediaries. Once predetermined conditions are fulfilled, smart contracts initiate prompt payments, streamlining the compensation process and ensuring timely rewards for sustainable practices. This automated system, guided by transparent and precise terms, promotes the adoption and enhancement of sustainable practices. The efficiency of smart contracts enhances the acknowledgment and incentivization of such practices, cultivating a more sustainable tourism ecosystem that benefits both businesses and destinations (Rashideh, 2020; Treiblmaier, 2022).

Tokenized Incentives

Token-based rewards present a novel method for incentivizing sustainable tourism practices. Travelers who participate in eco-friendly activities, responsible tours, and sustainable behaviors can earn tokens, establishing a positive feedback loop that reinforces sustainable choices. These rewards extend to activities involving local communities, culture, and businesses, promoting economic empowerment and ensuring that tourism has a positive impact on local areas. This system encourages a more sustainable and inclusive tourism ecosystem by fostering responsible traveler behavior and facilitating community engagement (Abdellah, 2023; Abdel Maksoud, 2023).

In addition, token-based incentives offer a means of encouraging responsible consumption among tourists. By making choices that prioritize sustainable transportation, local products, or low-impact activities, tourists can earn tokens, promoting environmentally and socially conscious decision-making. In addition to immediate benefits, these token-based rewards contribute to customer loyalty. Businesses can leverage this loyalty by enabling tourists to redeem accumulated tokens for discounts, exclusive experiences, or upgrades. This approach aligns with the growing demand for responsible tourism, creating a compelling value proposition that fosters long-term customer commitment to sustainable practices (Onder, 2023; Li et al., 2019).

Token-based rewards have the potential to cultivate collaboration among sustainable tourism businesses by establishing a shared ecosystem. This ecosystem allows for cross-rewards, enabling tourists to earn and redeem tokens across multiple providers who are committed to sustainability. By fostering such collaboration, not only does it foster a sense of community, but it also encourages businesses to collectively promote sustainable practices. The utilization of blockchain technology for these token ecosystems ensures transparency and traceability. Every transaction and token transfer is securely recorded on a decentralized and unalterable ledger, enhancing the integrity of the reward system. This transparency builds trust among participants, as both tourists and businesses can verify the validity and authenticity of token-based rewards (Lisi et al., 2021; AlKhader et al., 2023).

Token ecosystems within sustainable tourism provide a valuable avenue for gathering meaningful data by tracking token transactions and user behaviors. This data can play a crucial role in assessing the impact of sustainable activities, informing evidence-based strategies, and shaping future policies. By incentivizing sustainable practices, local engagement, and collaborative networks, token-based rewards have the potential to reshape both tourist behavior and business practices towards sustainability. The transparency and traceability offered by blockchain-based tokens foster trust among stakeholders and facilitate the collection of impactful data. Ultimately, through the implementation of token-based rewards, the tourism industry can promote a more sustainable, resilient, and fulfilling travel experience for both tourists and destinations (Treiblmaier, 2022; Kapil & Kapil, 2022).

Decentralized Platforms for Community Engagement

Blockchain technology enables direct and transparent transactions between tourists and local communities, offering advantages to both sides. By leveraging blockchain, peer-to-peer transactions are facilitated, eliminating intermediaries and allowing for direct interaction between tourists and local communities or service providers. This decentralized approach promotes authenticity and customization in tourist experiences while ensuring prompt and equitable compensation for local services. Financial transactions are recorded transparently on a decentralized ledger, enabling tourists and locals to verify transaction information. This transparency fosters trust, reduces the risk of fraud, and enhances the overall dependability of transactions for all participants involved (Abdel Maksoud, 2023).

Blockchain-powered payment systems enable tourists to make direct, secure, and immediate payments to local communities using digital currencies or tokens. This approach proves particularly beneficial in regions with limited access to traditional banking, as it facilitates worldwide economic engagement for local communities. Within blockchain, smart contracts can automate and enforce transaction terms and conditions between tourists and local communities. These contracts establish the scope of services, pricing, delivery conditions, and other parameters, minimizing the potential for misunderstandings or disputes by ensuring mutual agreement on terms (Tham & Sigala, 2020; Fiorentino & Bartolucci, 2021).

Blockchain-based platforms have the capability to implement transparent feedback and reputation systems, allowing tourists to provide reviews and ratings for services offered by local communities. These systems, being secure and accessible to all participants on the blockchain network, enhance trust and credibility. By leveraging these systems, tourists can make informed decisions based on the experiences of others, while local communities receive recognition for their high-quality offerings. Blockchain technology enables direct and transparent transactions, facilitating peer-to-peer interactions, decentralized payments, and the implementation of smart contracts. It ensures traceability, authenticity, and empowers local communities. Through this technology, fair and mutually beneficial interactions are promoted, contributing to sustainable tourism and fostering equitable economic growth (Bodkhe et al., 2019; Onder, 2023).

Challenges of Adopting Blockchain in Sustainable Tourism

The broad adoption of blockchain in sustainable tourism encounters various obstacles, including technological, regulatory, and cultural barriers. Blockchain technology remains relatively complex, demanding specialized expertise for development and implementation. Numerous tourism businesses, particularly smaller ones, may lack the technical knowledge and resources necessary to adopt and integrate blockchain solutions. The intricacy of blockchain systems can also present challenges concerning scalability, interoperability, and data privacy, necessitating substantial investments and infrastructure enhancements (Abdellah, 2023).

The regulatory landscape concerning blockchain and cryptocurrencies varies across jurisdictions, which can impede the adoption of blockchain in sustainable tourism. Uncertainty arises from the absence of clear regulations and legal frameworks in place. Addressing specific concerns, such as data protection, the enforceability of smart contracts, and cross-border transactions, is crucial to establish a favorable regulatory environment for blockchain technology (Prados-Castillo et al., 2023; Khan et al., 2021).

The successful implementation of blockchain in sustainable tourism necessitates collaboration and integration among diverse stakeholders, including tourism businesses, governments, local communities, and technology providers. Achieving interoperability between different blockchain platforms and legacy systems can prove challenging, requiring standardization, data sharing protocols, and consensus among multiple parties. Overcoming these technical obstacles is crucial for seamless integration and widespread adoption. Cultural acceptance and understanding of blockchain technology can also present barriers to adoption. Blockchain disrupts traditional business models and processes, potentially facing resistance from stakeholders who are unfamiliar with the technology or hesitant to embrace change. Building awareness, educating stakeholders, and demonstrating the benefits and potential of blockchain in sustainable tourism are vital for overcoming cultural barriers (Abdel Maksoud, 2023).

The implementation of blockchain solutions in sustainable tourism can entail substantial initial expenses, such as infrastructure development, data management, and training. This can be particularly challenging for tourism businesses, especially small and medium-sized enterprises (SMEs), who may hesitate to invest in blockchain technology without clear evidence of its long-term benefits and a favorable return on investment. Demonstrating the economic value and cost-effectiveness of blockchain solutions is crucial in encouraging adoption among SMEs and driving the wider adoption of blockchain in the tourism industry (Biswas & Gupta, 2019; Lin et al., 2023).

Blockchain networks, particularly those that utilize proof-of-work consensus mechanisms, require significant energy consumption. This energy usage raises concerns regarding the environmental impact and sustainability of blockchain technology. Achieving a balance between the advantages of blockchain and its environmental footprint poses a challenge that must be addressed to align with the objectives of sustainable tourism (Platt et al., 2021; Truby, 2018).

Despite the trust and security features associated with blockchain, the technology is not immune to vulnerabilities and risks. Smart contract bugs, hacking attempts, and data breaches can compromise the integrity of blockchain systems. To encourage the widespread adoption of blockchain in sustainable tourism, it is essential to address and mitigate these concerns through the implementation of robust security measures, auditing mechanisms, and adherence to industry best practices (Wang et al., 2021; Abdel Maksoud, 2023).

Conclusion and Recommendations

To fully leverage the capabilities of blockchain technology for sustainable tourism, it is imperative to provide a comprehensive set of strategic guidelines for stakeholders, policymakers, and businesses within the industry. These recommendations are designed to offer guidance in the effective integration and implementation of blockchain solutions, ensuring their positive contributions to environmental, social, and economic sustainability.

The successful implementation of blockchain-enabled sustainable tourism relies on the active participation and support of stakeholders, including local communities, tourism businesses, and tourists. It is crucial for stakeholders to actively engage with and endorse decentralized platforms that prioritize transparency and fair economic distribution. By doing so, stakeholders can contribute to decision-making processes and ensure that the benefits of sustainable tourism are distributed equitably among all involved parties. Educational initiatives and awareness programs are key components of stakeholder engagement. In community-driven blockchain projects, it is essential to prioritize efforts that educate stakeholders about the advantages and potential applications of blockchain technology in sustainable tourism. By promoting a comprehensive understanding of blockchain, stakeholders can make informed decisions and actively contribute to the development and implementation of blockchain solutions. Moreover, stakeholders should actively promote the adoption of smart contracts within the tourism ecosystem. Smart contracts are contractual agreements encoded on the blockchain, automatically executing predefined rules and conditions. Through the utilization of smart contracts, stakeholders can enforce sustainable practices and guarantee equitable compensation for all participants involved in the tourism value chain. For instance, smart contracts can be employed to automate carbon offset transactions, ensuring that tourism businesses fulfill their sustainability commitments and contribute to environmental conservation efforts. Through active participation in these endeavors, stakeholders can assume a pivotal role in driving the progress of sustainable tourism through blockchain technology. Their involvement and support are vital for fostering the adoption and implementation of blockchain solutions that prioritize transparency, fairness, and sustainability in the tourism industry. By collaborating with each other and advocating for the benefits of blockchain, stakeholders can make significant contributions to the development of a more sustainable and responsible tourism sector overall.

Policymakers have a crucial role in promoting the widespread adoption of blockchain technology in sustainable tourism. They can do so by creating an environment that supports its implementation. To achieve this, policymakers need to establish regulatory frameworks that are tailored to the unique challenges and opportunities posed by blockchain in the tourism sector. This requires close collaboration with industry experts, blockchain developers, and other stakeholders to develop legislation that not only encourages innovation but also safeguards consumer rights and privacy. By working together, policymakers can create a regulatory environment that fosters the responsible and secure use of blockchain technology in the tourism industry. Promoting ethical practices is a key element of regulatory frameworks for blockchain in sustainable tourism. Policymakers should strive to establish guidelines and standards that encourage the responsible and sustainable use of blockchain technology. This involves addressing important issues such as data privacy, security, and the ethical considerations surrounding blockchain implementation in tourism operations. To stay well-informed about the latest technological advancements and global best practices, policymakers should actively pursue partnerships with blockchain developers, academic institutions, and international organizations. By doing so, policymakers can ensure that their regulatory frameworks remain up-to-date and aligned with the evolving landscape of blockchain technology in the tourism sector. Through these collaborative efforts, policymakers can acquire valuable insights and knowledge that inform their regulatory decisions. Engaging with stakeholders enables policymakers to develop a deeper understanding of the potential benefits and risks associated with blockchain technology, empowering them to establish a regulatory landscape that promotes responsible and sustainable use of blockchain in the tourism industry. Moreover, policymakers should cultivate an environment that fosters innovation and experimentation with blockchain technology. This can be achieved by providing support mechanisms such as funding opportunities, incubation programs, and regulatory sandboxes. By creating these avenues of support, policymakers encourage the emergence of a vibrant innovation ecosystem. This, in turn, attracts blockchain startups and motivates existing tourism businesses to explore and adopt blockchain solutions. By driving the growth of sustainable tourism through the adoption of blockchain, policymakers contribute to the overall development and advancement of the industry.

Tourism businesses can strengthen their sustainability initiatives by strategically integrating blockchain technology into their operations. One effective approach is to adopt transparent supply chain systems built on blockchain. These systems can validate and showcase the authenticity of eco-friendly practices, thereby fostering trust among environmentally conscious consumers. By offering a transparent view of the entire supply chain, blockchain empowers businesses to demonstrate their dedication to sustainability and appeal to customers who prioritize environmentally friendly options. Apart from transparent supply chains, tourism businesses can explore the implementation of tokenized loyalty programs. By tokenizing loyalty rewards on the blockchain, businesses can incentivize tourists to opt for sustainable choices. This approach not only cultivates brand loyalty but also motivates customers to make environmentally conscious decisions. Through these tokenized loyalty programs, tourism businesses can align their sustainability objectives with customer engagement and environmental conservation efforts. To successfully adopt blockchain technology, it is advisable for tourism businesses to engage in collaboration with blockchain developers and experts. By closely partnering with these professionals, businesses can customize blockchain solutions to align with their distinct needs and goals. This collaboration guarantees that the implemented blockchain systems are optimized to address the specific requirements of each tourism business, thereby maximizing the advantages and value derived from blockchain technology in the realm of sustainability.

REFERENCES

Abdel Maksoud, W. (2023). *Opportunities and challenges of applying "Block Chain" Technology in the Tourism Industry* (Master Thesis, Faculty of Tourism and Hotels, University of Sadat City).

Abdellah, M. (2023). The impact of blockchain technology on the efficiency and effectiveness of tourism services. *Journal of the Faculty of Tourism and Hotels-University of Sadat City*, *7*(1/2), 27–48.

AlKhader, W., Jayaraman, R., Salah, K., Sleptchenko, A., Antony, J., & Omar, M. (2023). Leveraging blockchain and NFTs for quality 4.0 implementation in digital manufacturing. *Journal of Manufacturing Technology Management*, *34*(7), 1208–1234. doi:10.1108/JMTM-05-2023-0172

Ante, L. (2021). Smart contracts on the blockchain–A bibliometric analysis and review. *Telematics and Informatics*, *57*, 101519. doi:10.1016/j.tele.2020.101519

Banafa, A. (2022). *Blockchain technology and applications*. CRC Press. doi:10.1201/9781003337393

Baralla, G., Pinna, A., Tonelli, R., Marchesi, M., & Ibba, S. (2021). Ensuring transparency and traceability of food local products: A blockchain application to a Smart Tourism Region. *Concurrency and Computation*, *33*(1), e5857. doi:10.1002/cpe.5857

Bashir, I. (2017). *Mastering blockchain*. Packt Publishing Ltd.

Bhushan, B., Sinha, P., Sagayam, K., & Andrew, J. (2021). Untangling blockchain technology: A survey on state of the art, security threats, privacy services, applications and future research directions. *Computers & Electrical Engineering*, *90*, 106897. doi:10.1016/j.compeleceng.2020.106897

Biswas, B., & Gupta, R. (2019). Analysis of barriers to implement blockchain in industry and service sectors. *Computers & Industrial Engineering*, *136*, 225–241. doi:10.1016/j.cie.2019.07.005

Bodkhe, U., Bhattacharya, P., Tanwar, S., Tyagi, S., Kumar, N., & Obaidat, M. (2019). BloHosT: Blockchain enabled smart tourism and hospitality management. In 2019 international conference on computer, information and telecommunication systems (CITS) (pp. 1-5). IEEE.

Bodkhe, U., Tanwar, S., Parekh, K., Khanpara, P., Tyagi, S., Kumar, N., & Alazab, M. (2020). Blockchain for industry 4.0: A comprehensive review. *IEEE Access : Practical Innovations, Open Solutions*, *8*, 79764–79800. doi:10.1109/ACCESS.2020.2988579

Bouraga, S. (2021). A taxonomy of blockchain consensus protocols: A survey and classification framework. *Expert Systems with Applications*, *168*, 114384. doi:10.1016/j.eswa.2020.114384

Bulchand-Gidumal, J., & Melián-González, S. (2023). Fighting fake reviews with blockchain-enabled consumer-generated reviews. *Current Issues in Tourism*. Advance online publication. doi:10.1080/13683500.2023.2173054

Cao, S., Johnson, H., & Tulloch, A. (2023b). Exploring blockchain-based traceability for food supply chain sustainability: Towards a better way of sustainability communication with consumers. *Procedia Computer Science*, *217*, 1437–1445. doi:10.1016/j.procs.2022.12.342

Cao, S., Xu, H., & Bryceson, K. (2023a). Blockchain Traceability for Sustainability Communication in Food Supply Chains: An Architectural Framework, Design Pathway and Considerations. *Sustainability (Basel)*, *15*(18), 13486. doi:10.3390/su151813486

Cao, S., Xu, H., & Bryceson, K. (2023c). Blockchain Traceability for Sustainability Communication in Food Supply Chains: An Architectural Framework, Design Pathway and Considerations. *Sustainability (Basel)*, *15*(18), 13486. doi:10.3390/su151813486

Centobelli, P., Cerchione, R., Del Vecchio, P., Oropallo, E., & Secundo, G. (2022). Blockchain technology for bridging trust, traceability and transparency in circular supply chain. *Information & Management*, *59*(7), 103508. doi:10.1016/j.im.2021.103508

Chan, J., Marzuki, K., & Mohtar, T. (2021). Local community participation and responsible tourism practices in ecotourism destination: A case of lower Kinabatangan, Sabah. *Sustainability (Basel)*, *13*(23), 13302. doi:10.3390/su132313302

Chen, D. (2018). Utility of the blockchain for climate mitigation. *The Journal of The British Blockchain Association*, *1*(1), 75–80. doi:10.31585/jbba-1-1-(6)2018

Corrales, M., Fenwick, M., & Haapio, H. (2019). *Legal tech, smart contracts and blockchain*. Springer. doi:10.1007/978-981-13-6086-2

Di Vaio, A., & Varriale, L. (2020). Blockchain technology in supply chain management for sustainable performance: Evidence from the airport industry. *International Journal of Information Management*, *52*, 102014. doi:10.1016/j.ijinfomgt.2019.09.010

Dionysis, S. (2023). *Blockchain and distributed ledger technologies for supply chain traceability: industry considerations and consumer preferences* (Doctoral dissertation, University of Nottingham).

Dutta, P., Choi, T., Somani, S., & Butala, R. (2020). Blockchain technology in supply chain operations: Applications, challenges and research opportunities. *Transportation research part e: Logistics and transportation review, 142*, 102067.

Edgell, D. (2019). *Managing sustainable tourism: A legacy for the future*. Routledge.

Erol, I., Neuhofer, I., Dogru, T., Oztel, A., Searcy, C., & Yorulmaz, A. (2022). Improving sustainability in the tourism industry through blockchain technology: Challenges and opportunities. *Tourism Management*, *93*, 104628. doi:10.1016/j.tourman.2022.104628

Farahat, E. (2023). *Applications of artificial intelligence as a marketing tool and their impact on the competitiveness of the Egyptian tourist destination* (Doctoral Dissertation, Minia University).

Farahat, E., Mohamed, M., & Al-Romeedy, B. (2022). Artificial intelligence applications and its impact on the competitiveness of the Egyptian tourist destination. *Research Journal of the Faculty of Tourism and Hotels Mansoura University*, *11*(2), 57–93.

Filimonau, V., & Naumova, E. (2020). The blockchain technology and the scope of its application in hospitality operations. *International Journal of Hospitality Management*, *87*, 102383. doi:10.1016/j.ijhm.2019.102383

Fiorentino, S., & Bartolucci, S. (2021). Blockchain-based smart contracts as new governance tools for the sharing economy. *Cities (London, England)*, *117*, 103325. doi:10.1016/j.cities.2021.103325

Gayvoronskaya, T., & Meinel, C. (2021). *Blockchain*. Springer International Publishing. doi:10.1007/978-3-030-61559-8

Gomes, S., & Lopes, J. M. (2023). Insights for pro-sustainable tourist behavior: The role of sustainable destination information and pro-sustainable tourist habits. *Sustainability (Basel)*, *15*(11), 8856. doi:10.3390/su15118856

He, S., Tang, Q., Wu, C., & Shen, X. (2019). Decentralizing IoT management systems using blockchain for censorship resistance. *IEEE Transactions on Industrial Informatics*, *16*(1), 715–727. doi:10.1109/TII.2019.2939797

Holotescu, C. (2018). Understanding blockchain technology and how to get involved. *The 14thInternational Scientific ConferenceeLearning and Software for EducationBucharest,* 20.

Jabbar, S., Lloyd, H., Hammoudeh, M., Adebisi, B., & Raza, U. (2021). Blockchain-enabled supply chain: Analysis, challenges, and future directions. *Multimedia Systems*, *27*(4), 787–806. doi:10.1007/s00530-020-00687-0

Judmayer, A., Stifter, N., Krombholz, K., & Weippl, E. (2022). *Blocks and chains: introduction to bitcoin, cryptocurrencies, and their consensus mechanisms*. Springer Nature.

Kapil, S., & Kapil, K. (2022). Blockchain in hospitality and tourism industry way forward. *International Journal of Business and Economics*, *6*(2), 289–298.

Karamitsos, I., Papadaki, M., Al-Hussaeni, K., & Kanavos, A. (2023). Transforming Airport Security: Enhancing Efficiency through Blockchain Smart Contracts. *Electronics (Basel)*, *12*(21), 4492. doi:10.3390/electronics12214492

Karode, T., & Werapun, W. (2022). Robustness against fraudulent activities of a blockchain-based online review system. *Peer-to-Peer Networking and Applications*, *15*(1), 92–106. doi:10.1007/s12083-021-01225-z

Khan, S., Loukil, F., Ghedira-Guegan, C., Benkhelifa, E., & Bani-Hani, A. (2021). Blockchain smart contracts: Applications, challenges, and future trends. *Peer-to-Peer Networking and Applications*, *14*(5), 2901–2925. doi:10.1007/s12083-021-01127-0 PMID:33897937

Kotler, P., Kartajaya, H., & Setiawan, I. (2023). *Marketing 6.0: The Future Is Immersive*. John Wiley & Sons.

Lashkari, B., & Musilek, P. (2021). A comprehensive review of blockchain consensus mechanisms. *IEEE Access : Practical Innovations, Open Solutions*, *9*, 43620–43652. doi:10.1109/ACCESS.2021.3065880

Li, F., Yu, X., Ge, R., Wang, Y., Cui, Y., & Zhou, H. (2021). BCSE: Blockchain-based trusted service evaluation model over big data. *Big Data Mining and Analytics*, *5*(1), 1–14. doi:10.26599/BDMA.2020.9020028

Li, J., & Kassem, M. (2021). Applications of distributed ledger technology (DLT) and Blockchain-enabled smart contracts in construction. *Automation in Construction*, *132*, 103955. doi:10.1016/j.autcon.2021.103955

Li, K., Chen, X., Jiang, H., & Bertino, E. (2019). *Essentials of blockchain technology*. CRC Press. doi:10.1201/9780429674457

Lin, M., Sharma, A., Pan, B., & Quadri-Felitti, D. (2023). Information asymmetry in the innovation adoption decision of tourism and hospitality SMEs in emerging markets: A mixed-method analysis. *Tourism Management*, *99*, 104793. doi:10.1016/j.tourman.2023.104793

Lisi, A., De Salve, A., Mori, P., Ricci, L., & Fabrizi, S. (2021). Rewarding reviews with tokens: An ethereum-based approach. *Future Generation Computer Systems*, *120*, 36–54. doi:10.1016/j.future.2021.02.003

Modi, R. (2018). *Solidity Programming Essentials: A beginner's guide to build smart contracts for Ethereum and blockchain*. Packt Publishing Ltd.

Mohamed, M., Farahat, E., & Al-Romeedy, B. (2022). The Mediating Role of Marketing Effectiveness in the Relationship between Artificial Intelligence and Destination Competitiveness. *Minia Journal of Tourism and Hospitality Research*, *14*(1), 113–129.

Nguyen, C., Hoang, D., Nguyen, D., Niyato, D., Nguyen, H., & Dutkiewicz, E. (2019). Proof-of-stake consensus mechanisms for future blockchain networks: Fundamentals, applications and opportunities. *IEEE Access : Practical Innovations, Open Solutions*, *7*, 85727–85745. doi:10.1109/ACCESS.2019.2925010

Onder, I. (2023). Enhancing Sustainable Tourism through Blockchain-Based Tokenization: A Novel Approach for Revenue Generation. In *Blockchain for Tourism and Hospitality Industries*. Routledge. doi:10.4324/9781003351917-8

Peng, Y., & Huang, W. (2022). Using blockchain technology and sharing culture to promote sustainable forest management in tribal communities. *Journal of Environmental and Public Health*, *2022*, 1–12. Advance online publication. doi:10.1155/2022/1529407 PMID:35815248

Platt, M., Sedlmeir, J., Platt, D., Xu, J., Tasca, P., Vadgama, N., & Ibañez, J. (2021). The energy footprint of blockchain consensus mechanisms beyond proof-of-work. In *2021 IEEE 21st International Conference on Software Quality, Reliability and Security Companion (QRS-C)* (pp. 1135-1144). IEEE.10.1109/QRS-C55045.2021.00168

Politou, E., Alepis, E., Patsakis, C., Casino, F., & Alazab, M. (2020). Delegated content erasure in IPFS. *Future Generation Computer Systems*, *112*, 956–964. doi:10.1016/j.future.2020.06.037

Prados-Castillo, J., Guaita Martínez, J., Zielińska, A., & Gorgues Comas, D. (2023). A Review of Blockchain Technology Adoption in the Tourism Industry from a Sustainability Perspective. *Journal of Theoretical and Applied Electronic Commerce Research*, *18*(2), 814–830. doi:10.3390/jtaer18020042

Pranita, D., Sarjana, S., Musthofa, B., Kusumastuti, H., & Rasul, M. (2023). Blockchain Technology to Enhance Integrated Blue Economy: A Case Study in Strengthening Sustainable Tourism on Smart Islands. *Sustainability (Basel)*, *15*(6), 5342. doi:10.3390/su15065342

Rahman, M., Chamikara, M., Khalil, I., & Bouras, A. (2022). Blockchain-of-blockchains: An interoperable blockchain platform for ensuring IoT data integrity in smart city. *Journal of Industrial Information Integration*, *30*, 100408. doi:10.1016/j.jii.2022.100408

Rashideh, W. (2020). Blockchain technology framework: Current and future perspectives for the tourism industry. *Tourism Management*, *80*, 104125. doi:10.1016/j.tourman.2020.104125

Rehman Khan, S. A., Yu, Z., Sarwat, S., Godil, D. I., Amin, S., & Shujaat, S. (2022). The role of block chain technology in circular economy practices to improve organisational performance. *International Journal of Logistics*, *25*(4-5), 605–622. doi:10.1080/13675567.2021.1872512

Rejeb, A., Zailani, S., Rejeb, K., Treiblmaier, H., & Keogh, J. G. (2022). Modeling enablers for blockchain adoption in the circular economy. *Sustainable Futures : An Applied Journal of Technology, Environment and Society*, *4*, 100095. doi:10.1016/j.sftr.2022.100095

Sarfraz, M., Khawaja, K., Han, H., Ariza-Montes, A., & Arjona-Fuentes, J. (2023). Sustainable supply chain, digital transformation, and blockchain technology adoption in the tourism sector. *Humanities & Social Sciences Communications*, *10*(1), 1–13. doi:10.1057/s41599-023-02051-9

Stuit, A., Brockington, D., & Corbera, E. (2022). Smart, commodified and encoded: Blockchain technology for environmental sustainability and nature conservation. *Conservation & Society*, *20*(1), 12–23. doi:10.4103/cs.cs_41_21

Tham, A., & Sigala, M. (2020). Road block (chain): Bit (coin) s for tourism sustainable development goals? *Journal of Hospitality and Tourism Technology*, *11*(2), 203–222. doi:10.1108/JHTT-05-2019-0069

Treiblmaier, H. (2022). Blockchain and tourism. In *Handbook of e-Tourism* (pp. 475–495). Springer International Publishing. doi:10.1007/978-3-030-48652-5_28

Truby, J. (2018). Decarbonizing Bitcoin: Law and policy choices for reducing the energy consumption of Blockchain technologies and digital currencies. *Energy Research & Social Science*, *44*, 399–410. doi:10.1016/j.erss.2018.06.009

Tyan, I., Yagüe, M. I., & Guevara-Plaza, A. (2021). Blockchain technology's potential for sustainable tourism. In *Information and Communication Technologies in Tourism 2021: Proceedings of the ENTER 2021 eTourism Conference, January 19–22, 2021* (pp. 17-29). Springer International Publishing.

Vazquez Melendez, E. I., Bergey, P., & Smith, B. (2024). Blockchain technology for supply chain provenance: Increasing supply chain efficiency and consumer trust. *Supply Chain Management*. Advance online publication. doi:10.1108/SCM-08-2023-0383

Vigliotti, M., & Jones, H. (2020). *The Executive Guide to Blockchain: Using Smart Contracts and Digital Currencies in Your Business*. Springer Nature. doi:10.1007/978-3-030-21107-3

Wan, P., Huang, L., & Holtskog, H. (2020). Blockchain-enabled information sharing within a supply chain: A systematic literature review. *IEEE Access : Practical Innovations, Open Solutions*, *8*, 49645–49656. doi:10.1109/ACCESS.2020.2980142

Wang, Y., He, J., Zhu, N., Yi, Y., Zhang, Q., Song, H., & Xue, R. (2021). Security enhancement technologies for smart contracts in the blockchain: A survey. *Transactions on Emerging Telecommunications Technologies*, *32*(12), e4341. doi:10.1002/ett.4341

Zarrin, J., Wen Phang, H., Babu Saheer, L., & Zarrin, B. (2021). Blockchain for decentralization of internet: Prospects, trends, and challenges. *Cluster Computing*, *24*(4), 2841–2866. doi:10.1007/s10586-021-03301-8 PMID:34025209

Chapter 7
Revolutionizing Responsible Tourism:
Cutting-Edge Technological Advances in Geotagging

Muhammad Khizar Hayat Naeem
https://orcid.org/0009-0008-5996-5321
Harbin Institute of Technology, China

Munir Ahmad
https://orcid.org/0000-0003-4836-6151
Survey of Pakistan, Pakistan

Froilan Delute Mobo
https://orcid.org/0000-0002-4531-8106
Philippine Merchant Marine Academy, Philippines

ABSTRACT

The introduction of geotagging has transformed the tourism industry, impacting environmental conservation, social engagement, and economic development. Through GPS-enabled devices, augmented reality applications, social media networks, and more, tourists and stakeholders can enhance experiences and support responsible tourism practices. Geotagging can contribute significantly to biodiversity conservation, sustainable management, and environmental education, fostering a deeper appreciation for nature. Sociologically, it can promote community connection, cultural preservation, and tourist safety. Economically, it can boost revenue generation, marketing, and infrastructure development. Challenges include privacy concerns, data security risks, and the digital divide, which need addressing through prioritizing data protection and closing the digital gap. Future research should focus on enhancing geotagging tools, exploring blockchain and AI, and ensuring ethical use of geotagged data, thus propelling the tourism industry towards sustainable growth and innovation.

DOI: 10.4018/979-8-3693-3286-3.ch007

INTRODUCTION

The tourism industry has an important role in the world's economy and provides people with a variety of services, sites, and experience options, which makes millions of people decide to travel (Borysova et al. 2022; Lew 2011; Thommandru et al. 2023). From the unparalleled to the landmarks of civilization, tourism creates a mirror to the extremely diverse geography and human culture, a lifelong tapestry. Tourism, being a big economic force, in turn, employs a great number of people across all related sectors comprising hospitality, transport, entertainment, and retail, which makes it an essential part of economies in many countries throughout the globe.

These days, technology holds a key position overall with tourism being the biggest beneficiary of all. Location-based recommendation systems, fueled by advanced algorithms and huge databases not only help travelers to know about new places, attractions, accommodations, etc. but their preferences and interests can be also matched in these systems (Kumar et al. 2020; Al-Nafjan, Alrashoudi, and Alrasheed 2022; Bao et al. 2015). These recommendation systems basically take benefit from various sources of data, for example; user preferences, traveling history patterns, ratings and reviews, location information, and available real-time updates in order to deliver specific recommendations. These systems are enabled by the analysis of a vast amount of input data thus making it possible to generate insights about travelers' choices, behavior, and trends. These insights are useful in making their decisions and maximizing their experience.

The geo-tagged data and the use of GPS technologies are the backbone of the location-based recommendation systems that provide precise recommendations (Sun et al. 2015; Rehman, Khalid, and Madani 2016; AlBanna et al. 2016). Geo-tagging, a frequent feature of digital photography and community photo sharing, is the basis for this chain reaction. It is based on geolocation extraction and techniques such as geoparsing and geotagging, where images are geo-coordinated. With the automatic generation of hashtags and geotags, these platforms make searching for videos in the collections and communities online easier. The integration of geo-tagged data and GPS technology results in more accurate and impactful location-based recommendations which adds to the overall quality of travel experience for users.

Similarly, advanced medium-sharing services and mobile technologies create a large volume of geotagged data, providing a new method for tourism and hospitality researchers to analyze tourist movement and behavior (Wong, Law, and Li 2017). Geotags can improve visual image recognition by providing additional information such as raw values of latitude and longitude, and visual features of aerial photos around the location the geotag represents (Yaegashi and Yanai 2009). Significant growth and diversification in the use of tracking technologies in tourism research over the past decade have led to advancements in understanding tourist behavior and spatial dynamics (Shoval and Ahas 2016). GPS technology accurately tracks paths tourists take and provides a greater understanding of their spatial behavior, contributing to the management and planning of destinations (Edwards 2012). High-resolution data on tourist mobility, collected through tracking technologies like GPS and smartphones, can open up new lines of tourism research and help manage social, environmental, and cultural impacts (Shoval 2018).

In light of this context, the primary aim of this chapter is to examine technological advancement in geotagging in the context of tourism. To accomplish this goal, the chapter is organized as follows: The second section elucidates technological advancement in geotagging. The third section delves into the implications of geotagging on environmental, social, and economic aspects. The subsequent section explores challenges and potential solutions. The final section offers a concluding summary of the chapter with a window for future work.

METHODOLOGY

To thoroughly investigate the transformative effects of cutting-edge technological innovations in geotagging on responsible tourism, a comprehensive review of scholarly literature was conducted. The evaluation framework (Figure 1) comprises three key steps:

- **Technological Advancements in Geotagging**

This step involves the exploration of articles, research papers, and academic publications that delve into various technologies driving geotagging. The aim is to identify the advancements in geotagging technology.

- **Review of Geotagging Platforms**

This step involves the study of platforms that provide geotagging functionalities for their users. The aim is to make a comparison of different geotagging platforms.

- **Implications of Geotagging on Responsible Tourism**

Following the identification of geotagging-related technologies, this step focuses on examining the impact of geotagging on the environmental, socio-cultural, and economic aspects of responsible tourism. A detailed analysis is carried out to understand these implications.

- **Challenges and Potential Solutions**

Despite its support for responsible tourism, geotagging applications present challenges that need to be addressed. In this step, these challenges are identified and potential solutions are explored to mitigate them effectively.

Figure 1. Methodology

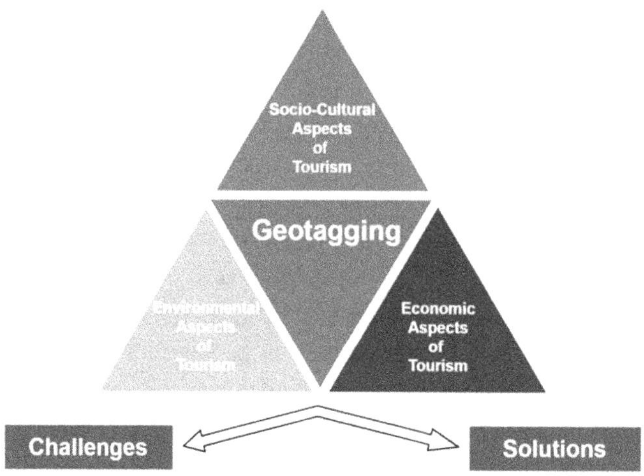

TECHNOLOGICAL ADVANCEMENTS IN GEOTAGGING

The development of geotagging technology (Figure 2) is fundamental for the introduction and implementation of responsible behaviors on the part of tourists which can lead to curious visitors, environmental awareness, and conservation support.

GPS Enabled Devices

Modern life has a touch of class with the use of gadgets like mobile phones and PDAs that may be expensive but normally come with a GPS receiver that allows the users to determine where the device is at any time. GPS-driven devices produce the individuals' location information that can be linked to a certain person. These devices can be used as sensors to measure activity on a city or regional scale, adding new layers of knowledge to urban studies (van der Spek et al. 2009). Moreover, GPS and geo-enabled technologies can help older people and the disabled by giving accurate addresses and location particulars for medical injuries and accident dispatch (Boulos et al. 2011). Furthermore, GPS-enabled cell phones can provide real-time location-based services like the travel assistance device (TAD) for transit riders with special needs, increasing their independence and security (Barbeau et al. 2010).

Modern geotagging technology offers high precision and accuracy in location tracking. GPS-enabled devices such as smartphones and dedicated GPS units enable tourists to navigate and explore locations with unparalleled accuracy thereby diminishing the dangers of getting lost. GPS offers advantages over traditional methods for tracking visitors, offering more reliable, accurate, and precise data, and allowing for more detailed analyses and investigations (Hallo et al. 2012). Thus, it can be said that GPS-enabled devices can empower tourism research to gain new insights into the development of the tourism industry (Shoval and Ahas 2016).

Geo-based technology contributes to the acquisition of geographic knowledge and behavior, enhancing tourism experiences (Tussyadiah and Zach 2012), and enabling real-time study of tourists' activities and movement patterns (Zakrisson and Zillinger 2012; Shoval, Isaacson, and Chhetri 2014). Smartphone-based GPS technology and a mobile app can help to track tourists through regional destinations and attractions over an extended period, contributing to tourism research (Hardy et al. 2017).

Augmented Reality (AR)

AR implies a technology that imposes virtual elements into a real environment to provide necessary and important data or instructional matter. AR is the display of cyber data on top of one's already existing environment, an environment in which viewing the objects, people and spaces around him/her provides high interactive customer engagement (Scholz and Smith 2016). AR is an artificial human-machine interaction as it vertically superimposes the visual, audible, or dynamic information on the physical or tangible environment of the industry (Ong, Yuan, and Nee 2008).

VR and AR in the tourism context have improved immersion and the feeling of telepresence (Loureiro, Guerreiro, and Ali 2020). AR applications in tourism contribute to marketing, information, and experience channels, with five value dimensions: marketing, economic, tourist, epistemic, and organizational (Cranmer, tom Dieck, and Fountoulaki 2020). For instance, AR applications in urban heritage tourism form a user experience based on the correlation of product features and tourists' perceptions and experiences (D. I. Han, tom Dieck, and Jung 2018). AR applications can enhance tourist experiences by translating psychological and behavioral indicators into relevant technical design elements for mobile AR tourism applications in the context of urban heritage tourism (D. I. D. Han, Jung, and Tom Dieck 2019). AR technology in tourism increases involvement in tourist activity and diversifies tourism experiences, positively impacting tourist satisfaction levels (Genç 2018).

AR applications feature geotagging capabilities to allow users to view digital content layered onto the actual world they are in. In the world of responsible tourism, AR can transform visitors 'experiences by offering them interactive directions, historical data, and ecological information about the places they are visiting. This deepens tourists' knowledge about the unique features of local culture, ecosystems, and conservation work. AR applications in tourism can enhance the experience by providing immersive and interactive content, making it possible to deliver augmented reality content (Kounavis, Kasimati, and Zamani 2012). Similarly, AR technology in water tourism, like the Maritime Educational Trip system, helps expand the tourism space by combining geolocation information with accumulated knowledge of a specific location (Kaźmierczak et al. 2021).

Social Media Networks

Social network platforms facilitate the geotagging element that enables users to tag the location when sharing their photos with friends and networks online. This way tourists do not only talk about their travel but also teach others responsible tourism practices. Marking places such as eco-friendly accommodations, sustainable attractions, and conservation sites, can be an inspiration for other people to also make environmentally conscious choices.

Social media platforms play a significant role in facilitating geotagging, enabling users to tag locations when sharing content. Geotagged social media data can be used to construct tourism hotspot networks, improve understanding of tourism activities, and provide decision support for resource development (Wu et al. 2018). GPS devices can enable tourists to record trajectories and upload geo-tagged photos on social media, which can be used to generate tourism paths by matching user-uploaded photos to tourism sites and identifying places of interest inside tourism sites (Zeng et al. 2012).

Geofencing

Geofencing platforms provide context-relevant information to mobile customers who enter a geospatial "fence" or area limit around the area where the message is bound, which in turn makes location-oriented advertising possible (Greenwald et al. 2011). Geofencing technology develops digital boundaries around identified geographical points and users receive alerts when they cross the boundaries. Responsible tourism can benefit from geofencing technology due to its provision of location-based messages, safety tips, and conservation guidelines to tour visitors. For instance, geofencing can guide visitors on areas to avoid such as protected wildlife zones, disallowed activity areas, or ecofriendly activities that exist in the destination. Similarly, geofencing technology can help tourists navigate specific areas by providing detailed maps, multimedia content, and interactive exhibits using smartphones (Park 2014). For example, geofencing technology affects Egyptian travel agencies' marketing campaigns and customers' purchasing intention for tour services as noted by Taha and El-Mawardy (2022). Similarly, the Contextual Geofencing Mobile Tourism Service proactively can send notifications to visitors' mobile devices according to their context, providing personalized information (Martin, Alzua, and Lamsfus 2011).

Mobile Applications

Mobile phones are increasingly integral to daily life, offering numerous applications catering to various aspects, including tourism and travel. Mobile computing devices equipped with GPS receivers can be used to build mobile information technology for tourist support at all stages of their journey, with Google Maps services playing a role in this support (Savchuk and Pasichnyk 2015).

Personal travel applications that are geotagging ready are the new buzz in the application stores. Moreover, these kinds of applications provide customized travel experiences based on user interests and preferences. These applications can make riding, lodgings, and adventure operators' recommendations sustainable. Mobile devices play a significant role in receiving efficient and effective recommendations. The machine-learning approach can be harnessed to offer coffee shop recommendations based on geotagging (Shukla et al. 2023).

Supporting valuable choices in tourism such as mobile applications can open the door to sustainable and responsible travel for users who do not want to harm the environment. Mobile-based geotagging applications offer real-time monitoring and ensure data authenticity. Moreover, open-source technology facilitates AI-enabled geotagging and the dissemination of ground information through WebGIS applications (Paul, Chauhan, and Dutta 2023).

Crowdsourced Mapping

Crowdsourcing is a business model that utilizes Human Intelligence Tasks (HITs) and involves the crowd, the crowdsourcer, the crowdsourced task, and the crowdsourcing platform (Hosseini et al. 2015). Crowdsourcing is an online, distributed problem-solving and production model that uses a crowd of innovators for problem-solving and production (Brabham 2008). Crowdsourcing in mapping involves humans actively involved in tasks like data collection, which are spatiotemporal and must be completed at a specific location and time (Tong et al. 2020). Crowdsourcing Mapping involves workers performing tasks related to a location using their mobile phones and subscribing to distance, live travel time, and detailed route information from web mapping services (Zhang et al. 2020).

The crowdsourced mapping platforms allow users to share their geotagged data including points of interest, hiking trails, and environmental hazards. These web-based sites aided collective endeavors of mapping that benefit both tourists and the locals who live there. Moreover, geo-tagged information can be crowdsourced for travelers to gain current and useful maps as well as recent insights, which ultimately helps to strengthen community-led conservation programs. For example, crowdsourcing geospatial data from Weibo in Shanghai provides insights into the relationship between crowdedness and popularity of tourist attractions, potentially aiding in the sustainable development of tourist attractions (Shi, Zhao, and Chen 2017). Furthermore, crowdsourced mapping can help businesses identify trends and tailor recommendations, increasing user confidence and empowering sustainable tourism (Leal et al. 2020). For Example, UbiAsk, a mobile crowdsourcing application, assists foreign visitors by involving local residents to answer image-based questions in a timely fashion, with reliable response speed and response quantity (Liu et al. 2012). Crowdsourcing tools can amplify public participation in the tourism market by allowing members of different ages to contribute to territorial knowledge, engage in activities, and collaborate through digital tools (Ribeiro et al. 2020).

Figure 2. Technological advancements in geotagging
Source: Scopus AI

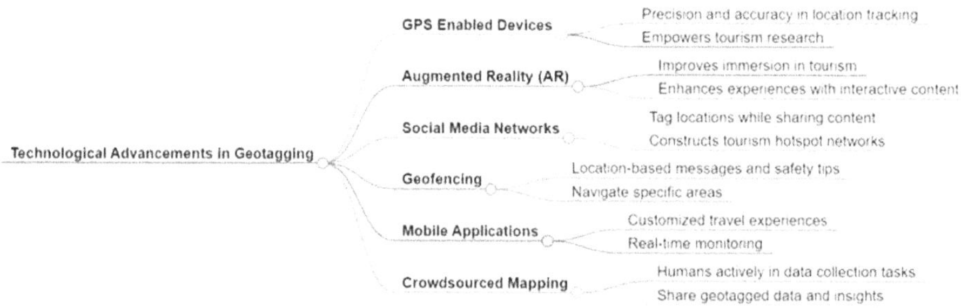

REVIEW OF GEOTAGGING PLATFORMS

In today's digital age, numerous platforms can empower users to geotag their locations, enhancing their online experiences and interactions. This section offers an exhaustive overview of various geotagging platforms, providing users with diverse options for tagging their locations.

Instagram

Widely used not only by visual content creators, Instagram lets users attach location tags to the photos and videos that they share. This feature enables individuals to share their travel experiences, discover new destinations, and engage with location-specific content within the Instagram community (Gunter and Önder 2021; Krisna, Handayani, and Azzahro 2019).

Facebook

Facebook enables geotagging by leveraging its popular user base as well as integrated social mapping functions such as status updates, check-ins, or photo uploads. Users can share their whereabouts with friends and followers, explore nearby attractions, and participate in location-based discussions and events (BAIK 2020).

Twitter

Twitter is not just about microblogging; users can also put location tags in their tweets to provide context and relevance to their posts. Geotagged tweets allow individuals to share real-time updates, discover trending topics in specific locations, and engage in location-based conversations (Chua et al. 2016; Issa et al. 2017).

Google Maps

As a leading mapping service, Google Maps offers robust geotagging capabilities, allowing users to mark and share locations with precision. Whether creating custom maps, reviewing businesses, or sharing directions, Google Maps empowers individuals to contribute geospatial information and enrich the mapping ecosystem.

Yelp

One of the primary goals of Yelp is the use of reviews of local businesses and recommendations, and the geotag function to add more depth to user-generated content. Users can geotag their reviews, photos, and check-ins, helping others discover and assess nearby establishments based on location, ratings, and reviews.

TripAdvisor

A popular platform for travel planning and advice, TripAdvisor features geotagging as a search tool to help visitors explore destinations and make the right choices. Users can geotag their reviews, upload location-specific photos, and access geolocated recommendations and insights from fellow travelers.

OpenStreetMap

Unlike proprietary mapping services, OpenStreetMap is an interactive collaborative mapping project that gives users a chance to contribute, edit, and download free-of-charge geographical data. Geotagging on OpenStreetMap allows individuals to map and annotate locations worldwide, fostering community-driven mapping initiatives and promoting spatial data accessibility.

Flickr

Recognized as a photo-sharing platform, Flickr supports geotagging to geographically reference uploaded photos. Users can directly interact with geotagged photos, make their travel maps as well as share their photo content with geospatial groups displaying geospatially curated photo collections in the Flickr community (Sun et al. 2015; Yuan and Medel 2016).

Pinterest

Pinterest is a visual discovery and bookmarking platform where users can discover and save ideas for various interests, including travel, home decor, fashion, and more. Users can create boards to organize and save images, including those with geotags and share them with others. Geotagging on Pinterest allows users to curate collections of location-specific inspiration, such as travel destinations, local attractions, or dream vacation spots.

The array of geotagging platforms and websites, including Instagram, Facebook, Twitter, Google Maps, Yelp, TripAdvisor, OpenStreetMap, Pinterest, and Flickr, offer users diverse avenues for tagging and sharing their locations. Whether documenting travel experiences, reviewing local businesses, or contributing to collaborative mapping efforts, these platforms play a pivotal role in enriching online interactions and fostering geospatial awareness in the digital realm as described in Table 1.

Table 1 outlines the key features of the geotagging platforms. It highlights the distinct features, user bases, and primary uses of each geotagging platform, aiding in the understanding of their respective functionalities and purposes in the digital landscape.

Table 1. Key features of geotagging platforms

Platform/Website	Key Features	User Base	Primary Use
Instagram	Geotagging photos and videos, location-based content discovery	Over 2 billion monthly active users (https://datareportal.com/)	Sharing travel experiences, visual storytelling
Facebook	Status updates, check-ins, photo uploads with location tags	Over 3 billion monthly active users (https://datareportal.com/)	Sharing personal experiences, connecting with friends
Twitter	Geotagging tweets for real-time updates and location-based discussions	Over 600 million monthly active users (https://datareportal.com/)	Sharing news, opinions, and updates
Google Maps	Marking and sharing locations, reviewing businesses, creating custom maps	Widely used globally	Navigation, local business discovery
Yelp	Geotagging reviews, photos, and check-ins for local businesses	Over 178 million unique monthly visitors (https://www.enterpriseappstoday.com/)	Discovering and reviewing local establishments
TripAdvisor	Geotagging reviews, photos, and accessing location-based recommendations	Over 463 million unique monthly visitors (https://www.semrush.com/)	Planning and researching travel destinations
OpenStreetMap	Collaborative mapping, editing, and accessing open-source geospatial data	A global community of contributors	Creating and maintaining open-source maps
Flickr	Geotagging photos and exploring geospatially curated photo collections	Millions of users worldwide	Sharing and discovering photography
Pinterest	Pinning and organizing images, boards, and locations	Over 480 million monthly active users (https://datareportal.com/)	Collecting and sharing visual inspiration

IMPLICATIONS OF GEOTAGGING ON ENVIRONMENTAL, SOCIAL, AND ECONOMIC ASPECTS

Modern developments related to geo-tagging have a great impact on the environmental, socio-cultural, and economic dimensions of tourism, which is the agenda of this section as depicted in Figure 3.

Environmental Implications

Conservation and Preservation

Small-scale geo-sensor networks, wireless communication networks, low-power, short-range radio-based communication networks, miniaturization of computing and storage platforms, and novel microsensors and sensor materials have changed dynamic environmental monitoring (Nittel 2009). Advances in microsensor technology and miniaturized computing platforms enable geo-sensor networks to continuously collect geospatial information in unprecedented spatial and temporal scales (Nittel, Labrinidis, and Stefanidis 2008). Whereas, GPS is especially advantageous in monitoring and preserving the environment. Scientists and canaonlies people use mobilized devices in conjunction with GPS to study wildlife's routes, habitat changes, and the environmental effects caused by tourism. Geotagged information collected through remote sensors, drones, and satellite monitoring, presents manifold uses for wildlife conservation, and to reduce human-wildlife conflicts.

Through geo-tracking and geotagging techniques, hikers and conservationists can jointly monitor, protect, and save unique ecosystems and biodiversity centers. Through precise tracking of visitors' movements and identification of sensitive environmental areas, geotagging facilitates minimizing tourism's carbon footprint. For example, Geo-tagger application encourages people to observe the natural world around them, document their observations, share them, and discuss them with peers and colleagues, potentially promoting conservation and preservation (Fails et al. 2015). Similarly, geotagging of natural history collections can help in obtaining valuable information in biological and ecological research by assigning geographical coordinates and spatial attributes to specimens (Nowak, Słupecka, and Jackowiak 2021).

Sustainable Management

Geographical information systems (GIS) enhance understanding of relationships between the environment and the economy, providing information on environmental opportunities and constraints in a visible and understandable form to decision-makers (Scott and Rajabifard 2017; Kurowksa et al. 2021; Manning 1990). Geotagging is one form of spatial data that can be fueled in GIS to do insightful analysis.

Geotagging has the potential to make the execution of sustainable tourism easy and effective by giving real-time information on tourist movements, environmental degradation, and habitat disturbances. Through this data, destination managers can be able to formulate useful conservation measures as well as control human activities in these ecosystems to reduce ecological footprints. For example, geotagged photos can help address challenges in capturing and understanding tourist travel behaviors, providing useful practical implications for destination development, transportation planning, and impact management (Vu et al. 2015).

Environmental Education

Geotagged data provides valuable information for spatial analysis in geography., whereas, geography can facilitate environmental education by allowing students to explore spatial patterns and interconnections of environmentally related topics at different scales (McKeown-Ice 1994). Incorporating geospatial technologies in environmental curricula can allow students to take the lead in their research, view field data interactively, and analyze multiscale data during field data collection (Norton et al. 2019). Students can learn physical geography concepts by analyzing environmental data collected in contrasting environments through an innovative pedagogical approach (Roberts et al. 2010). Geo-tagging applications can enhance environmental consciousness and education among tourists through knowledge maps, ecological data, and conservation ideas. As geotagged content becomes increasingly popular, travelers progressively find out more about the local ecosystems, cultural heritage, and sustainable living ways.

Social Implications

Community Engagement

Geotagging can facilitate community engagement and involvement in tourism planning activities. Local people can use photos tagged with location details about cultural sites, traditional practices, and community-based enterprises to make tours richer and more profitable for residents. Moreover, geotagged social media data from Twitter yields more detailed spatial, temporal, and demographic information on tourist movements, potentially aiding in collaborative tourism planning (Chua et al. 2016). Furthermore, geotagging social media data helps construct tourism hotspot networks, improving understanding of tourism activities and providing decision support for resource development (Wu et al. 2018).

Geotagged photos from social media can help address challenges in capturing and understanding tourist travel behaviors, providing practical implications for destination development, transportation planning, and impact management (Vu et al. 2015). Geotagging on social media helps predict tourists' preferences in new or unknown cities and generates better recommendations compared to traditional landmark recommendation methods (Majid et al. 2013). Additionally, geotagging helps in collaborative tourism planning by adding location metadata to existing databases and accessing information stored in these databases via geospatial services (Dickinger et al. 2008).

Cultural Preservation

Geotagging leverages the principles and techniques of geomatics to accurately determine and record the geographic location of objects or events. Geoinformatics contributes to the achievement of Cultural Heritage SDGs by enhancing protection and promoting sustainable tourism (Xiao et al. 2018). Geomatics tools and techniques, such as high-resolution data recording and integrating information from different sensors, can help document and preserve cultural heritage (Tucci and Bonora 2015). Geotagging of historical locations, archaeological sites, and indigenous lands can help to protect and promote cultural heritage sites. Through cultural diversity and heritage conservation advocacy, geotagging strengthens intercultural awareness and is a building block toward respect for indigenous rights, customs, and cultures. Additionally, geotagging on social media helps identify and rank cultural heritage resources, enabling smart cultural tourism services based on user context and location (Nguyen, Camacho, and Jung 2017). Geotagging historical and cultural texts enables location-based searching and new geographically related discoveries, enhancing access to heritage data (Kew et al. 2019).

Safety and Security

The geotagging benefits the safety and security of tourists by providing location-based emergency services, navigation guidance, and risk warnings. Visitors can find geolocated data on nearby hazards, and medical facilities, as well as emergency contacts on their smartphones to help them make prompt, informed decisions and meet unpredictable situations. Moreover, geofencing enables remote monitoring of geographic areas and automatic detections when tracked tourists enter or exit these areas, contributing to the safety and security of tourists. For example, geotagged data from Twitter can provide situational and public safety awareness by helping to better understand the situation and detect local geospatial threats (Amirkhanyan and Meinel 2016).

Economic Implications

Tourism Revenue Generation

Geotagged photos on Flickr can reveal tourists' preferences for landmarks and routings, providing users with the most popular destinations and suitable routes in a city (Sun et al. 2015). Geotagging of the tourist places can increase the inflow of visitors, promote attractions in the area, and improve the comfort of tourists. Also, geotagged content like online reviews, photos, and travel blogs advocate the potential tourist's decision to visit a tourist site.

Destination Marketing

Location-based advertising on mobile devices offers valuable opportunities for companies to target consumers individually, based on their current location, and in real-time, while allowing content to be replaced quickly by remote access (Bauer and Strauss 2016). Platforms with geotagging functions can be highly effective in tourism marketing, where they can be used to create unique and personalized experiences for visitors, promote specific products and services, track consumers' preferences, and measure marketing performance. Geotagged data analytics of the destinations can amplify their marketing campaigns, including visitor satisfaction, and thereby add to destination competitiveness. Furthermore, geotagging helps in searching for specific landmarks and provides a good overview of the destination in navigation applications and travel guides (Mu et al. 2020).

Infrastructure Development

Geotagged data can facilitate destination managers to determine which projects to start/complete at the right time and hence use their resources wisely. Geotagging data about tourists and their destinations, routes, or ways of transportation is important for planning and investing money into infrastructure because it is a way to find out the most popular places, routes, or transport means. Moreover, geotagged data can be used to provide services and amenities that visitors need. Geodata that includes info on tourists and their destinations and routes using which modes of transport contribute profoundly to the planning and investment that go into infrastructural projects. Employing geotagged information allows for a determination of locations, paths, and means of public transportation most attractive to travelers.

Figure 3. Implications of geotagging on environmental, social, and economic aspects

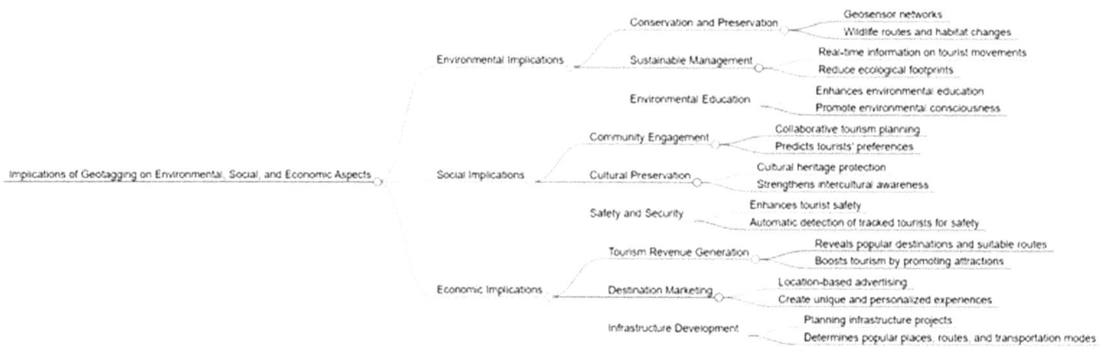

CHALLENGES AND POTENTIAL SOLUTIONS

The application of up-to-date geotagging tools in the sphere of responsible tourism potentially carries advantages and impediments. Whilst these tools are capable of showing the advantages, they also raise questions about privacy, data security, and the digital divide.

Privacy Concerns

The geotagging tools collect and transfer geo-based data; that is why the privacy problem among tourists arises because of the possibility of unauthorized use or disclosure of personal information. To allay privacy issues, the platforms that implement geotagging should employ rigorous data protection measures such as anonymization techniques, encryption protocols, and user consent mechanisms. The adoption of openness in data gathering practices and compliance with the existing regulations relating to privacy, for instance, GDPR (General Data Protection Regulation), can reassure users of the trustworthiness and reliability of the geotagging technologies.

Data Security Risks

Security threats, which include data breaches, malware attacks, and unauthorized access, can undermine existing geotagging tools and reveal confidential information about geotagged data. Implementing cybersecurity industry best practices, like encryption, authentication, and data access controls, helps geotagging tools to ensure that data security risks are minimized. Implementing regular security audits and threat assessments, and constantly training staff are required for building a resilient organization with the current threats in mind.

Digital Divide

Gaps of socioeconomic status and digital divide can be the causes of the fact that less privileged communities and remote destinations have limited access to cutting-edge geotagging tools, thus showing social inequality in tourism development and participation. The promotion of digital literacy programs, the offering of low-cost mobile technology facilities, and the investment in telecommunication infrastructure can be the tools to shorten the digital divide and make geotagging tools available to the marginalized parts of the world. Community-oriented efforts and alliances with local organizations provide oppressed civilians with enough resources to employ geotagging technologies for sustainable tourism.

Ethical Use of Geotagged Data

The responsible use of geotagged data raises questions about data ownership, privacy, and transparency of data collection and control, processing, and sharing. The establishment of transparent ethics accumulations and frameworks for geotagged data capture, storage, and use will ensure responsible data governance and will also protect user rights. Engaging stakeholders in conversations, performing privacy impact assessments, as well as safeguarding accountability through legal oversight can on the other hand support ethical standards and can also lead to trust in geotagging technologies.

CONCLUSION

The introduction of geotagging has revolutionized the tourism industry, affecting environmental conservation, social engagement, and economic development. GPS-enabled devices, augmented reality applications, social media networks, geofencing platforms, mobile applications, and crowdsourced mapping are available to tourists and stakeholders alike for the purpose of better experiences and making decisions to support responsible tourism practices.

The environmental impacts of geotagging on biodiversity are significant as it provides an opportunity for the conservation of species, sustainable environmental management, and environmental education. Stakeholders can analyze geotagged data to monitor wildlife, protect ecosystems, educate visitors about environmental matters, and thus, foster a deeper appreciation for nature and motivate sustainable behavior by using the power of geotagged data.

Sociologically, geotagging emphasizes community connection, cultural heritage, and security matters. Communities can participate in tourism planning as a core activity, ensure the preservation of cultural heritage, and enhance the safety of tourists through location-based services. Simply said, geotagging facilitates people to interact with local communities, investigate cultural areas, and move through unknown areas safely.

Geotagging from the economic side has a contribution to tourism revenue generation, destination marketing, and infrastructure development. With the use of geotagged data businesses can grab consumers on time, advertise tourist destinations, and invest in projects that are essential to tourists resulting in an economic boost and enhancing visitor experiences.

The incorporation of geotagging tools comes with obvious challenges like privacy concerns, data security risks, and the digital divide. In order to address these issues, stakeholders should aim at giving priority to data protection, cybersecurity measures, and efforts to stop the digital gap as this will make sure that geotagging technologies benefit both members of society and ethical standards are upheld while user privacy is respected.

Moving forward, geotagging technology along with responsible tourism is an area where new research can be done and further innovation can be implemented. The first issue is the need to develop the tools used to geotag information with the aim of improving privacy, data safety, and the level of user transparency. It can include the advancement of impenetrable encryption protocols, privacy-ensuring algorithms, and interfaces that make it easy for users so that they can be able to control their geotagged information.

Furthermore, the opportunities that cutting-edge innovations like blockchain and artificial intelligence provide to build up and improve geotagging for the tourism industry can be considered in future studies. As an illustration, blockchain technology provides a distributed and unalterable system for keeping and sharing attributively geotagged data so that its validity can be guaranteed.

REFERENCES

Al-Nafjan, A., Alrashoudi, N., & Alrasheed, H. (2022). Recommendation System Algorithms on Location-Based Social Networks: Comparative Study. *Information (Basel)*, *13*(4), 188. Advance online publication. doi:10.3390/info13040188

AlBanna, B., Sakr, M., Moussa, S., & Moawad, I. (2016). Interest Aware Location-Based Recommender System Using Geo-Tagged Social Media. *ISPRS International Journal of Geo-Information*, *5*(12), 245. Advance online publication. doi:10.3390/ijgi5120245

Amirkhanyan, A., & Meinel, C. (2016). Analysis of the Value of Public Geotagged Data from Twitter from the Perspective of Providing Situational Awareness. In. Lecture Notes in Computer Science: Vol. 9844. *Including Subseries Lecture Notes in Artificial Intelligence and Lecture Notes in Bioinformatics)*. LNCS. doi:10.1007/978-3-319-45234-0_48

Baik, J. (2020). The Geotagging Counterpublic: The Case of Facebook Remote Check-Ins to Standing Rock. *International Journal of Communication*, 4.

Bao, J., Zheng, Y., Wilkie, D., & Mokbel, M. (2015). Recommendations in Location-Based Social Networks: A Survey. *GeoInformatica*, *19*(3), 525–565. Advance online publication. doi:10.1007/s10707-014-0220-8

Barbeau, S. J., Winters, P. L., Georggi, N. L., Labrador, M. A., & Perez, R. (2010). Travel Assistance Device: Utilising Global Positioning System-Enabled Mobile Phones to Aid Transit Riders with Special Needs. IET Intelligent Transport Systems, 4. doi:10.1049/iet-its.2009.0028

Bauer, C., & Strauss, C. (2016). Location-Based Advertising on Mobile Devices: A Literature Review and Analysis. *Management Review Quarterly*, *66*(3), 159–194. Advance online publication. doi:10.1007/s11301-015-0118-z

Borysova, O., Gryniuk, O., Mykhailenko, T., Shparaha, T., & Hryniuk, T. (2022). Global Trends in Tourism and Imperatives of the Circular Economy in the Context of Sustainable Environmental Development of States. *Revue d'Economie Financiere*, *20*(1), 196–202. Advance online publication. doi:10.55365/1923.x2022.20.23

Boulos, M. N., Anastasiou, A., Bekiaris, E., & Panou, M. (2011). Geo-Enabled Technologies for Independent Living: Examples from Four European Projects. *Technology and Disability*, *23*(1), 7–17. Advance online publication. doi:10.3233/TAD-2011-0300

Brabham, D. C. (2008). Crowdsourcing as a Model for Problem Solving: An Introduction and Cases. *Convergence (London)*, *14*(1), 75–90. Advance online publication. doi:10.1177/1354856507084420

Chua, A., Servillo, L., Marcheggiani, E., & Vande Moere, A. (2016). Mapping Cilento: Using Geotagged Social Media Data to Characterize Tourist Flows in Southern Italy. *Tourism Management*, *57*, 295–310. Advance online publication. doi:10.1016/j.tourman.2016.06.013

Cranmer, E. E., tom Dieck, M. C., & Fountoulaki, P. (2020). Exploring the Value of Augmented Reality for Tourism. *Tourism Management Perspectives*, *35*, 100672. Advance online publication. doi:10.1016/j.tmp.2020.100672

Dickinger, A., Scharl, A., Stern, H., Weichselbraun, A., & Wöber, K. (2008). Acquisition and Relevance of Geotagged Information in Tourism. In Information and Communication Technologies in Tourism 2008. doi:10.1007/978-3-211-77280-5_48

Edwards, D. (2012). Using GPS to Track Tourists Spatial Behaviour in Urban Destinations. SSRN *Electronic Journal*. doi:10.2139/ssrn.1905286

Fails, J. A., Herbert, K., Hill, E., De Stefano, A., Hesse, B., Cushman, P., Gant, T., & ... 2015). Geotagger: A Collaborative Environmental Inquiry Platform. *2015 International Conference on Collaboration Technologies and Systems, CTS 2015*. 10.1109/CTS.2015.7210453

Genç, R. (2018). *The Impact of Augmented Reality*. Technology on Tourist Satisfaction., doi:10.1007/978-3-319-64027-3_8

Greenwald, A., Hampel, G., Phadke, C., & Poosala, V. (2011). An Economically Viable Solution to Geofencing for Mass-Market Applications. *Bell Labs Technical Journal*, *16*(2), 21–38. Advance online publication. doi:10.1002/bltj.20500

Gunter, U., & Önder, I. (2021). An Exploratory Analysis of Geotagged Photos From Instagram for Residents of and Visitors to Vienna. *Journal of Hospitality & Tourism Research (Washington, D.C.)*, *45*(2), 373–398. Advance online publication. doi:10.1177/1096348020963689

Hallo, J. C., Beeco, J. A., Goetcheus, C., McGee, J., McGehee, N. G., & Norman, W. C. (2012). GPS as a Method for Assessing Spatial and Temporal Use Distributions of Nature-Based Tourists. *Journal of Travel Research*, *51*(5), 591–606. Advance online publication. doi:10.1177/0047287511431325

Han, D. I., tom Dieck, M. C., & Jung, T. (2018). User Experience Model for Augmented Reality Applications in Urban Heritage Tourism. *Journal of Heritage Tourism*, *13*(1), 46–61. Advance online publication. doi:10.1080/1743873X.2016.1251931

Han, D. I. D., Jung, T., & Claudia Tom Dieck, M. (2019). Translating Tourist Requirements into Mobile AR Application Engineering Through QFD. *International Journal of Human-Computer Interaction*, *35*(19), 1842–1858. Advance online publication. doi:10.1080/10447318.2019.1574099

Hardy, A., Hyslop, S., Booth, K., Robards, B., Aryal, J., Gretzel, U., & Eccleston, R. (2017). Tracking Tourists' Travel with Smartphone-Based GPS Technology: A Methodological Discussion. *Information Technology & Tourism*, *17*(3), 255–274. Advance online publication. doi:10.1007/s40558-017-0086-3

Hosseini, M., Shahri, A., Phalp, K., Taylor, J., & Ali, R. (2015). Crowdsourcing: A Taxonomy and Systematic Mapping Study. *Computer Science Review*, *17*, 43–69. Advance online publication. doi:10.1016/j.cosrev.2015.05.001

Issa, E., Tsou, M. H., Nara, A., & Spitzberg, B. (2017). Understanding the Spatio-Temporal Characteristics of Twitter Data with Geotagged and Non-Geotagged Content: Two Case Studies with the Topic of Flu and Ted (Movie). *Annals of GIS*, *23*(3), 219–235. Advance online publication. doi:10.1080/19475683.2017.1343257

Kaźmierczak, R., Szczepańska, A., Kowalczyk, C., Grunwald, G., & Janowski, A. (2021). Using Ar Technology in Tourism Based on the Example of Maritime Educational Trips—A Conceptual Model. *Sustainability (Basel)*, *13*(13), 7172. Advance online publication. doi:10.3390/su13137172

Kew, T., Shaitarova, A., Meraner, I., Goldzycher, J., Clematide, S., & Volk, M. (2019). Geotagging a Diachronic Corpus of Alpine Texts: Comparing Distinct Approaches to Toponym Recognition. *Proceedings of the Workshop on Language Technology for Digital Historical Archives*, 11–18. 10.26615/978-954-452-059-5_003

Kounavis, C. D., Kasimati, A. E., & Zamani, E. D. (2012). Enhancing the Tourism Experience through Mobile Augmented Reality: Challenges and Prospects. *International Journal of Engineering Business Management*, *4*(1), 10. Advance online publication. doi:10.5772/51644

Krisna, D. F., Handayani, P. W., & Azzahro, F. (2019). The Antecedents of Hashtag and Geotag Use in Smart Tourism: Case Study in Indonesia. *Asia Pacific Journal of Tourism Research*, *24*(12), 1141–1154. Advance online publication. doi:10.1080/10941665.2019.1665559

Kumar, A. P., Shankar, S. G., Praveen, K. R. M., Gadekallu, T. R., Al-Ahmari, A., & Abidi, M. H. (2020). Location Based Business Recommendation Using Spatial Demand. *Sustainability (Basel)*, *12*(10), 4124. Advance online publication. doi:10.3390/su12104124

Kurowksa, K., Marks-Bielska, R., Bielski, S., Aleknavičius, A., & Kowalczyk, C. (2021). Geographic Information Systems and the Sustainable Development of Rural Areas. *Land (Basel)*. Advance online publication. doi:10.3390/land10010006

Leal, F., Malheiro, B., Veloso, B., & Burguillo, J. C. (2020). Responsible Processing of Crowdsourced Tourism Data. *Journal of Sustainable Tourism*. Advance online publication. doi:10.1080/09669582.2020.1778011

Lew, A. A. (2011). Tourism's Role in the Global Economy. *Tourism Geographies*, *13*(1), 148–151. Advance online publication. doi:10.1080/14616688.2010.531046

Liu, Y., Lehdonvirta, V., Alexandrova, T., & Nakajima, T. (2012). Drawing on Mobile Crowds via Social Media. *Multimedia Systems*, *18*(1), 53–67. Advance online publication. doi:10.1007/s00530-011-0242-0

Loureiro, S. M. C., Guerreiro, J., & Ali, F. (2020). 20 Years of Research on Virtual Reality and Augmented Reality in Tourism Context: A Text-Mining Approach. *Tourism Management*, *77*, 104028. Advance online publication. doi:10.1016/j.tourman.2019.104028

Majid, A., Chen, L., Chen, G., Mirza, H. T., Hussain, I., & Woodward, J. (2013). A Context-Aware Personalized Travel Recommendation System Based on Geotagged Social Media Data Mining. *International Journal of Geographical Information Science*, *27*(4), 662–684. Advance online publication. doi:10.1080/13658816.2012.696649

Manning, E. W. (1990). Geographic Information Systems and Sustainable Development. *Government Information Quarterly*, *7*(3), 329–342. Advance online publication. doi:10.1016/0740-624X(90)90028-M

Martin, D., Alzua, A., & Lamsfus, C. (2011). A Contextual Geofencing Mobile Tourism Service. Information and Communication Technologies in Tourism 2011. doi:10.1007/978-3-7091-0503-0_16

McKeown-Ice, R. (1994). Environmental Education: A Geographical Perspective. *The Journal of Geography*, *93*(1), 40–42. Advance online publication. doi:10.1080/00221349408979684

Mu, S., Piwowarczyk, M., Kutrzyński, M., Trawiński, B., & Telec, Z. (2020). Comparative Analysis of Selected Geotagging Methods. Communications in Computer and Information Science, 1178. doi:10.1007/978-981-15-3380-8_29

Nguyen, T. T., Camacho, D., & Jung, J. E. (2017). Identifying and Ranking Cultural Heritage Resources on Geotagged Social Media for Smart Cultural Tourism Services. *Personal and Ubiquitous Computing*, *21*(2), 267–279. Advance online publication. doi:10.1007/s00779-016-0992-y

Nittel, S. (2009). A Survey of Geosensor Networks: Advances in Dynamic Environmental Monitoring. *Sensors (Basel)*, *9*(7), 5664–5678. Advance online publication. doi:10.3390/s90705664 PMID:22346721

Nittel, S., Labrinidis, A., & Stefanidis, A. (2008). Introduction to Advances in Geosensor Networks. In. Lecture Notes in Computer Science: Vol. 4540. *Including Subseries Lecture Notes in Artificial Intelligence and Lecture Notes in Bioinformatics)*. LNCS. doi:10.1007/978-3-540-79996-2_1

Norton, E., Li, Y., Mason, L. R., & Washington-Allen, R. A. (2019). Assessing the Impact of a Geospatial Data Collection App on Student Engagement in Environmental Education. *Education Sciences*, *9*(2), 118. Advance online publication. doi:10.3390/educsci9020118

Nowak, M. M., Słupecka, K., & Jackowiak, B. (2021). Geotagging of Natural History Collections for Reuse in Environmental Research. *Ecological Indicators*, *131*, 108131. Advance online publication. doi:10.1016/j.ecolind.2021.108131

Ong, S. K., Yuan, M. L., & Nee, A. Y. C. (2008). Augmented Reality Applications in Manufacturing: A Survey. *International Journal of Production Research*, *46*(10), 2707–2742. Advance online publication. doi:10.1080/00207540601064773

Park, Y.-J. (2014). Tourism Information System Using Geofencing and NFC Technology. *Journal of Digital Convergence*, *12*(8), 211–217. Advance online publication. doi:10.14400/JDC.2014.12.8.211

Paul, A., Chauhan, S., & Dutta, D. (2023). Mobile-Based Image Interpretation and Geotagging Using Artificial Intelligence and Open-Source Geospatial Technology. *Applied Geomatics*, *15*(4), 795–805. Advance online publication. doi:10.1007/s12518-023-00522-x

Rehman, F., Khalid, O., & Madani, S. A. (2016). A Comparative Study of Location-Based Recommendation Systems. *The Knowledge Engineering Review*, *32*, e7. Advance online publication. doi:10.1017/S0269888916000308

Ribeiro, V., Pereira, M., Matos, O., Alves, J., Remoaldo, P., Gôja, R., & Freitas, I. (2020). Geo-Crowdsourcing Contributions for Cultural Mapping. *Journal of Information Science Theory and Practice*, *8*(1). Advance online publication. doi:10.1633/JISTaP.2020.8.1.5

Roberts, D., Bradley, E., Roth, K., Eckmann, T., & Still, C. (2010). Linking Physical Geography Education and Research through the Development of an Environmental Sensing Network and Project-Based Learning. *Journal of Geoscience Education*, *58*(5), 262–274. Advance online publication. doi:10.5408/1.3559887

Savchuk & Pasichnyk. (2015). Modern Tendention in the Use of Gps Technology in Tourism Industry. *Journal on Economics of Technology*.

Scholz, J., & Smith, A. N. (2016). Augmented Reality: Designing Immersive Experiences That Maximize Consumer Engagement. *Business Horizons*, *59*(2), 149–161. Advance online publication. doi:10.1016/j.bushor.2015.10.003

Scott, G., & Rajabifard, A. (2017). Sustainable Development and Geospatial Information: A Strategic Framework for Integrating a Global Policy Agenda into National Geospatial Capabilities. *Geo-Spatial Information Science*, *20*(2), 59–76. doi:10.1080/10095020.2017.1325594

Shi, B., Zhao, J., & Po, J. C. (2017). Exploring Urban Tourism Crowding in Shanghai via Crowdsourcing Geospatial Data. *Current Issues in Tourism*, *20*(11), 1186–1209. Advance online publication. doi:10.1080/13683500.2016.1224820

Shoval, N. (2018). Sensing Tourists: Geoinformatics and the Future of Tourism Geography Research. *Tourism Geographies*, *20*(5), 910–912. Advance online publication. doi:10.1080/14616688.2018.1437768

Shoval, N., & Ahas, R. (2016). The Use of Tracking Technologies in Tourism Research: The First Decade. *Tourism Geographies*, *18*(5), 587–606. Advance online publication. doi:10.1080/14616688.2016.1214977

Shoval, N., Isaacson, M., & Chhetri, P. (2014). GPS, Smartphones, and the Future of Tourism Research. The Wiley Blackwell Companion to Tourism. doi:10.1002/9781118474648.ch20

Shukla, A., Choudhury, T., Benara, N., Garg, P., Tiwari, A., & Um, J. S. (2023). Geo-Based Recommendation System Utilising Geo Tagging and K-Means Clustering. *Spatial Information Research*, *31*(3), 253–263. Advance online publication. doi:10.1007/s41324-022-00495-w

Sun, Y., Fan, H., Bakillah, M., & Zipf, A. (2015). Road-Based Travel Recommendation Using Geo-Tagged Images. *Computers, Environment and Urban Systems*, *53*, 110–122. Advance online publication. doi:10.1016/j.compenvurbsys.2013.07.006

Taha & El-Mawardy. (2022). *The effect of applying geofencing technology on the Egyptian customers' purchasing intention to tour services.* . doi:10.21608/mkaf.2022.254115

Thommandru, A., Espinoza-Maguiña, M., Ramirez-Asis, E., Ray, S., Naved, M., & Guzman-Avalos, M. (2023). Role of Tourism and Hospitality Business in Economic Development. *Materials Today: Proceedings*, *80*, 2901–2904. Advance online publication. doi:10.1016/j.matpr.2021.07.059

Tong, Y., Zhou, Z., Zeng, Y., Chen, L., & Shahabi, C. (2020). Spatial Crowdsourcing: A Survey. *The VLDB Journal*, *29*(1), 217–250. Advance online publication. doi:10.1007/s00778-019-00568-7

Tucci, G., & Bonora, V. (2015). Geomatics and Management of At-Risk Cultural Heritage. *Rendiconti Lincei. Scienze Fisiche e Naturali*, *26*(S1), 105–114. Advance online publication. doi:10.1007/s12210-015-0427-0

Tussyadiah, I. P., & Zach, F. J. (2012). The Role of Geo-Based Technology in Place Experiences. *Annals of Tourism Research*, *39*(2), 780–800. Advance online publication. doi:10.1016/j.annals.2011.10.003

van der Spek, S., van Schaick, J., de Bois, P., & de Haan, R. (2009). Sensing Human Activity: GPS Tracking. *Sensors (Basel)*, *9*(4), 3033–3055. Advance online publication. doi:10.3390/s90403033 PMID:22574061

Vu, H. Q., Li, G., Law, R., & Ye, B. H. (2015). Exploring the Travel Behaviors of Inbound Tourists to Hong Kong Using Geotagged Photos. *Tourism Management*, *46*, 222–232. doi:10.1016/j.tourman.2014.07.003

Wong, E., Law, R., & Li, G. (2017). Reviewing Geotagging Research in Tourism. In Information and Communication Technologies in Tourism 2017. doi:10.1007/978-3-319-51168-9_4

Wu, X., Huang, Z., Peng, X., Chen, Y., & Liu, Y. (2018). Building a Spatially-Embedded Network of Tourism Hotspots from Geotagged Social Media Data. *IEEE Access : Practical Innovations, Open Solutions*, *6*, 21945–21955. Advance online publication. doi:10.1109/ACCESS.2018.2828032

Xiao, W., Mills, J., Guidi, G., Rodríguez-Gonzálvez, P., Barsanti, S. G., & González-Aguilera, D. (2018). Geoinformatics for the Conservation and Promotion of Cultural Heritage in Support of the UN Sustainable Development Goals. *ISPRS Journal of Photogrammetry and Remote Sensing*, *142*, 389–406. Advance online publication. doi:10.1016/j.isprsjprs.2018.01.001

Yaegashi, K., & Yanai, K. (2009). Can Geotags Help Image Recognition? In. Lecture Notes in Computer Science: Vol. 5414. *Including Subseries Lecture Notes in Artificial Intelligence and Lecture Notes in Bioinformatics)*. LNCS. doi:10.1007/978-3-540-92957-4_32

Yuan, Y., & Medel, M. (2016). Characterizing International Travel Behavior from Geotagged Photos: A Case Study of Flickr. *PLoS One*, *11*(5), e0154885. Advance online publication. doi:10.1371/journal.pone.0154885 PMID:27159195

Zakrisson, I., & Zillinger, M. (2012). Emotions in Motion: Tourist Experiences in Time and Space. *Current Issues in Tourism*, *15*(6), 505–523. Advance online publication. doi:10.1080/13683500.2011.615391

Zeng, Z., Zhang, R., Liu, X., Guo, X., & Sun, H. (2012). Generating Tourism Path from Trajectories and Geo-Photos. In. Lecture Notes in Computer Science: Vol. 7651. *Including Subseries Lecture Notes in Artificial Intelligence and Lecture Notes in Bioinformatics)*. LNCS. doi:10.1007/978-3-642-35063-4_15

Zhang, D., Wen, S., Chen, F., Li, Z., & Zhao, L. (2020). Spatial Crowdsourcing Based on Web Mapping Services. *World Wide Web (Bussum)*, *23*(1), 631–648. Advance online publication. doi:10.1007/s11280-019-00708-7

Chapter 8
Smart Tourism Experience and Responsible Travelers' Behavior:
A Systematic Literature Review

Mourad Aarabe
https://orcid.org/0009-0003-9772-6683
National School of Business and Management of Fez, Morocco

Nouhaila Ben Khizzou
National School of Business and Management of Fez, Morocco

Lhoussaine Alla
https://orcid.org/0000-0002-7238-1792
National School of Applied Sciences of Fez, Morocco

Ahmed Benjelloun
National School of Business and Management of Fez, Morocco

ABSTRACT

The research on smart tourism experiences that take into account the responsible behavior of travelers is booming. The relationship between the development of memorial consumer experiences and the promotion of responsible consumer behavior is a highly sought-after area of research. This is especially true regarding the potential use of emerging technology solutions to enhance tourism services, improve customer smartness, predict responsible behavior, and increase customer satisfaction and loyalty. The purpose of this research is to investigate how the advancement of smart tourism experiences can promote responsible behavior among customers of tourism businesses and destinations. This chapter presents a systematic review of the literature on the interaction between tourism experience and responsible traveler behavior in the context of the transformation of responsible tourism through digital platforms.

DOI: 10.4018/979-8-3693-3286-3.ch008

INTRODUCTION

In the contemporary context, social and environmental issues are receiving increasing attention. The global appeal of the Western consumerist trend has been criticized for the strain it places on our planet (Séré de Lanauze et al., 2021). The tourism sector, which plays a major role in these issues, is at a critical juncture and must fundamentally review its model in light of sustainability imperatives (Adongo et al., 2018; Park et al., 2022).

The rapid transformation of tourism raises important questions about how to integrate ethical and responsible considerations into travel planning. It is important to consider environmental degradation, carbon footprints, and social concerns in developing these considerations (Zhang et al., 2023). Research on responsible tourism has become predominant due to the rapid development of technology and the globalization of the travel industry. Promoting social responsibility and sustainability is a top priority while offering enjoyable and enriching travel experiences. It is important to take into account the community and the environment (Acharya et al., 2023).

Our contribution aims to provide an innovative approach to integrating smart tourism experiences to promote responsible behavior among customers of tourism establishments and destinations (Chuang, 2023). This approach allows travelers to participate in co-creating their experience, gain a better understanding of customer preferences, use technology to predict and influence responsible behavior, and explore the design and evaluation of sustainable tourism experiences to maximize their effectiveness. This study fills a gap in the literature (Bhuiyan et al., 2021; Chuang, 2023).

In this context, the tourism sector is leveraging emerging technologies to transform the way travelers experience their trips (Sustacha et al., 2023). In this shift, artificial intelligence plays a critical role, enabling personalized recommendation systems to offer experiences based on individual preferences, resulting in an enjoyable and memorable travel experience (Bhaskar & Sharma, 2022; Lanquar, 2023). Virtual reality and augmented reality technologies have the potential to transform tourism experiences and destination management by providing an immersive dimension (Hoang & Trang, 2023). According to Yan & Jia, (2021), this immersive experience can stimulate tourists' awe in religious contexts and positively influence their pro-environmental behavior. IoT technologies enable devices to connect and collect data in real time, creating a smooth travel experience (Hoang & Trang, 2023). Big data analytics is crucial for improving the marketing effectiveness of tourism companies (Alla et al., 2022). The use of this technology helps predict trends, improve services, and anticipate customer needs, resulting in a more efficient travel experience for tourists (Vecchio et al., 2018).

Environmentally conscious travelers are acting responsibly by becoming increasingly aware of the environmental impact of their travel and stay (Baloch et al., 2022; Falcão & Roseira, 2022). This translates into choosing environmentally friendly modes of transportation (Li et al., 2023) and supporting local communities through ethical consumption (Vermeir & Verbeke, 2006). Responsible traveler behavior promotes positive relationships between tourists and destinations, resulting in more authentic and meaningful experiences (Ulusoy, 2016). The combination of smart tourism and responsible travel practices provides a significant opportunity for a more ethical and responsible tourism industry (SONKO & DEAC, 2020). Tourism stakeholders can leverage technological advances to promote responsible behavior among travelers and enhance their overall experience (T. H. Lee & Jan, 2023).

In this study, we will conduct a systematic literature review using the PRISMA protocol to address this topic. Our goal is to synthesize existing contributions, identify research limitations and gaps, and propose perspectives for future research. We will present a rigorous approach to consolidate the current theoretical foundations of the field and provide guidance for future research, trends, and challenges for both theorists and practitioners. This chapter has three objectives. Firstly, it reviews the literature on smart tourism experience and responsible traveler behavior. Secondly, it analyzes the strengths and limitations of existing research. Finally, it identifies avenues for future research and proposes directions for future studies. To accomplish these goals, we will divide our work into five sections: a review of the literature, a presentation of the studied methodology, a descriptive analysis of current literature, a diagnosis of research strengths and gaps, and a conclusion with a prospective analysis and research perspectives on future challenges and opportunities in the field.

1. LITERATURE REVIEW

In the age of the digital revolution and emerging technologies, consumers are becoming more autonomous, connected, and engaged in increasingly digitized environments. They are constantly creating, sharing, and demanding smarter consumer experiences. This engagement is justified by the search for and sharing of information, the quest for a sense of belonging, the search for connections, and the building of social networks (Andrews, 2002; Zhang & Hiltz, 2003). The smart consumer is the central figure in the consumer environment, acting as the primary agent of transaction and interaction. They create consumer culture, evaluate products and services, and provide brand advice as well as being active, informed buyers.

Kim et al., (2014), identified three fundamental aspects that characterize the smart consumer experience: " *a focus on the co-creation, sharing and transformation of value between consumers, companies and environments*", "*the promotion of integrated relational capital*" and "*the mobilization of cognitive capital through the combination of megadata analysis and omnichannel distribution*".

Several authors (Buhalis & Amaranggana, 2015; Kumar et al., 2023; T. H. Lee & Jan, 2023; Pai et al., 2020; Sustacha et al., 2023, 2023) have attempted to conceptualize the notion of smart tourism experience from a multidisciplinary perspective. This includes the fields of tourism, marketing, technology, and sustainability. Sustacha et al., (2023), describe smart tourism experience as the utilization of technological advancements to enhance travel quality and personalization. Buhalis & Amaranggana, (2015) argue that the tourism experience includes pre- and post-stay aspects, extending beyond the stay itself. In their study, T. H. Lee & Jan, (2023) analyzed the attributes of smart tourism experiences, including utility and hedonic experiences, provided by emerging technologies like virtual and augmented reality.

The literature highlights the significant impact of emerging technologies on smart tourism experiences and destination management (Hoang & Trang, 2023). Indeed, the fusion of technology and data is crucial in the tourism industry. From this perspective, the Internet of Things (IoT) plays an essential role as an effective tool for collecting real-time data on visitor flows, weather conditions, and more (Rosário & Dias, 2024). From a proactive perspective, the automated processing of data through the use of artificial intelligence algorithms can customize tourist offers according to individual preferences, improving satisfaction and loyalty (Hoang & Trang, 2023). For its part, augmented reality offers immersive and interactive experiences, allowing travelers to virtually explore destinations and attractions before physically visiting them (Kounavis et al., 2012). According to the sustainable approach of Bec et al., (2021), immersive technologies such as virtual and mixed reality offer a second life to tourist destinations through the preservation of world heritage and the revitalization of tourist destinations.

Big data is an essential component of smart tourism experiences as it improves the marketing effectiveness of tourism businesses (Lhoussaine et al., 2022 ; Alla et al, 2022 ; Bouhtati et al, 2023a,b) and smart tourism destinations (Vecchio et al., 2018). It facilitates decision making by providing real-time information on tourism trends and visitor preferences, enabling personalized tourism offers, and promoting the development of innovative (Shamim et al., 2021) and sustainable (Agrawal et al., 2022) business models and services.

From a complementary perspective, AI also offers opportunities to improve customer experience and optimize operations in the tourism and hospitality sectors (Bhaskar & Sharma, 2022). Technological practices such as chatbots for customer service, personalized recommendation systems, predictive analytics for dynamic pricing, facial and biometric recognition, and customer sentiment and feedback (NLP) analysis are increasingly being used in these sectors. Khan et al., (2024) discuss the potential of AI to transform sustainable tourism by improving operational efficiency, providing personalized experiences for travelers, and contributing to the preservation of natural and cultural resources.

The adoption of blockchain technology in tourism and hospitality also has the potential to improve transaction security and manage payments, as noted by (Bodkhe et al., 2019) . However, there are still challenges to be addressed, such as data security, privacy, and ethical concerns related to human dependence on these technologies (Bhaskar & Sharma, 2022; Rosário & Dias, 2024).

Previous research (Baloch et al., 2022; T. H. Lee & Jan, 2023; Park et al., 2022; Samaddar & Mondal, 2023; Shen et al., 2020; SONKO & DEAC, 2020) has highlighted the importance of taking ethical and sustainable considerations into account in the travel planning process. The tourism industry faces challenges in adapting to the demands of sustainability. Marchi et al., (2023) emphasize the importance of integrating economic, social, and environmental considerations into responsible tourism.

Research on responsible travel behavior has explored various factors that influence tourists' sustainability decisions and behavioral intentions, including personal values (Yan & Jia, 2021), social norms (Yoo et al., 2017), and perceived image (Tavitiyaman et al., 2021). However, there are still gaps in our understanding of the underlying mechanisms that motivate travelers to behave responsibly (Wider et al., 2023). The effectiveness of interventions to promote sustainable travel practices is also unclear (Hoang & Trang, 2023).

Many studies have been conducted on smart tourism experiences and responsible traveler behavior. Few have explored the interaction between these two concepts, and most have adopted a fragmented approach. The purpose of this work is to address this gap by conducting a thorough analysis of the current literature on the topic.

2. METHODOLOGY

To provide the answers to our problem, we conducted a systematic literature review. The purpose of this review is to summarize and synthesize the existing literature on the smart tourist experience and its correlation with responsible traveler behavior (Donthu et al., 2021). To ensure transparency, reproducibility, and quality of our work, we followed the PRISMA (Preferred Reporting Items for Systematic Reviews and Meta-Analyses) protocol. We rigorously collected and selected data, as outlined by (Mateo, 2020). The protocol was chosen due to its relevance and robustness. It is widely recognized and recommended for conducting a systematic review of the literature in a rigorous and transparent manner, which enhances the quality and reliability of the results (Page et al., 2021).

We began by defining our search equation, which included all relevant keywords related to our research topic: 'Smart tourism' or 'smart tourism experience' and 'responsible traveler behavior' or 'responsible tourism'. This equation was used to search various academic databases, such as SCOPUS, Springer, Cairn, and MDPI. We filtered the identified documents by field of research to include only social sciences and management sciences research, resulting in 1,168 resources. We eliminated 292 duplicates, leaving us with 876 unique references to evaluate.

We then proceeded to a multi-stage selection based on the eligibility of the references to our research object. Relevance was assessed by examining titles, keywords, and abstracts, resulting in a corpus of 362 potentially relevant references. After a full-text review, 65 sources were deemed relevant to the research topic and included in the review.

To analyze our corpus, we utilized NVIVO and VOSVIEWER. Using NVIVO, we organized, coded, and analyzed textual data extracted from selected references. Vosviewer was used to visualize co-occurrence network mappings of key terms in the literature to understand trends and relationships between concepts.

Figure 1. Information flow in the different phases of the systematic review (Moher et al., 2009)

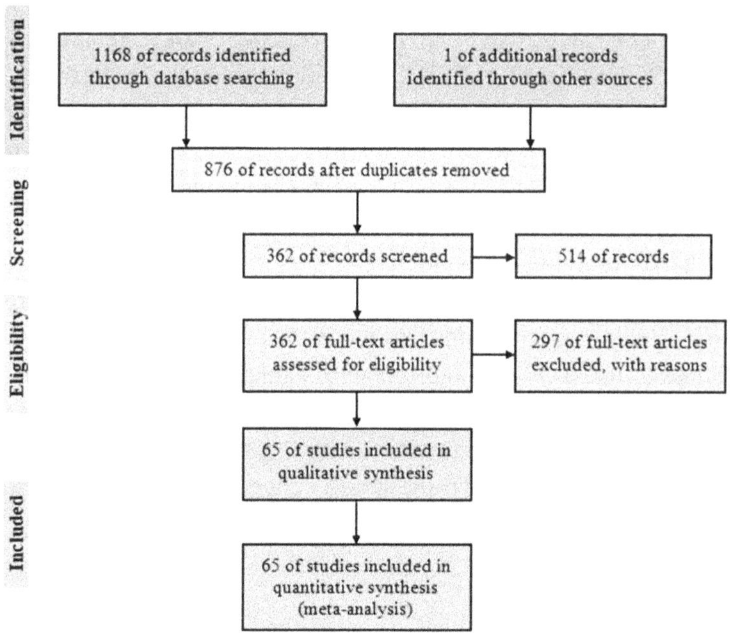

3. A DESCRIPTIVE ANALYSIS OF THE INCLUDED STUDIES

The characteristics of the studies included in our systematic literature review will be presented using Nvivo version 12 software. This will include the type of references, years of publication, journals, and scientific databases in which the studies were published, as well as a word cloud. An interpretation of the results will also be provided.

3.1. Reference Types

Figure 2 displays the distribution of the 65 references by type. Articles are the primary source of information for our literature review, accounting for 88% of the total. The remaining references consist of book chapters (8%) and conference papers (5%).

Figure 2. Distribution by reference type (Authors)

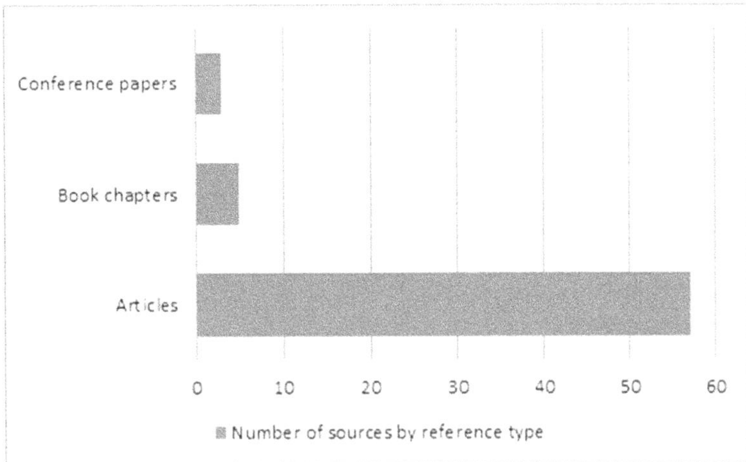

3.2. Publication Years

Figure 3 displays the distribution of our references by year of publication. The curve indicates a rising trend in the number of publications over time, reflecting the growing interest in the subject in recent years. The growth in publications was gradual from 2012 to 2017, with a significant increase in 2019. Starting in 2022, the number of publications increased, peaking at 19 in 2023 and 4 in 2024, despite this study being conducted at the beginning of 2024.

This analysis shows the recent and dynamic evolution of research on the subject, with a focus on the last three years, which account for over 50% of publications. This highlights the relevance of technological advances and emerging trends in the field of smart tourism and the responsible tourism experience.

3.3. Publication Sources

The table below displays the main journals in which our sources were published. The majority of sources were published in nine journals, with 'Sustainability' and 'Journal of Hospitality and Tourism Management' each accounting for 9%, and 'Tourism Management Perspectives' 6%. This indicates that these journals are prime venues for the publication of research on the theme of smart tourism and responsible travel behavior. According to the data, 57% of research on smart tourism and responsible travel behavior is published in 37 different journals classified in the 'other' category. This indicates that research on the subject is disseminated across a wide range of journals specializing in tourism, sustainable development, hotel management, and information science.

Figure 3. Distribution of references by year of publication (Authors)

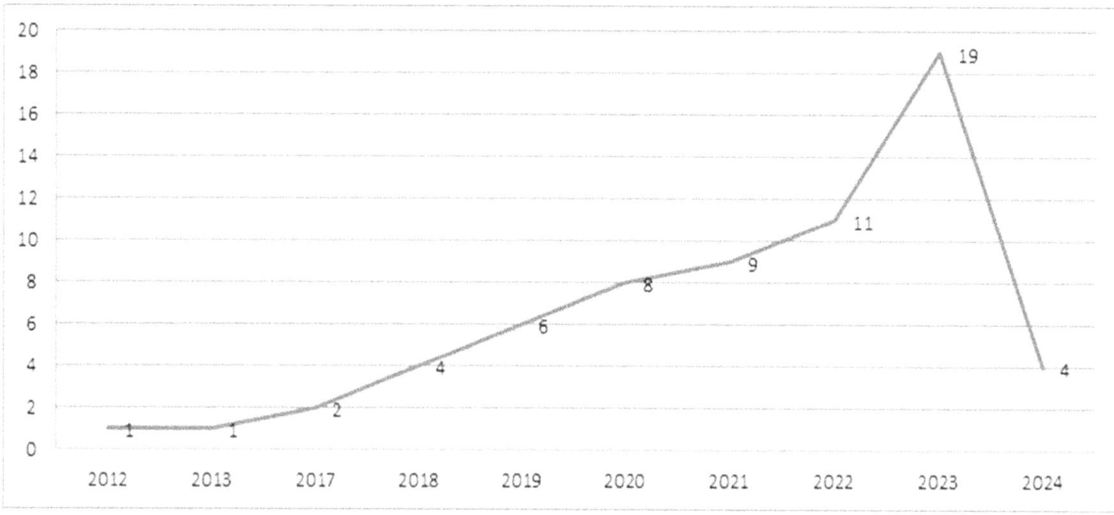

Table 1. Distribution of references by journal title

Journal Title	Number of Corresponding Sources	%
Sustainability	6	9%
Journal of Hospitality and Tourism Management	6	9%
Tourism Management Perspectives	4	6%
International Journal of Information Management Data Insights	2	3%
Journal of Destination Marketing & Management	2	3%
Alternatives Managerialisms Economiques	2	3%
Information Technology & Tourism	2	3%
Tourism Management	2	3%
Administrative Sciences	2	3%
Others	37	57%
Total	**65**	**100%**

Source: Authors

The analysis demonstrates the wide range of publication sources and the extensive research conducted in this field.

The listed journals are available in various scientific databases. Table 2 illustrates that 60% of our resources are accessible through two major databases (46% in Sciendirect and 14% in MDPI). Springer is also a significant source, with 7 resources accounting for 11% of the total. Other databases, such as SCOPUS, Cairn Info, Emerald Insight, and IMIST, also contribute to providing relevant references in this field.

Table 2. The breakdown of references by database

Databases	Number of Corresponding Sources	%
ScienceDirect	30	46%
MDPI	9	14%
springer	7	11%
SCOPUS	5	8%
cairn info	2	3%
Emerald Insight	2	3%
IMIST	2	3%
Others	8	12%
Total	**65**	**100%**

Source: authors

3.4. Keyword Co-Occurrence

The 65 articles analyzed revealed the most recurrent themes in the field. Figure 4 presents a mapping of 82 items grouped into eleven clusters, classified according to the following main keywords: "Sustainable tourism" (purple), "pro-environmental behavior" (dark blue), "smart tourism", "eco-friendly behavior", "socially responsible destination", and "biospheres value" (red), "tourism experience" and "digital tourism" (green), "big data" and "business analytics" (orange), "technology adoption" (light blue), "destination management", and "smart destination" (brown). The map highlights important themes in current tourism research, such as sustainable tourism, the tourism experience, and the use of data and emerging technologies. These trends reflect the concerns of the industry and provide new research opportunities.

Figure 4. Co-occurrence of keywords (Authors)

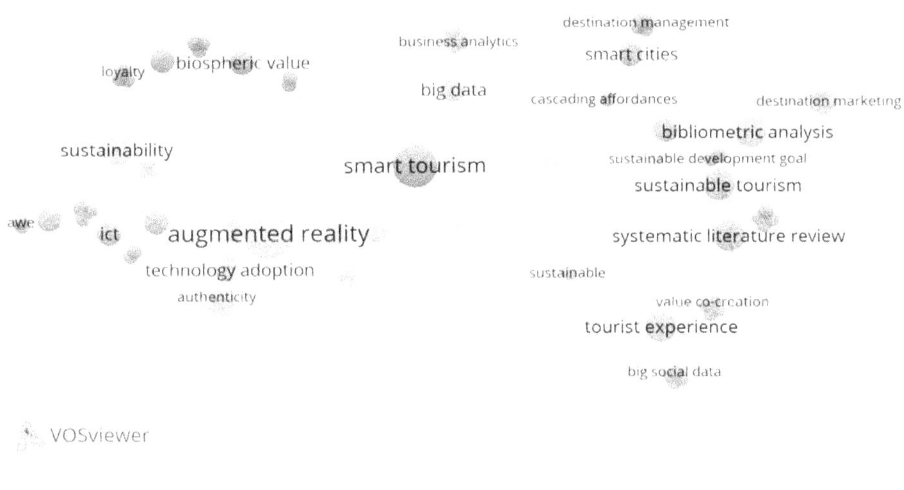

An analysis of keyword trends reveals the dominant topics in literature. This research focuses on tourism, which is cited 13515 times, confirming its importance. Sustainable tourism is gaining interest as it aims to minimize environmental and socio-cultural impacts while ensuring long-term viability. Data analysis, including big data and statistical modeling, is increasingly utilized. Experience is important for understanding tourist satisfaction and loyalty. The tourism industry is utilizing 'smart' technologies such as the Internet of Things, augmented reality, and artificial intelligence to enhance the tourist experience. Tourists prioritize 'value' over cost (Lhoussaine et al., 2022; Alla et al, 2022; Bouhtati et al, 2023a,b). Sustainable development necessitates effective destination planning and management. These trends reflect the concerns and opportunities of the tourism industry. Table 3 displays the 20 most commonly used keywords in the literature.

Table 3. Top 20 most frequently used keywords

Rank	Word	Number	Rank	Word	Number
1	tourism	13515	11	experience	5039
2	use	6717	12	positive	5006
3	issues	6661	13	destination	4972
4	sustainable	5970	14	place	4936
5	study	5937	15	technology	4348
6	data	5563	16	research	4253
7	creation	5558	17	value	4096
8	making	5511	18	tourist	4031
9	development	5392	19	based	3632
10	management	5181	20	smart	3428

Source: authors

Figure 5. Breakdown of references by methodological approach (Authors)

3.5. Methodological Approaches

The literature on smart tourism experience and responsible traveler behavior employs a variety of methodologies, including qualitative, quantitative, and mixed approaches. This diversity enhances the relevance and reliability of the study and promotes a comprehensive understanding of the phenomenon under study.

While there are various approaches to improve the relevance and reliability of a study, it is crucial to consider methodological challenges. One of the most significant challenges is ensuring the reliability of the collected data and its impact on the validity of the results. Additionally, other sectoral and spatial factors may affect the generalizability of the findings.

4. TRENDS AND FINDINGS

This chapter summarizes the main trends and findings of the literature review on responsible travel behavior and the smart tourism experience. It identifies areas requiring further study or development and guides future research.

4.1. Smart Tourism Experience

This work focuses on various aspects of tourism, including the integration of emerging technologies to enhance interaction with tourism sites and services (Shin et al., 2022) and optimizing the customer journey (Cuomo et al., 2021). The antecedents of tourism, such as expectations, behavioral intentions (Shin et al., 2022), emotions (Doyle & Kelliher, 2023) and quality (Kounavis et al., 2012) are studied in depth. The following table summarizes the main studies on the smart tourism experience.

Table 4. Summary of the main studies on the smart tourism experience

Authors	Key Findings
(Ribes & Baidal, 2018)	Smartness and sustainability share common elements, such as long-term planning, creativity, and public-private collaboration.
(Rocha, 2020)	Tourist destinations that integrate information and communication technology (ICT) into their management can enhance the tourist experience, stimulate economic development, and promote sustainability.
(Shen et al., 2020)	Social networking sites have a significant impact on smart tourists at all stages of the travel cycle (before, during and after). The most noticeable impact occurs during the first two stages.
(Stankov & Gretzel, 2020)	Tourism 4.0 technologies provide opportunities to enhance the tourism experience, but also pose risks such as information overload, technology-induced stress, and the dehumanization of travel experiences.
(C.-K. Lee et al., 2021)	Visitor behavior towards the environment is influenced by sustainable smartness, biospheres values, DSR, and the tourism experience.
(Cuomo et al., 2021)	Big Social Data has the potential to enhance the tourism experience and inform management decisions by analyzing tourist perceptions, online discussions, attractive site features, and topics discussed by influencers.
(Shin et al., 2022)	Consumers' technological experience is part of their overall hotel and tourism experience.

Table 4 continued

Authors	Key Findings
(Hoang & Trang, 2023)	The integration of technology and data in smart tourism can enhance the sector's operational efficiency, sustainability, inclusiveness, quality of experience, and competitiveness.
(Li et al., 2023)	An intelligent transport system that prioritizes both the tourist experience and operating costs can improve transport efficiency at tourist attractions.
(Nicola-Gavrilă, 2023)	The tourism industry is being impacted by virtual experiences provided by artificial intelligence, augmented reality, and blockchain. These technologies are also having an effect on the environment.
(Sustacha et al., 2023)	Smart technology features, such as interaction and information, are key to enhancing the tourism experience. However, it is important to note that privacy and security concerns may compromise these benefits.
(T. H. Lee & Jan 2023)	Smart tourism experiences, such as aesthetics, ease of experience, and learning, can have a significant impact on visitors' biosphere values, which in turn influence their environmentally responsible behavior.
(Bhaskar & Sharma, 2022; Khan et al., 2024)	The use of AI can improve visitor satisfaction and optimize operations in the travel and hospitality sectors.
(Agrawal et al., 2022; Alla et al., 2022; Lhoussaine et al, 2022; Del Vecchio et al., 2018; Shamim et al., 2021; Vecchio et al., 2018; Zeng et al., 2020; Bouhtati et al, 2023a,b))	Big Data Analytics (BDA) can revolutionize sustainable tourism by improving productivity, offering personalized experiences, and helping to preserve natural resources. BDA supports the transformation of a city into a smart destination. In the tourism industry, BDA promotes sustainable tourism by selecting destinations, enhancing tourism experiences, and managing resources and environmental impact. BDA can enhance customer relationship management, customer intelligence, campaign optimization, and return on investment (ROI) in tourism marketing. Social data analysis can improve decision-making in smart tourism destinations. It also enables personalized marketing strategies, greater transparency and trust with customers and stakeholders, and the emergence of new business models. Social Big Data is a valuable source of open innovation in tourism, as it supports the definition of sustainable tourism experiences.
(Buhalis & Amaranggana, 2015; Hoang & Trang, 2023; Rocha, 2020; Rosário & Dias, 2024)	Integrating the Internet of Things into smart tourism can improve the operational efficiency of businesses in the sector by using recommendation systems and smart city technologies, thereby enhancing personalized and safe experiences for tourists. Integrating Internet of Things (IoT) and smart city technologies into tourism destinations can enhance the tourist experience and help destinations become more competitive.
(Bec et al., 2021; Cranmer et al., 2020, 2023; Kounavis et al., 2012)	Augmented reality mobile applications have the potential to enhance the tourism experience, but they face challenges related to platform interoperability and internet dependency. Smart tourism with augmented reality (AR) provides relevant information about destinations, tourist attractions, activities, and services, promoting responsible decision-making and consumer behavior. Immersive technologies offer opportunities for revitalizing declining tourist destinations.

Source: authors

4.2. Responsible Travelers Behavior

The authors have demonstrated a growing interest in research on responsible behavior. T. H. Lee & Jan, (2023) highlight the impact of tourism on society, the economy, and the environment, as well as the behavioral intentions of tourists in the face of ethical and sustainable considerations (Samaddar & Mondal, 2023; Shen et al., 2020). The literature shows that there are several issues related to promoting responsible tourism (Baloch et al., 2022). To promote ethical and fair tourism, it is important to educate visitors on sustainable practices and community-serving activities (SONKO & DEAC, 2020).

Table 5. Summary of main studies on responsible travel behavior

Authors	Key Findings
(Samaddar & Mondal, 2023)	The critical dimensions that influence behavioral intentions towards responsible travel behaviors include performance anticipation, effort anticipation, reference power, facilitating factors, perceived value, and perceived risk.
(Shen et al., 2020)	Social networking sites can be utilized to promote sustainable and responsible behavior among tourists. To encourage responsible and smart behavior, smart tourist destinations should employ a variety of tactics. Smart technologies can assist in making tourists more responsible and mitigating the effects of over tourism. Social networking sites influence the behavior of smart tourists throughout their trip, including before, during, and after.
(Baloch et al., 2022)	Factors such as institutional policies, environmentally-friendly infrastructure, tourism carrying capacity, and community involvement impact responsible tourism behavior.
(SONKO & DEAC, 2020)	Adopting sustainable tourism practices can help minimize negative impacts on the environment and local communities. Environmental management policies are essential to promote more responsible tourism.
(T. H. Lee & Jan, 2023)	Smart tourism experiences have a significant impact on visitors' biosphere values, particularly in terms of aesthetics, ease of experience and learning. These experiences enhance visitors' appreciation of the value of the biosphere, which, in turn, influences their ecologically responsible behavior.

Source: authors

5. PROSPECTIVE ANALYSIS AND RESEARCH PROSPECTS

The authors examine several emerging research perspectives related to the smart tourism experience and responsible traveler behavior. These perspectives have managerial implications and offer opportunities for future research. The table below summarizes the main perspectives raised by the authors:

This analysis emphasizes the necessity for additional research on smart tourism experiences and responsible traveler behavior to address current and future challenges and promote sustainable and ethical tourism.

Table 6. Summary of main perspectives addressed by authors

Authors	Main Research Perspectives
(Agrawal et al., 2022; Benbba et al., 2022; Gretzel et al., 2015; Kounavis et al., 2012; Shin et al., 2022; Stankov & Gretzel, 2020; Sustacha et al., 2023; Yin et al., 2023; Zheng & Wu, 2023)	This paper aims to explore the implications of emerging technologies, such as artificial intelligence, augmented reality, and blockchain, on the tourism experience.
(Baloch et al., 2022; Lanquar, 2023; Nicola-Gavrilă, 2023)	Additionally, it seeks to deepen our understanding of the factors that influence responsible behavior among travelers, including their perception of sustainability, commitment to environmentally friendly practices, and contribution to local communities. This text presents an in-depth analysis of the motivations, values, and barriers that shape travelers' choices.
(Benbba et al., 2022; Cranmer et al., 2023; Cuomo et al., 2021; Hallem et al., 2017; Khan et al., 2024; C.-K. Lee et al., 2021; Li et al., 2023; Nicola-Gavrilă, 2023; Pan et al., 2018; SONKO & DEAC, 2020)	It also considers the socio-economic and environmental implications of smart tourism. Further studies are necessary to evaluate the impact of emerging technologies on destination sustainability, as well as to identify best practices in tourism governance and management.

Source: **Authors**

6. DISCUSSION AND MODELING SKETCH

This literature review highlights several mechanisms influencing responsible traveler behavior and the smart tourism experience. According to Stankov & Gretzel, (2020), technological advancements present both opportunities and challenges in the tourism industry. While emerging technologies can enhance the customer experience, they can also lead to a dehumanization of travel experiences and information overload. Therefore, the integration of these technologies should aim to promote ecologically responsible behavior among travelers (T. H. Lee & Jan, 2023).

The results of our study are in accordance with those of several recent studies (Femenia-Serra & Neuhofer, 2018; Gretzel et al., 2015). In recent years, there has been a global trend towards using IoT to connect the physical and digital environments, exploiting big data and analytics to personalize experiences, and integrating artificial intelligence to predict tourist behaviors and needs. This is achieved through the use of augmented reality and virtual reality, intelligent interfaces such as chatbots, and platforms for sharing and co-creation with tourists. These trends have been noted in various studies (Karasakal & Albayrak, 2022; Sarmento & Loureiro, 2021; Shen et al., 2020). For these studies, responsible visitor behavior includes adopting pro-environmental behaviors such as saving water and energy, reducing waste, and engaging in sustainable tourism activities like eco-responsible tours and volunteering. It also involves respecting the environment and local culture, sharing positive experiences and responsible advice on social media, and actively participating in the co-creation of smart tourism experiences.

Responsible traveler behavior can be achieved through a combination of mechanisms, tools, and technological solutions (T. H. Lee & Jan, 2023; Shen et al., 2020). According to Cranmer et al.,(2023), augmented reality technologies provide relevant information about destinations, tourist attractions, and services, promoting responsible decision-making and consumption behavior. Buhalis & Amaranggana, (2015) suggest that integrating Internet of Things and smart city technologies in tourism destinations can enhance the tourist experience and strengthen destination competitiveness. Furthermore, mobile applications and social networking platforms have the potential to promote sustainable tourism by providing tips and encouraging responsible practices (Kounavis et al., 2012). According to Lanquar, (2023), artificial intelligence can significantly contribute to tourism innovation and creativity by facilitating advanced data analysis, personalizing tourism offers, predicting traveler preferences, and creating unique experiences.

T. H. Lee & Jan, (2023) have identified the aspects of smart tourism experiences, such as aesthetics, usability, and learning, that have a significant impact on visitors' biosphere values. Enriched experiences can positively influence visitors' ecologically responsible behavior by increasing their awareness of environmental preservation and the adoption of sustainable practices (Shen et al., 2020). Technological innovations can encourage travelers to adopt environmentally-friendly behaviors, promoting responsible and ethical tourism (Lanquar, 2023).

Figure 6 presents the conceptual model that illustrates the relationship between responsible travel behavior and the smart tourism experience. The model demonstrates how technological advancements in tourism can promote responsible behavior among travelers.

It is hypothesized that the smart tourism experience can promote sustainable tourism by studying the interaction between responsible traveler behavior and the experience. This can raise travelers' awareness of environmental issues and encourage them to adopt more responsible practices throughout their trip. Two hypotheses can be envisaged in this context:

H1: Enhancing the smart tourism experience increases the likelihood that travelers will adopt responsible behaviors.
H2: Smart tourism experiences promote the development of sustainable and responsible destinations.

Figure 6. Research conceptual model sketched (Authors)

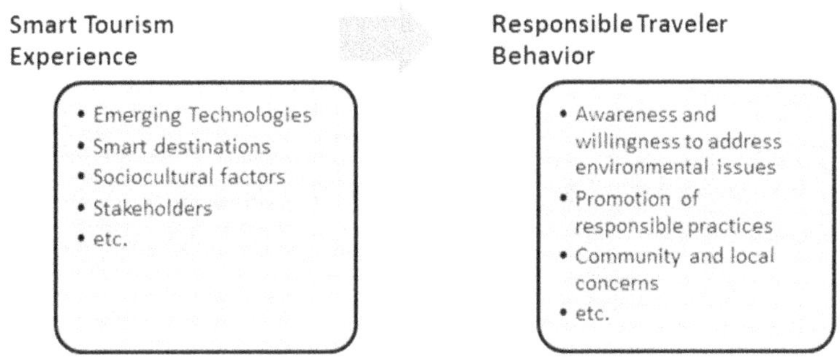

Our conceptual model aims to explore the relationship between smart tourism experience and responsible traveler behavior. The two hypotheses formulated following the literature review serve as a starting point to open the debate and guide future research. They may be used to deepen or broaden the subject through possible empirical validation of these findings (see research perspectives). Our model aims to identify the mechanisms and elements that foster the creation of responsible and ethical tourism destinations by focusing on the interaction between technology and traveler behavior. We take a holistic approach to examining the relationships between these factors, contributing to a better understanding of the complex dynamics shaping the future of sustainable tourism.

7. CONCLUSION, CONTRIBUTIONS, AND FUTURE PERSPECTIVES

Recent technological advancements have led to the emergence of innovative smart technology solutions that offer more opportunities in the tourism sector. These solutions can help develop smart and memorable consumer experiences while promoting responsible customer behavior.

In response to the question, our literature review allows us to draw conclusions about the smart tourism experience and responsible traveler behavior. Key themes such as sustainability, technological innovation and social and community responsibility have a certain consensus among researchers and industry professionals. The increasing prevalence of these themes highlights their significance for both the scientific community and the tourism industry.

It is essential to comprehend the correlation between sustainability and Smartness in the current tourism context. New technologies have the potential to personalize travelers' experiences, optimize interactions between tourists and destinations, and promote sustainability. However, achieving responsible tourism for destinations and local communities requires a balance between technological innovations and environmental, social, and cultural concerns.

However, it is important to note that this study has limitations. The results may not be generalizable to other cultural, geographical, and economic contexts beyond those considered in the sample. Additionally, the study was limited to analyzing secondary data from a literature review. It is also worth noting that customer experience and behavior are complex and evolving concepts that are influenced by various temporal, spatial, and contextual factors. Therefore, further research using a holistic longitudinal approach is necessary.

In conclusion, it is essential to conduct further research on smart tourism experiences and responsible traveler behavior. Future studies should focus on understanding the requirements and expectations of travelers who prioritize ethical and sustainable considerations. Additionally, the impact of technology on tourism and the environment should be studied. The purpose of this analysis is to provide a foundation for researchers and practitioners to conduct empirical studies on the subject. Empirical validation of the proposed model will be conducted as part of future research. The goal is to identify ideas and practical implications for promoting responsible tourism in the digital landscape of the future.

REFERENCES

Acharya, S., Mekker, M., & De Vos, J. (2023). Linking travel behavior and tourism literature : Investigating the impacts of travel satisfaction on destination satisfaction and revisit intention. *Transportation Research Interdisciplinary Perspectives*, *17*, 100745. doi:10.1016/j.trip.2022.100745

Adongo, C. A., Taale, F., & Adam, I. (2018). Tourists' values and empathic attitude toward sustainable development in tourism. *Ecological Economics*, *150*, 251–263. doi:10.1016/j.ecolecon.2018.04.013

Agrawal, R., Wankhede, V. A., Kumar, A., Luthra, S., & Huisingh, D. (2022). Big data analytics and sustainable tourism : A comprehensive review and network based analysis for potential future research. *International Journal of Information Management Data Insights*, *2*(2), 100122. doi:10.1016/j.jjimei.2022.100122

Alla, L., Hmioui, A., & Taghzouti, A. (n.d.). Impact de la creation de valeur pour le client sur la performance des entreprises touristiques marocaines : Etude exploratoire des pratiques des hotels de fes. *Les Cahiers Du Cedimes*, 75.

AllaL.KamalM.BouhtatiN. (2022). Big data et efficacité marketing des entreprises touristiques : Une revue de littérature. *Alternatives Managériales Economiques, 4*(0). doi:10.48374/IMIST.PRSM/ame-v1i0.36928

Andrews, D. C. (2002). Audience-specific online community design. *Communications of the ACM*, *45*(4), 64–68. doi:10.1145/505248.505275

Baloch, Q. B., Shah, S. N., Maher, S., Irshad, M., Khan, A. U., Kiran, S., & Shah, S. S. (2022). Determinants of evolving responsible tourism behavior : Evidences from supply chain. *Cogent Social Sciences*, *8*(1), 2099565. doi:10.1080/23311886.2022.2099565

Bec, A., Moyle, B., Schaffer, V., & Timms, K. (2021). Virtual reality and mixed reality for second chance tourism. *Tourism Management*, *83*, 104256. doi:10.1016/j.tourman.2020.104256

BenbbaB.IsmailiF. E. A.ArchiY. E. (2022). Smart tourism destination et développement durable : Quels apports à l'expérience touristique? *Alternatives Managériales Economiques, 4*(0), Article 0. doi:10.48374/IMIST.PRSM/ame-v1i0.36926

Bhaskar, P., & Sharma, K. (2022). A Critical Insight into the Role of Artificial Intelligence (AI) in Tourism and Hospitality Industries. *A Refereed Monthly. International Journal of Management*, *15*, 76–85.

Bhuiyan, M., Darda, M. A., & Hasan, M. R. (2021). Tourist perception and satisfaction on safari tourism at Bangabandhu Sheikh Mujib Safari Park in Bangladesh. *International Journal of Geoheritage and Parks*, *9*(4), 430–440. doi:10.1016/j.ijgeop.2021.11.005

Bodkhe, U., Bhattacharya, P., Tanwar, S., Tyagi, S., Kumar, N., & Obaidat, M. S. (2019). BloHosT : Blockchain enabled smart tourism and hospitality management. *2019 international conference on computer, information and telecommunication systems (CITS)*, 1-5. https://ieeexplore.ieee.org/abstract/document/8862001/

Buhalis, D., & Amaranggana, A. (2015). Smart Tourism Destinations Enhancing Tourism Experience Through Personalisation of Services. In I. Tussyadiah & A. Inversini (Eds.), *Information and Communication Technologies in Tourism 2015* (pp. 377–389). Springer International Publishing. doi:10.1007/978-3-319-14343-9_28

Chuang, C.-M. (2023). The conceptualization of smart tourism service platforms on tourist value co-creation behaviours : An integrative perspective of smart tourism services. *Humanities & Social Sciences Communications*, *10*(1), 1–16. doi:10.1057/s41599-023-01867-9

Cranmer, E. E., tom Dieck, M. C., & Fountoulaki, P. (2020). Exploring the value of augmented reality for tourism. *Tourism Management Perspectives*, *35*, 100672. doi:10.1016/j.tmp.2020.100672

Cranmer, E. E., tom Dieck, M. C., & Jung, T. (2023). The role of augmented reality for sustainable development : Evidence from cultural heritage tourism. *Tourism Management Perspectives*, *49*, 101196. doi:10.1016/j.tmp.2023.101196

Cuomo, M. T., Tortora, D., Foroudi, P., Giordano, A., Festa, G., & Metallo, G. (2021). Digital transformation and tourist experience co-design : Big social data for planning cultural tourism. *Technological Forecasting and Social Change*, *162*, 120345. doi:10.1016/j.techfore.2020.120345

Del Vecchio, P., Mele, G., Ndou, V., & Secundo, G. (2018). Open innovation and social big data for sustainability : Evidence from the tourism industry. *Sustainability (Basel)*, *10*(9), 3215. doi:10.3390/su10093215

Donthu, N., Kumar, S., Mukherjee, D., Pandey, N., & Lim, W. M. (2021). How to conduct a bibliometric analysis : An overview and guidelines. *Journal of Business Research*, *133*, 285–296. doi:10.1016/j.jbusres.2021.04.070

Doyle, J., & Kelliher, F. (2023). Bringing the past to life : Co-creating tourism experiences in historic house tourist attractions. *Tourism Management*, *94*, 104656. doi:10.1016/j.tourman.2022.104656

Falcão, D., & Roseira, C. (2022). Mapping the socially responsible consumption gap research : Review and future research agenda. *International Journal of Consumer Studies*, *46*(5), 1718–1760. doi:10.1111/ijcs.12803

Femenia-Serra, F., & Neuhofer, B. (2018). *Smart tourism experiences : Conceptualisation, key dimensions and research agenda*. Academic Press.

Gretzel, U., Sigala, M., Xiang, Z., & Koo, C. (2015). Smart tourism : Foundations and developments. *Electronic Markets*, *25*(3), 179–188. doi:10.1007/s12525-015-0196-8

Hallem, Y., Hikkerova, L., & Sahut, J.-M. (2017). Commercialisation de la durabilité dans le tourisme au travers de l'expérience client : Le cas d'un parc naturel: *Management & Avenir, 96*(6), 187-205. doi:10.3917/mav.096.0187

Hoang, H., & Trang, P. H. (2023). Navigating the Rise of Smart Tourism : Implications of Technology and Data for Sustainable Industry Growth. *Brawijaya Journal of Social Science*, *3*(1), 1–18. doi:10.21776/ub.bjss.2023.003.01.1

Karasakal, S., & Albayrak, T. (2022). How to create flow experience during travel : The role of destination attributes. *Journal of Vacation Marketing*, *28*(3), 303–318. doi:10.1177/13567667211053386

Khan, M. N., Gul, A., Khan, F., Khan, M. W., & Naz, A. (2024). Unlocking the potential : Artificial intelligence (AI) applications in sustainable tourism. *Remittances Review, 9*(1). https://remittancesreview.com/menu-script/index.php/remittances/article/view/1258

Kim, H., Ahn, S.-K., & Forney, J. A. (2014). Shifting paradigms for fashion : From total to global to smart consumer experience. *Fashion and Textiles*, *1*(1), 1. Advance online publication. doi:10.1186/s40691-014-0015-4

Kounavis, C. D., Kasimati, A. E., & Zamani, E. D. (2012). Enhancing the Tourism Experience through Mobile Augmented Reality : Challenges and Prospects. *International Journal of Engineering Business Management*, *4*, 10. doi:10.5772/51644

Kumar, N. R., Pavithra, R., Yuvaraj, V., & Kumar, T. M. (2023). The role of smart tourism in India for optimizing visitor experiences through technology. *JIMS8I International Journal of Information Communication and Computing Technology, 11*(2), 692-701.

Lanquar, R. (2023). Tourisme, créativité et intelligence artificielle. *Marché et organisations, 47*, 185-203. doi:10.3917/maorg.047.0185

Lee, C.-K., Olya, H., Ahmad, M. S., Kim, K. H., & Oh, M.-J. (2021). Sustainable intelligence, destination social responsibility, and pro-environmental behaviour of visitors : Evidence from an eco-tourism site. *Journal of Hospitality and Tourism Management*, *47*, 365–376. doi:10.1016/j.jhtm.2021.04.010

Lee, T. H., & Jan, F.-H. (2023). How do smart tourism experiences affect visitors' environmentally responsible behavior? Influence analysis of nature-based tourists in Taiwan. *Journal of Hospitality and Tourism Management*, *55*, 1–10. doi:10.1016/j.jhtm.2023.02.016

Lhoussaine, A., Kamal, M., & Bouhtati, N. (2022). Big data et efficacité marketing des entreprises touristiques : Une revue de littérature. *Alternatives Managériales Economiques*, *1*, 39–58.

Li, C., Zheng, W., Zhuang, X., & Chen, F. (2023). Intelligent transport design with a dual focus : Tourist experience and operating cost. *Annals of Tourism Research*, *101*, 103597. doi:10.1016/j.annals.2023.103597

Marchi, V., Marasco, A., & Apicerni, V. (2023). Sustainability communication of tourism cities : A text mining approach. *Cities (London, England)*, *143*, 104590. doi:10.1016/j.cities.2023.104590

Mateo, S. (2020). Procédure pour conduire avec succès une revue de littérature selon la méthode PRISMA. *Kinésithérapie, la Revue*, *20*(226), 29–37. doi:10.1016/j.kine.2020.05.019

Moher, D., Liberati, A., Tetzlaff, J., & Altman, D. G. (2009). Preferred Reporting Items for Systematic Reviews and Meta-Analyses : The PRISMA Statement. *PLoS Medicine*, *6*(7), e1000097. doi:10.1371/journal.pmed.1000097 PMID:19621072

Nicola-Gavrilă, L. (2023). Empirical Approaches Regarding Interdependency between Technology and Sustainable Tourism. *Journal of Environmental Management and Tourism*, *14*(4 (68)), 2140–2147. doi:10.14505/jemt.v14.4(68).25

Page, M. J., McKenzie, J. E., Bossuyt, P. M., Boutron, I., Hoffmann, T. C., Mulrow, C. D., Shamseer, L., Tetzlaff, J. M., Akl, E. A., & Brennan, S. E. (2021). The PRISMA 2020 statement : An updated guideline for reporting systematic reviews. *BMJ (Clinical Research Ed.)*, *372*. https://www.bmj.com/content/372/bmj.n71.short PMID:33782057

Pai, C.-K., Liu, Y., Kang, S., & Dai, A. (2020). The Role of Perceived Smart Tourism Technology Experience for Tourist Satisfaction, Happiness and Revisit Intention. *Sustainability (Basel)*, *12*(16), 16. Advance online publication. doi:10.3390/su12166592

Pan, S.-Y., Gao, M., Kim, H., Shah, K. J., Pei, S.-L., & Chiang, P.-C. (2018). Advances and challenges in sustainable tourism toward a green economy. *The Science of the Total Environment*, *635*, 452–469. doi:10.1016/j.scitotenv.2018.04.134 PMID:29677671

Park, C., Lee, S., Lee, C.-K., & Reisinger, Y. (2022). Volunteer tourists' environmentally friendly behavior and support for sustainable tourism development using Value-Belief-Norm theory : Moderating role of altruism. *Journal of Destination Marketing & Management*, *25*, 100712. doi:10.1016/j.jdmm.2022.100712

Ribes, J. F. P., & Baidal, J. I. (2018). Smart sustainability : A new perspective in the sustainable tourism debate. *Investigaciones Regionales-Journal of Regional Research*, *42*, 151–170.

Rocha, J. (2020). Smart Tourism and Smart Destinations for a Sustainable Future. In W. Leal Filho, A. M. Azul, L. Brandli, A. Lange Salvia, & T. Wall (Eds.), *Decent Work and Economic Growth* (pp. 1–10). Springer International Publishing. doi:10.1007/978-3-319-71058-7_88-1

Rosário, A. T., & Dias, J. C. (2024). Exploring the Landscape of Smart Tourism : A Systematic Bibliometric Review of the Literature of the Internet of Things. *Administrative Sciences*, *14*(2), 22. doi:10.3390/admsci14020022

Samaddar, K., & Mondal, S. (2023). AR and VR-based travel : A responsible practice towards sustainable tourism. *International Journal of Tourism Cities*. https://www.emerald.com/insight/content/doi/10.1108/IJTC-05-2022-0135/full/html

Sarmento, E. M., & Loureiro, S. M. C. (2021). Exploring the role of norms and habit in explaining pro-environmental behavior intentions in situations of use robots and AI agents as providers in tourism sector. *Sustainability (Basel)*, *13*(24), 13928. doi:10.3390/su132413928

Séré de Lanauze, G., Lallement, J., & de Ferran, F. (2021). Représentations sociales du consommateur responsable : Les autres consommateurs ont-ils vraiment envie de le suivre? *Management & Avenir*, *125*(5), 141–160. doi:10.3917/mav.125.0141

Shamim, S., Yang, Y., Zia, N. U., & Shah, M. H. (2021). Big data management capabilities in the hospitality sector : Service innovation and customer generated online quality ratings. *Computers in Human Behavior*, *121*, 106777. doi:10.1016/j.chb.2021.106777

Shen, S., Sotiriadis, M., & Zhou, Q. (2020). Could smart tourists be sustainable and responsible as well? The contribution of social networking sites to improving their sustainable and responsible behavior. *Sustainability (Basel)*, *12*(4), 1470. doi:10.3390/su12041470

Shin, H. H., Jeong, M., So, K. K. F., & DiPietro, R. (2022). Consumers' experience with hospitality and tourism technologies : Measurement development and validation. *International Journal of Hospitality Management*, *106*, 103297. doi:10.1016/j.ijhm.2022.103297

Sonko, S. M., & Deac, A. L. (2020). From simple tourism to smart tourism : The bet for the development of a responsible and sustainable tourism. cases in the Ziguinchor Region (Senegal). *Annals of the University of Oradea, Geography Series/Analele Universitatii Din Oradea, Seria Geografie, 30*(1). https://www.academia.edu/download/103324740/auog.301102-838.pdf

Stankov, U., & Gretzel, U. (2020). Tourism 4.0 technologies and tourist experiences : A human-centered design perspective. *Information Technology & Tourism*, *22*(3), 477–488. Advance online publication. doi:10.1007/s40558-020-00186-y

Sustacha, I., Baños-Pino, J. F., & Del Valle, E. (2023). The role of technology in enhancing the tourism experience in smart destinations : A meta-analysis. *Journal of Destination Marketing & Management*, *30*, 100817. doi:10.1016/j.jdmm.2023.100817

Tavitiyaman, P., Qu, H., Tsang, W. L., & Lam, C. R. (2021). The influence of smart tourism applications on perceived destination image and behavioral intention : The moderating role of information search behavior. *Journal of Hospitality and Tourism Management*, *46*, 476–487. doi:10.1016/j.jhtm.2021.02.003

Ulusoy, E. (2016). Experiential responsible consumption. *Journal of Business Research*, *69*(1), 284–297. doi:10.1016/j.jbusres.2015.07.041

Vecchio, P. D., Mele, G., Ndou, V., & Secundo, G. (2018). Creating value from Social Big Data : Implications for Smart Tourism Destinations. *In (Big) Data we trust: Value creation in knowledge organizations*, *54*(5), 847-860. doi:10.1016/j.ipm.2017.10.006

Vermeir, I., & Verbeke, W. (2006). Sustainable Food Consumption : Exploring the Consumer "Attitude – Behavioral Intention" Gap. *Journal of Agricultural & Environmental Ethics*, *19*(2), 169–194. doi:10.1007/s10806-005-5485-3

Wider, W., Gao, Y., Chan, C. K., Lin, J., Li, J., Tanucan, J. C. M., & Fauzi, M. A. (2023). Unveiling trends in digital tourism research : A bibliometric analysis of co-citation and co-word analysis. *Environmental and Sustainability Indicators*, *100308*, 100308. Advance online publication. doi:10.1016/j.indic.2023.100308

Yan, A., & Jia, W. (2021). The influence of eliciting awe on pro-environmental behavior of tourist in religious tourism. *Journal of Hospitality and Tourism Management*, *48*, 55–65. doi:10.1016/j.jhtm.2021.05.007

Yin, D., Li, M., & Qiu, H. (2023). Do customers exhibit engagement behaviors in AI environments? The role of psychological benefits and technology readiness. *Tourism Management*, *97*, 104745. doi:10.1016/j.tourman.2023.104745

Yoo, C., Kwon, S., Na, H., & Chang, B. (2017). Factors affecting the adoption of gamified smart tourism applications : An integrative approach. *Sustainability (Basel)*, *9*(12), 2162. doi:10.3390/su9122162

Zeng, D., Tim, Y., Yu, J., & Liu, W. (2020). Actualizing big data analytics for smart cities : A cascading affordance study. *International Journal of Information Management*, *54*, 102156. doi:10.1016/j.ijinfomgt.2020.102156

Zhang, X., Xi, W., Xu, F. Z., & Wu, R. (2023). Deconstructing consumers' low-carbon tourism promotion preference and its consequences : A heuristic-systematic model. *Journal of Hospitality and Tourism Management*, *57*, 48–60. doi:10.1016/j.jhtm.2023.08.020

Zhang, Y., & Hiltz, S. R. (2003). Factors that influence online relationship development in a knowledge sharing community. *AMCIS 2003 Proceedings*, 53.

Zheng, Y., & Wu, Y. (2023). An investigation of how perceived smart tourism technologies affect tourists' well-being in marine tourism. *PLoS One*, *18*(8), e0290539. doi:10.1371/journal.pone.0290539 PMID:37624780

Chapter 9
Leveraging Digital Platforms for Responsible Sports Tourism:
Budapest's Role in the 2020 European Football Championship

Nikolett Ágnes Tóth
University of Public Service, Hungary

Szabolcs Mátyás
University of Public Service, Hungary

Gergő József Tóth
University of Miskolc, Hungary

Zoltán Bujdosó
Hungarian University of Agriculture and Life Sciences, Hungary

Gábor Mátrai
Association of the European Digital Technology Industry, Hungary

Tamara Gajić
https://orcid.org/0000-0003-3016-8368
Serbian Academy of Sciences and Arts, Serbia

Lóránt Dénes Dávid
Hungarian University of Agriculture and Life Sciences, Hungary

ABSTRACT

The goal of the study is to present the safety and security systems for sporting events, with a special focus on the theoretical and practical tools used during the 2020 European Football Championship, which are of particular importance for ensuring security. The authors have analysed the policing statistics regarding the 2020 European Football Championship and described the latest sustainable technological solutions in use, with a particular focus on artificial intelligence, drone technology, and digitalization. It was pointed out how technological development, innovation, and digitization affect the sustainable operation and activity of law, sports law, sport tourism, law enforcement and sports police.

DOI: 10.4018/979-8-3693-3286-3.ch009

INTRODUCTION

By the end of the 20th century sport had become a strategic sector of major importance. Sport is an extremely complex phenomenon, affecting physical education, education policy, and healthcare, and its economic and socio-political significance is also growing. Today, it is legitimate to talk of a ‘sports industry’. Sport is becoming increasingly important in domestic and foreign policy and is also a very important tool for enhancing the image of a country.

This is particularly true of football, which is the most popular team sport in the world, which presents numerous examples in sports law and sports policing, the area concerned with the security of sporting events. This study will address the issue of legal innovation and innovative solutions adopted by police organizations and the links between them, which are of particular importance for sports policing. It can be concluded that in the modern world, sporting events, being events that attract large crowds, cannot exist nor can they be organized without guaranteeing order and security. The importance of event sustainability is also very important. Sporting events, particularly football tournaments, are increasingly focusing on sustainability initiatives to minimize their environmental impact. From reducing carbon emissions to implementing recycling programs and utilizing eco-friendly materials in infrastructure, such as stadiums, there’s a concerted effort to promote sustainability throughout all aspects of these events. Additionally, initiatives like promoting public transportation, encouraging fans to offset their carbon footprint, and implementing renewable energy sources contribute to a more sustainable future for sports. By integrating sustainability practices into football events, organizers aim to inspire fans and stakeholders to adopt environmentally responsible behaviors both during and beyond the games, ultimately fostering a more sustainable sporting culture worldwide. The authors referred to the phenomenon of football hooliganism and the measures taken to prevent and suppress it, during the 2016 European Football Championship, with notes of all the negative consequences of the pandemic, which later had an even stronger impact on the stalling of the 2020 European Football Championship.

Literature Review

A Brief Introduction to the Concept of Sports Policing

The creation, maintenance, and improvement of order, security, and the sense of security of the population and athletes have become increasingly important, primarily because the top events of the most popular team sports and individual sports have become mass events in all parts of the world, including Hungary (Imre, 2018.) Many events of elite, top-level sports attract tens of thousands of people not only during the event, but also before and after the event. Extraordinary events and acts of terrorism lead to a constant rethinking and revision of the definition of security, so it is also essential to create a synthesis of theory and practice. The publication primarily focuses on the Hungarian aspects, comparing them with the experiences of other countries organizing the European Championships (Tóth, 2021.)

The concept of sports policing in Hungary can be placed in the system of administrative law, and is also a specific area of sports administration, which can be clearly defined as a specialized area of law enforcement; a set of laws regulates everything related to sports policing, supplemented by the regulations of sports federations, which also serve as guidelines for keeping sports policing records and the providing sustainability and security for sports events (Tóth, 2019.; Belas et al., 2024.)

The primary responsibility of each state is to maintain and/or restore public safety at sporting events and related venues. Sports organizations, as organizers, are obliged to take, or to initiate with the competent authorities, all measures necessary to protect the personal safety and the protection of the property of the attendees/participants in sporting events and also to prevent infringements (Mátyás et al., 2021.)

The Effects of the COVID-19 Pandemic on the Security of Sports Events: Epidemic Emergency Situations and Sports Events

In the context of safety and security, the impact of the coronavirus epidemic on the safe running of sporting events should be mentioned (Watson et al., 2021.) The pandemic that broke out at the end of 2019 created an emergency situation and had an unforeseen, and far-reaching negative impact on the functioning of the global economy and highlighted more than ever before the vulnerabilities and weaknesses of the global economy, which is interconnected in all its elements (Watson et al., 2021.)

The effects of the health closures and quarantines were also particularly felt in the field of sports and sporting activities, and the closures and quarantines typically resulted in the banning of mass gatherings, including sporting events (Bogoch et al., 2020.) Figure 1 shows the cumulative number of COVID-19 infections between May 2020 and May 2022.

Figure 1. The total number of confirmed COVID-19 cases in countries around the world (1 May through 16 May 2022), by Johns Hopkins University

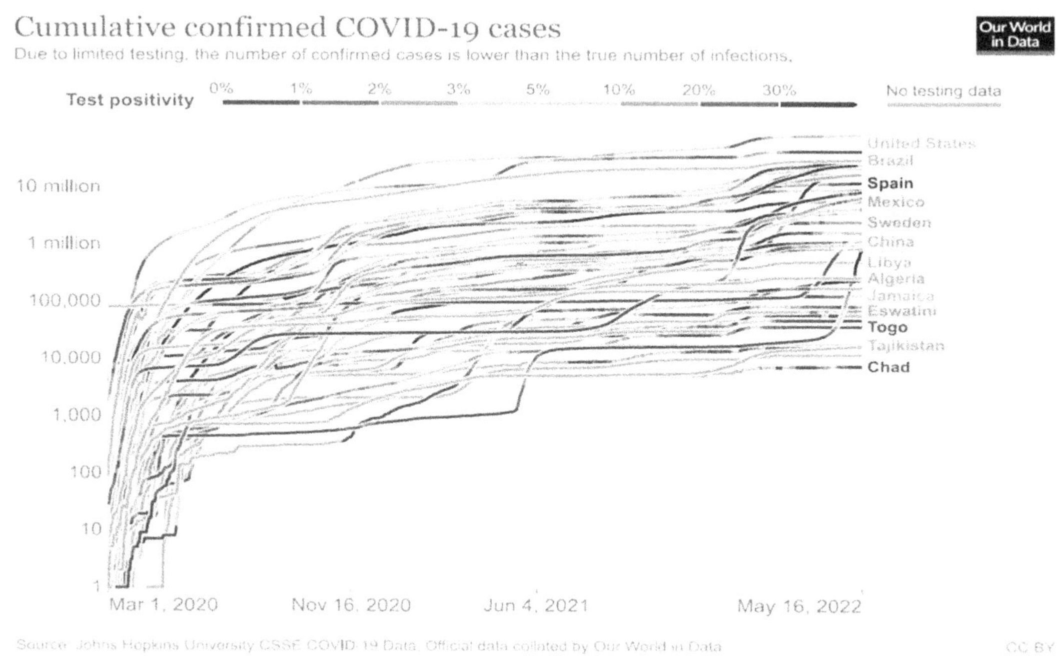

Similarly, to most countries in the world, Hungary also introduced measures restricting people's daily lives, which were only gradually lifted as the first wave of the epidemic subsided. Sporting life came to a complete standstill for months (Boyle & Haggerty, 2009.)

By March 2020, the pandemic had "brought football to its knees", and no one knew what would happen to the European Football Championship, scheduled for 12 June to 12 July 2020 (Tóth, 2020.)At that time, the Union of European Football Association (UEFA) denied press reports that the continental tournament, to be held in 12 countries, had been postponed. UEFA was also forced to take an immediate decision on certain series, postponing all international cup matches to be held the following week (Šuput, 2021.) It then became clear that it would not be enough to exclude spectators from stadiums to control the spread of the coronavirus epidemic. However, the national federations of other team sports had already acted: they had ceased all handball, basketball, volleyball, and water polo competitions, in all age groups. It was up to the clubs to organize training sessions, but in most places, this was also abandoned (Close et al., 2006.)

Football tournaments were also suspended in several countries. In Italy, for example, the decision to ban not only football matches but all sporting events was taken on 10 March 2020, and later, of the major championships, the Spanish, French, English, and German championships were also suspended (Martinez-Ferran et al., 2020.) Football continued in five of the 55 UEFA member countries, including Hungary, albeit behind closed gates. Later, it was decided to postpone the Summer Olympics and the European Football Championship (Kamyuka et al., 2020.)

During the COVID-19 epidemic, therefore, only football matches were played in Europe, with no tournaments or matches being organized in the other spectator team sports (Tóth, 2021.)

The policing aspect of UEFA EURO 2020 European Football Championship – "during a pandemic."

The success and results of sports policing can only be evaluated after the organization of sporting events, and even the considerable practical experience and theoretical knowledge acquired in the past is not always a guarantee for the management and solution of new challenges (Tóth, 2021.)

The postponement of UEFA EURO 2020 European Football Championship due to COVID-19 pandemic.

The UEFA EURO 2020 European Football Championship, or simply Euro 2020, was the 16th European Championship organized by UEFA for men's national football teams. The tournament was held in 11 cities across Europe. The original plan was for the matches to take place in the summer of 2020 (Tóth, 2020.) For the first time in the history of the European Championships, there was no specific host for the event. Of all the venues, London's Wembley Stadium, the largest capacity venue, hosted the two semi-finals and the final (McGuine et al., 2021.) The opening match was played at the Olympic Stadium in Rome, with Italy taking part, and for the first time in the history of the tournament, a video assistant referee (VAR) was used (Mátyás et al., 2021.)The 13 host cities were announced on 19 September 2014, but after the withdrawal of two cities, 11 cities - London, Munich, Bilbao, Baku, Copenhagen, Amsterdam, Budapest, Rome, St Petersburg, Bucharest, and Glasgow - were awarded the right to host the event· On 17 March 2020, UEFA held a video conference with the leaders of the 55 member countries, including representatives of FIFPro, the UEFA Europa League and the WHO, where a decision was taken on all national and international competitions, including the European Championship. The conference ultimately decided to postpone the tournament, with the date being moved to summer 2021 (11 June - 11 July) (McNiven & Harris, 2021.)

Figures 2 and 3 show the evolution of COVID infections in January 2020 and July 2021 respectively (Weekly epidemiological update on COVID-19 - 3 August 2021 (who.int)

Figure 2. Number of confirmed COVID-19 cases and total deaths, by date of report and WHO region (January through July 2021), WHO*

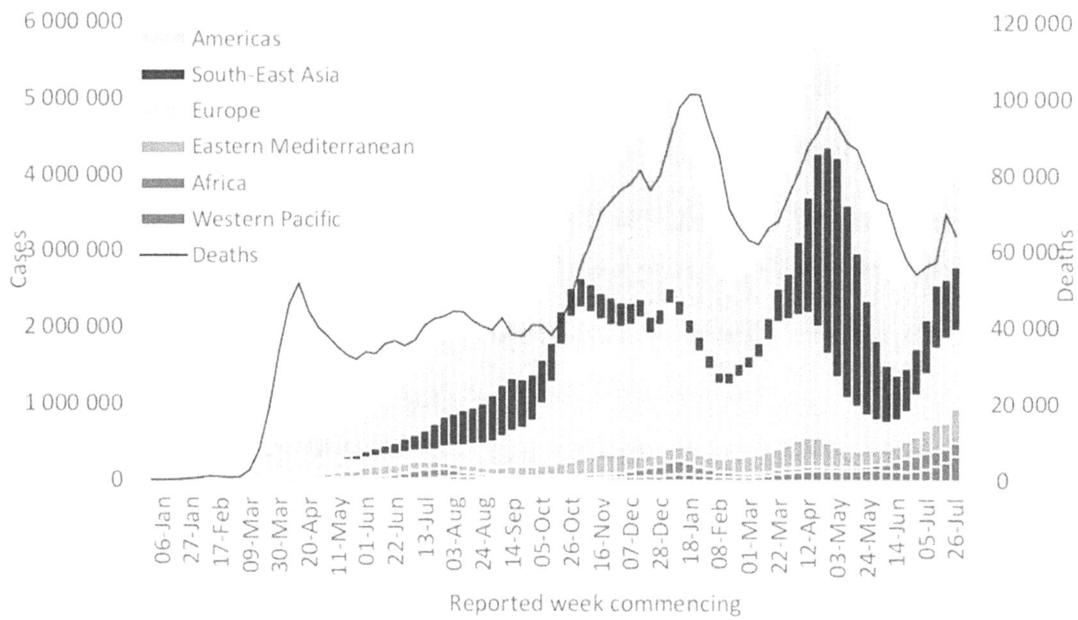

Figure 3. Number of confirmed COVID-19 cases, by WHO region (December 2019 through July 2020), WHO

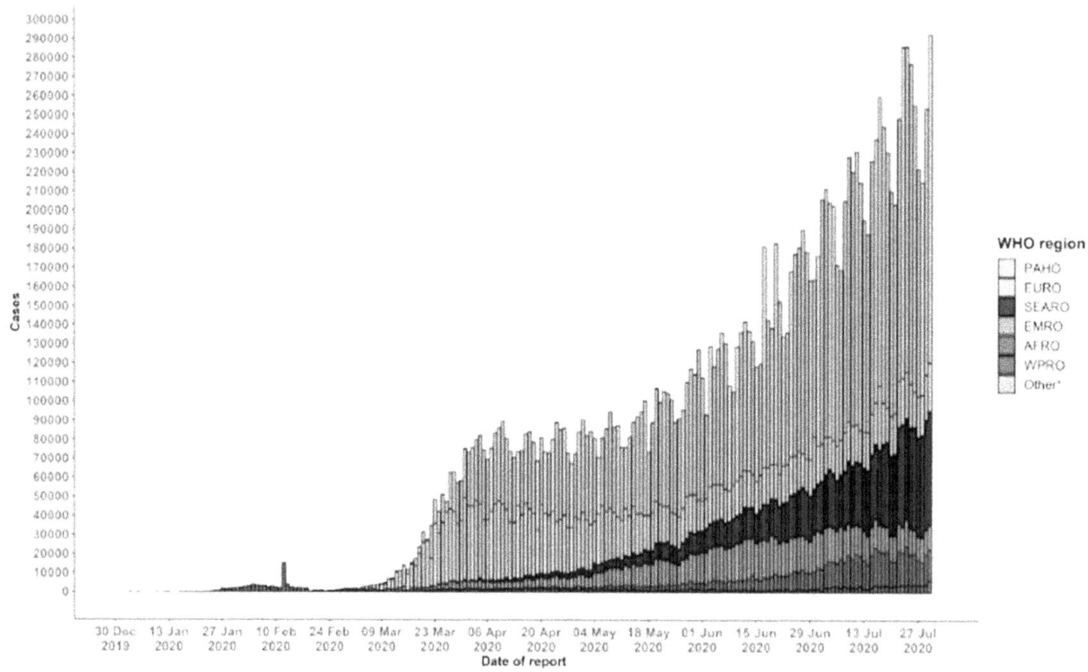

However, due to the COVID-19 pandemic, restrictions on spectator numbers were imposed at many venues (Drezner et al., 2021.) UEFA asked the organizers to work with their local or national governments and health authorities to develop a plan for health restrictions, but UEFA also wanted guarantees that matches would be played in the presence of spectators (Schreiber et al., 2022.) The deadline for communicating the plan was 7 April 2021. On 9 April, UEFA announced that eight cities had granted that their capacities be between 25% and 100%, with the remaining four (Bilbao, Dublin, Munich, and Rome) was granted an extension until the end of April. In mid-April, Rome's plan was accepted, followed by Munich at the end of April, while Seville was replaced by Bilbao and Dublin was stripped of its hosting rights since Bilbao and Dublin could not guarantee spectator attendance (Mátyás et al., 2021.)

Based on the above, it can be clearly stated that the organization and conduct of the 16th European Football Championship posed serious, unprecedented challenges and coordination tasks in many respects, especially regarding guaranteeing security, observing epidemiological measures, and maintaining and ensuring order.

Methodology

The theoretical basis of the present study is provided by the review and analysis of sports law, the organization of sports events, the literature on sports policing in Hungary and abroad, and scientific publications on the subject. Of course, for the practical presentation of the subject matter, it was important to process and evaluate relevant journals, statistical databases, summary reports, reportage, and news coverages (Gonda, 2013.)

Particular attention was also paid to the analysis of UEFA security regulations and risk management, evaluation reports on international events in the field of sporting policing, as well as the presentation of activities relating to the training and operation of police and law enforcement organizations and departments (Coaffee & Wood, 2006.; Turek et al., 2023).

From the point of view of the presentation of the subject, the prominent part of the study is the presentation of epidemiological data and measures taken in the context of the COVID pandemic and the analysis of their impact on security and the safe conduct of sporting events (Tóth, 2021.) To better understand and present the security and sports policing measures during the UEFA EURO 2020 European Football Championship, it was necessary to review the rules, regulations, practical standards, and steps used in domestic and international football matches during the pandemic (Steenee-Johannessen et al., 2020.) The above activities were greatly supported by the pieces of information and databases provided by the Hungarian Football Federation (MLSZ), as well as reports, data, and statistics prepared by the police forces involved in the security of the matches in Budapest (Tóth, 2021.)

In order to understand the impact of the latest innovative digital solutions on security and safe conduct at mass gatherings, it was also important to analyze the latest technology trends, developments and the conditions and possibilities of their practical application (Hyland, 2000.) Before preparing the study, the authors carried out extensive data collection and data analysis, and the data and statistics presented were compiled in such a way as to ensure comparability and provide an opportunity for further analyses.

Results and Discussion

Hungary's bid to host the tournament became a certainty when the 2015 UEFA decision in Geneva was announced. Several factors contributed to the success of the bid. According to UEFA's report, the new

Puskás Arena, with a capacity of 68,156 spectators, met UEFA's most stringent requirements in terms of both accessibility and capacity. The geographical location of Budapest, its easy accessibility, its forward-looking vision, and its commitment to sports were convincing arguments in favor of the Hungarian capital.

Organizing UEFA EURO 2020 European Football Championship in Hungary

In the summer of 2021, the new Puskás Ferenc Stadium hosted the European Championship matches in Hungary, where private security and police forces were responsible for ensuring order and security at sporting events. The 60th anniversary of the 2020 European Championships was also a major focus of attention (Tóth, 2021.) Puskás Ferenc Stadium is located in the inner urban area of Budapest, surrounded by two metro lines and a high-capacity tram line, but embedded in the residential area of the city. There is no car parking at and around the stadium (Imre, 2018.) Taking all this into account, the mobility plan was based on public transport links and the priority given to pedestrian traffic, so from a safety point of view, the main priorities were as follows: to protect pedestrians, separating them as far as possible from road traffic, without crossings where possible and without obstacles, and also to avoid the risk of tripping; to regulate and manage road traffic at critical points, including with the support of traffic police; to reinforce public transport safety measures. Close cooperation between border policing organizations was also necessary during the European Championship, as it was impossible to predict which teams would perform well, how far they would progress in the competition, how much their fan base would grow with a good performance, and whether they would have to move from one host country to another (Tóth, 2019.)

Puskás Arena also welcomed many foreign fans and their number per match was as follows:

- Hungary vs Portugal group stage: 16,000 spectators
- Hungary vs France group stage: 17,600 spectators
- Portugal vs France group stage: 24,780 spectators
- Netherlands vs Czech Republic Round of 16: 25,000 spectators

Figures 4-7 show the entry statistics for the above matches.

The European Championship football matches in Hungary were attended by a total of 83,380 foreign fans, which required special attention and preparation, as well as the possibility to intervene quickly and flexibly in the pre-arranged system, according to the situation at the time. Here, particular attention had to be paid to the issue of entry rules (for example, where appropriate, the possible requirement for epidemiological documents, vaccination certificates, and their verification) and the transmission, processing, and storage of data, particularly taking into account the fact that the United Kingdom was, at the time of the EC, no longer a member of the European Union.

It is also worth drawing attention to the different safety protocols for the VIP sector, where there had not typically been any screening of persons and baggage in the past. At the European Championships, all this finally changed, for example, no luggage larger than A/4 size was allowed, there was also a clothing and luggage search for those entering this sector, and a bomb squad inspection was also carried out in this sector, using magnetic auto control gates. Figure 8 shows the number of spotters by country during the 2020 EC:

Figure 4. Number of total entries per minute, HUN-POR

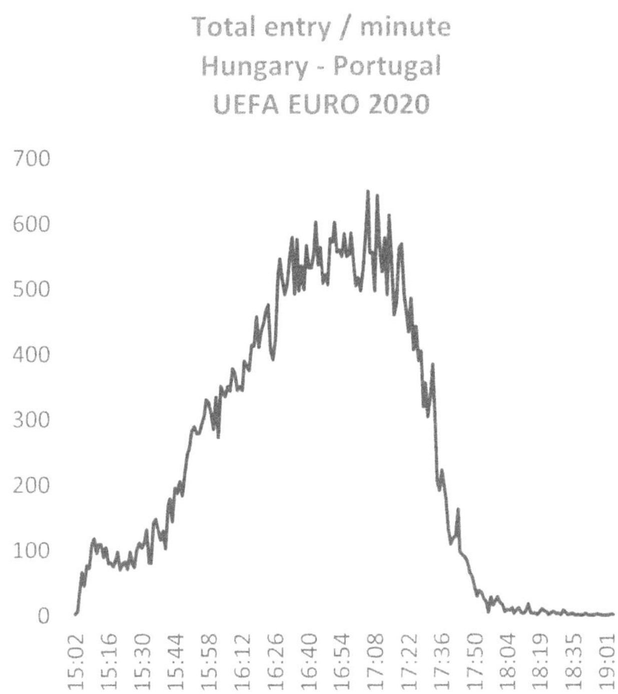

Figure 5. Number of total entries per minute HUN-FRA

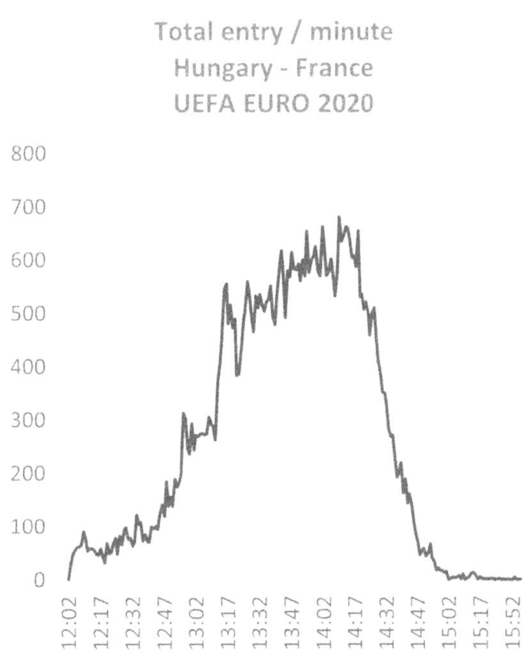

Figure 6. Number of total entries per minute, POR-FRA

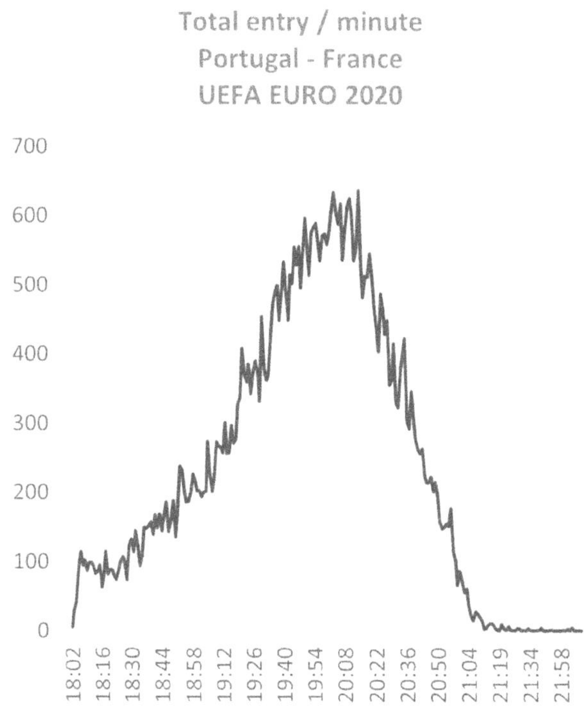

Figure 7. Number of total entries per minute, NED-CZ

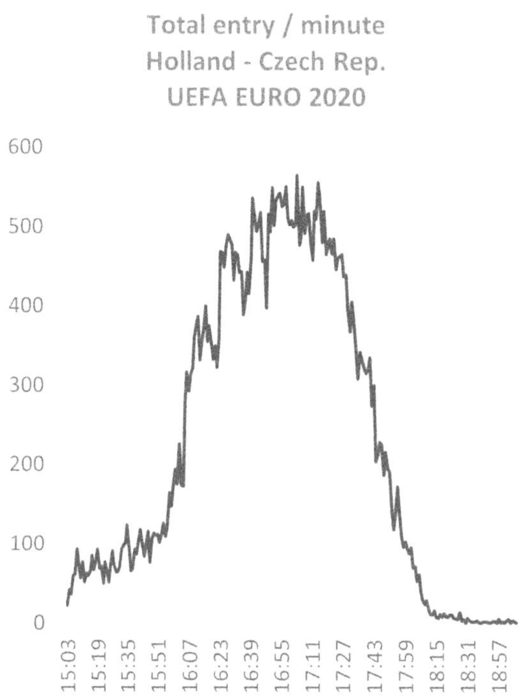

Figure 8. Number of spotters during EURO 2020, broken down by country

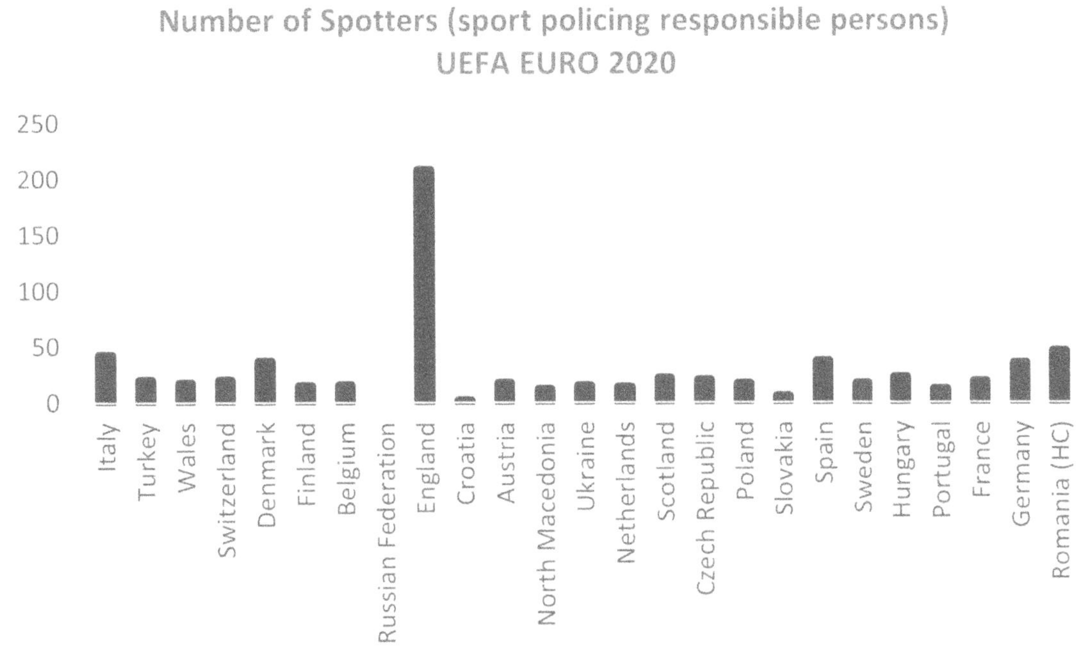

The Functioning of the Command-and-Control Center, the Data and Information Analyzed, and Their Evaluation

Securing the Venue for the Matches

Based on all the data and information received at the command-and-control center in Puskás Arena during the European Football Championships, the majority of the infringements committed were disorderly conduct by fans sympathetic to the players, who ran onto the pitch, usually after the match had finished, in order to make physical contact with their favorite players, talk to them or take a photo with them.

The effectiveness of the bio-secure bubble system set up in response to the COVID-19 outbreak worked as expected, as those confirmedly not infected were only allowed to enter the various locations (accommodation, event areas, etc.) where their movements and contacts could be monitored. The staff and liaison officers were continuously tested and, as a result, there were no cases of COVID-19 infection, which is a tremendous achievement and confirms the effectiveness of the epidemiological measures taken and their compliance and enforcement. Figure 9 illustrates the COVID-19 entry protocol.

Admission and entry of spectators was smooth and continuous, and congestion was rare. The private security service was adequately staffed to secure the entrance gates. A total of 219,391 people were admitted to the four matches in Budapest. Police measures were taken against nine persons all for attempts of unauthorized access to the Arena (persons without a valid ticket jumped over a fence). According to the rules established by the current UEFA stadium regulations, the following were forbidden:

Figure 9. COVID-19 entry protocol during EURO-2020, Budapest

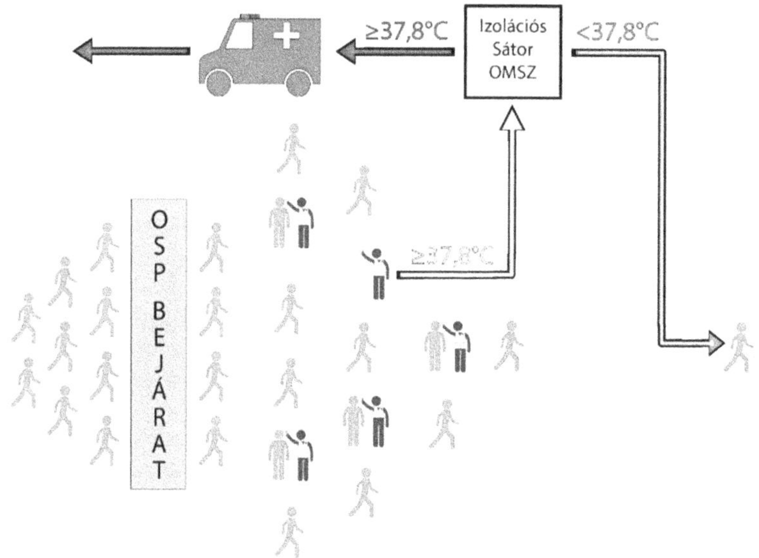

- a video camera or professional camera, unless authorized by the organizer,
- a club membership sign, flag, banner, or flagpole, or wear clothing that does not clearly indicate such membership,
- an animal, except an assistance dog in the case of a disabled spectator,
- food, drink, drugs, heavy objects, objects that cannot fit under a chair, or objects that may endanger the conduct of the sports event, the safety of others, the safety of persons or property, the safety of players or officials, the entertainment of other spectators, be used for violent acts, or the possession of which is prohibited by law or the entry of which into the sports event is prohibited by the organizer,
- display or wear any sign, flag, or symbol of authoritarianism prohibited by law, or any such clothing, which incites hatred towards others, has political content, or is offensive to others,
- any promotional or commercial object, unless authorized by the organizer,
- a video camera or professional camera, unless authorized by the organizer,
- a club membership sign, flag, banner or flagpole, or clothing which does not clearly indicate such membership,
- animals, except assistance/service dogs in the case of a disabled spectator (Mátyás et. al., 2022.).

The largest supporters' zone in the country, with a maximum capacity of 11,000 people, was created at the City Park Ice Rink. It was sold out during the Hungarian national team matches, and on other days it had a capacity of a few thousand people. There was one case of a major offence in the supporters' zone, and the perpetrators were apprehended and arrested. In order to perform image documentation tasks, a total of 97 cameras were installed in 38 installation positions (87 cameras in the capital and 10 cameras in rural locations), and the necessary equipment for their operation was installed.

Figures 10 to 13, based on data provided by UEFA, show the security statistics for all UEFA EURO 2020 matches and the number of offences committed, broken down by host country and by week.

Figure 10. Match security statistics, UEFA 2020, by county (11 through 16 June 2021)

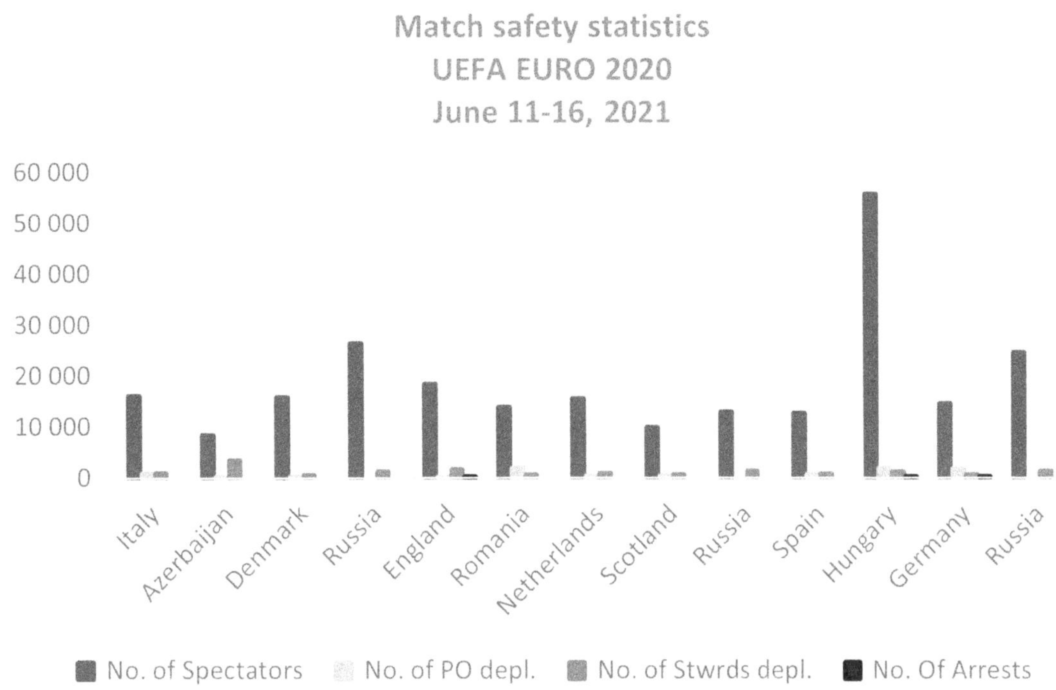

Figure 11. Match security statistics, UEFA 2020, by county (11 through 16 June 2021)

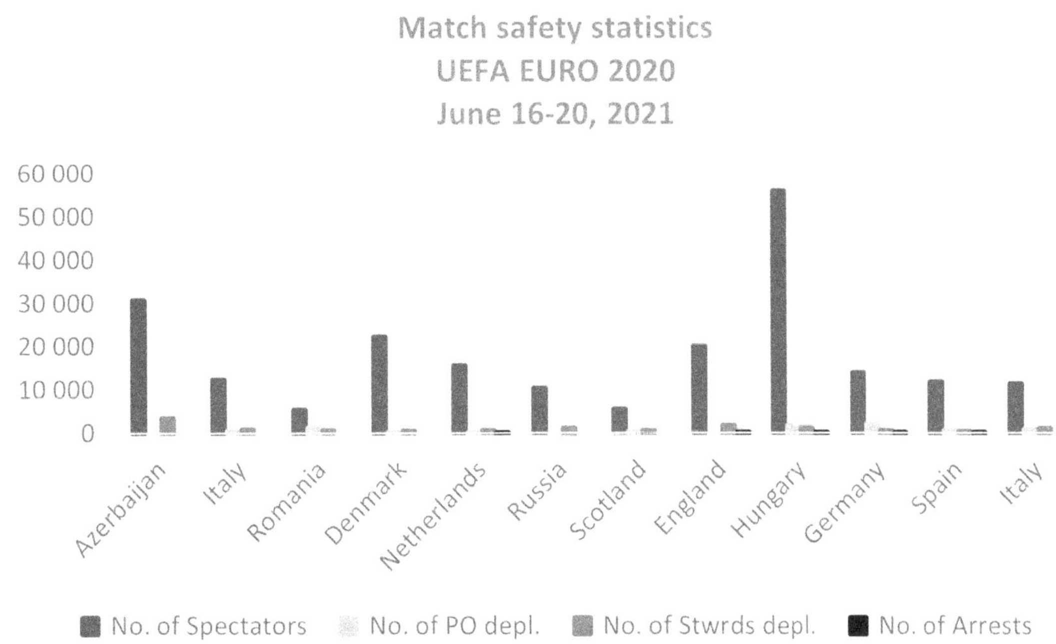

Figure 12. Match security statistics, UEFA 2020, by county (20 through 23 June 2021

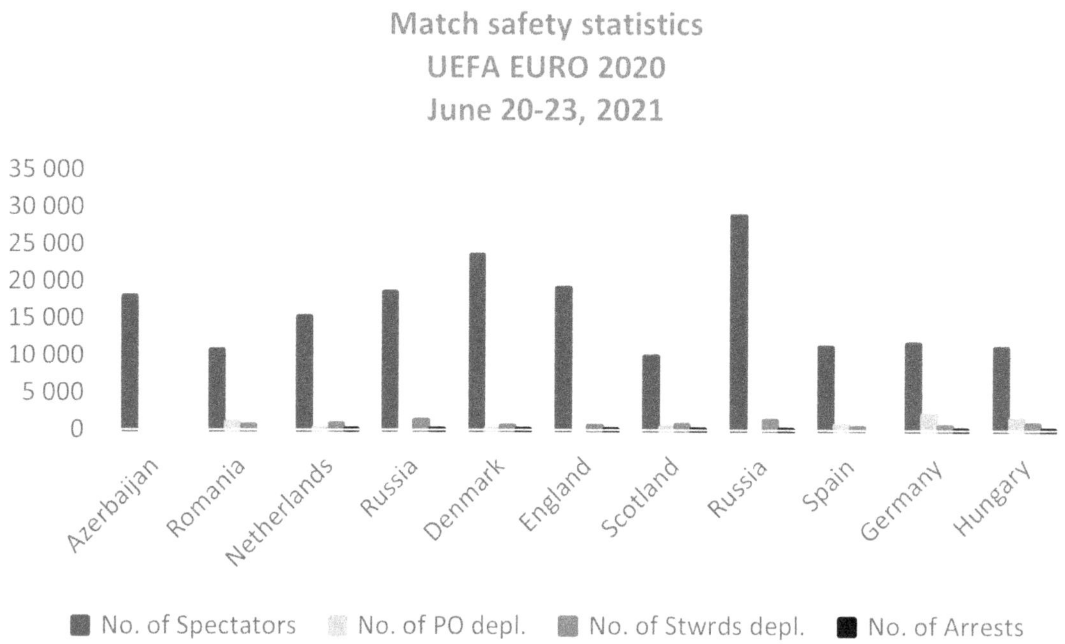

Figure 13. Match security statistics, UEFA 2020, by county (26 June through 11 July 2021)

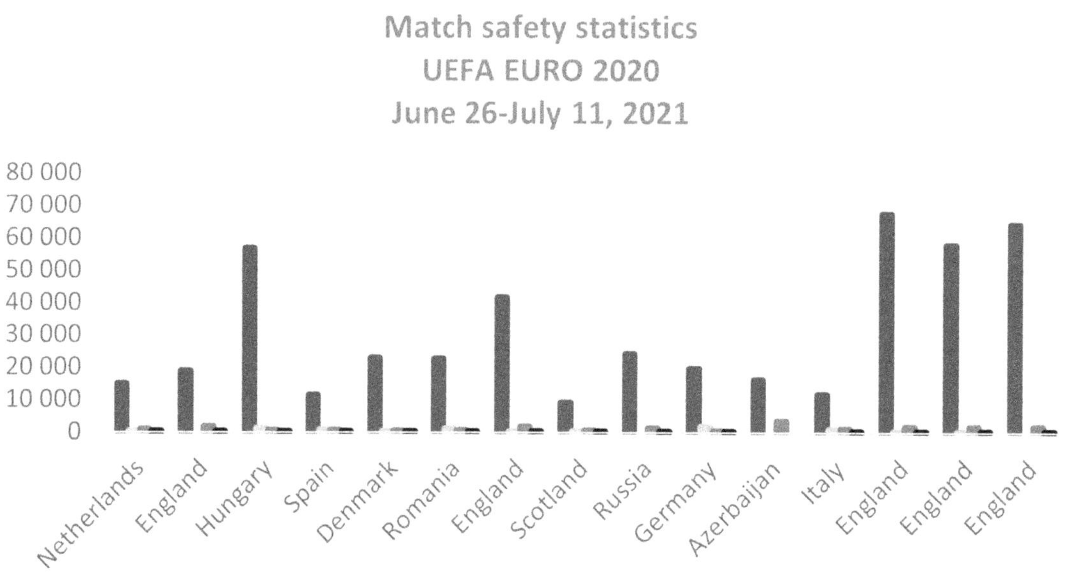

The statistics on the security of the matches clearly show which stadiums had committed to what capacity, and what capacity had been allowed by UEFA. This does not, of course, mean that the maximum permitted spectator quota was used in all venues. The statistics also show that, although the number of spectators is an important factor in determining the size of the police force, a basic number is necessary for a safe operation and, if and when the previous behavior and aggressiveness of the supporters' camps so requires, additional forces may be necessary. Of course, the 'threat' to or the vulnerability of a host country is also a determining factor. It should also be highlighted here that the participation of volunteers is an important factor in the safe conduct of a match and that their coordinated action is becoming increasingly important. The statistics alone show that the conduct of matches, and therefore the UEFA EURO 2020 European Football Championship, was excellent from a security point of view, with very few illegal acts and arrests.

Figure 14, based on data provided by UEFA, shows the proportion of offences broken down to host cities. The pattern of offences varies from city to city, but the data shows that, for example, the use of pyrotechnic articles was almost universally present and that ticket fraud is on the decline due to strict rules of purchase and entry.

Figure 14. Ratio of infringements/offences during UEFA EURO 2020, by country

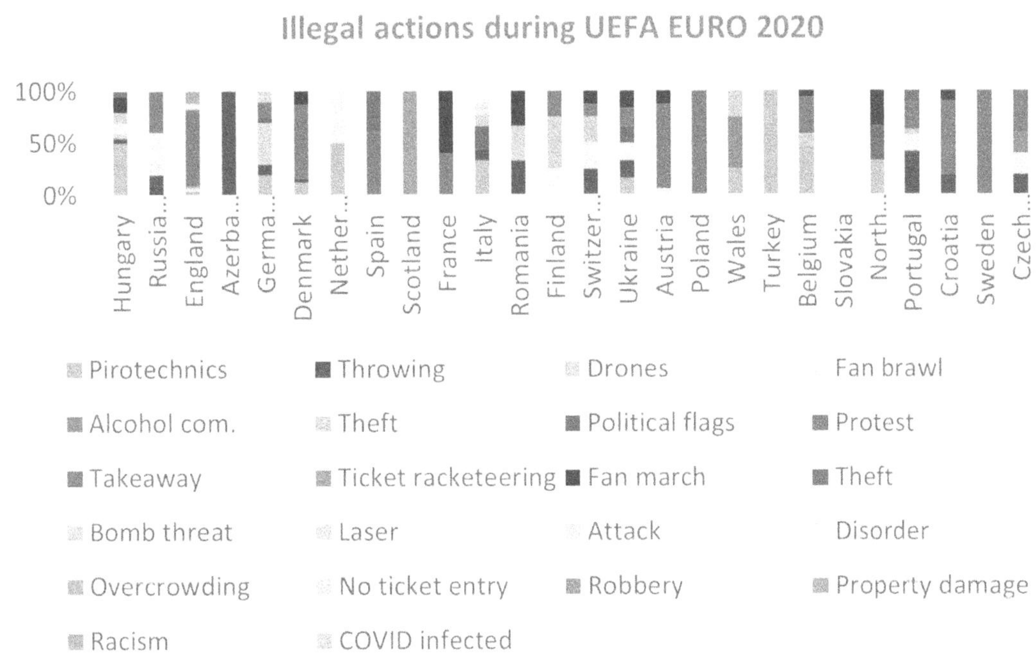

It can be said that the use of the access control systems and the consistent application of the sports policing register played a role in reducing the number of infringements/offences committed in stadiums, in parallel with the introduction of the CCTV systems (Hagemann, 2008.)

As regards the technical aspects of innovation, it is important to underline that over the past decades, IT storage capacities, data processing, and data transmission systems, IT tools, and technologies have developed and grown enormously (Hall, 2006.) We only need to look at the spread and everyday use of IoT, cloud-based solutions, supercomputers, and artificial intelligence. In the 21st century, our lives are clearly shaped by technology in everyday life and, among other things, in the world of sport and, increasingly, in the world of sports policing. High performance requires high technology, as does guaranteeing safety (Walters & Tacon, 2010.)

The Role of New, Innovative, and Digital Solutions in Sports Events Security

The establishment, preservation, and improvement of order, security, safety and security for the public and athletes have also become increasingly important in recent years, mainly because of the mass attendance at team sports, spectator sports, and individual sports events, and the mass participation in recreational sports events (Lenskyj, 2002.; Müller et al. 2017) The several events in elite and grassroots sports events attract tens of thousands of people, not only inside and outside sports facilities, but also before, during and after. Increasingly, sporting events also have a show element, which raises new security requirements. Increasingly, sporting events also have their show elements, which raise new security requirements (Whitson & Horne, 2006.)

The Relationship Between Artificial Intelligence and Law Enforcement

Artificial intelligence is based on structured data, databases, algorithms, and its growing use is becoming a dominant part of an increasingly data-driven society. The most promising application of AI for law enforcement is its ability to identify patterns, patterns of criminal behavior, to recognize and identify objects, persons, behaviors, manifestations and behaviors based on the analysis of databases by algorithms, thus enabling better prediction, prevention and, of course, detection of crime, better planning and organization of crime prevention and security for law enforcement tasks, and, where appropriate, for high-profile events (Tulloch, 2000.) The ability to predict crime before it happens is also called predictive police measures or predictive policing. However, the use of AI in predictive policing also requires the regulation of ethical and legal issues. Artificial intelligence is a disruptive technology, radically changing society and security issues. Security is a constant focus, also in the everyday use of AI.

Drone Detection: Drones

Drones pose an increasing threat, as professional forecasts show that their number is increasing and within seven years there will be nearly 30,000 self-guiding drones in Hungary, whose regulation, control, and monitoring will pose a serious professional and safety challenge. In this context, I think it is important to note a quote from Yuval Noah Harari: Humanity is developing powerful new technologies that are so effective that they could destroy mankind or, in contrast, even perfect it. The inherent difficulty of drone technology is that it is incredibly difficult to pinpoint the exact location of drones, but there is already a device that can alert you when they are hovering nearby. Detectors are now being produced that will signal when a drone is detected. This includes both voice and radio drone identification. Drone detection was used in Hungary during the European Championships, and the detection system in use in Puskás Ferenc Arena was able to locate the source of the signal of the drone and thus its operator. The protection against drones is also set out in detail by UEFA and drone detection was an important point in the detailed requirements for the organization of the 2020 European Championship.

Biometric Identification: Facial Recognition Systems

Biometric identification is a type of personal identification in which the identification process is based on biometrics, i.e., the measurable biological, physiological, or behavioral characteristics of a person, which differ from one individual to another. Biometrics in law enforcement identification (Mátyás et al., 2022.): DNA-based identification, identification based on hand geometry analysis, identification based on signature verification, identification based on retinal scanning, identification based on facial thermal imaging, voice-based identification, facial recognition- based identification, fingerprint-based identification, iris scan-based identification, finger, or palm vein pattern recognition (Samatas, 2007.)

The method identifies the person, and the sample is taken from a real live person, significantly reducing the possibility of deception. Biometric identifiers are unique bodily characteristics that are specific to an individual and therefore it is almost impossible for two individuals to have identical biometric identifiers (Stott et al., 2001.) Compared to other biometric systems using fingerprint/palmprint and iris diagnostics, facial recognition has particular advantages as it is a "physically" contactless process. Facial images can be captured remotely without the need to learn about the person to be identified, and no human interaction is required for identification. In addition, facial recognition also serves the purpose of "deterring" crime, as a captured and archived facial image can later help to identify a person with high accuracy. This chapter should also address the issue of mandatory wear of face masks and facial recognition in the context of the COVID-19 coronavirus outbreak, as masks cover a significant part of the face. However, as technology has improved, this has not been a problem, with a US study and survey already showing that 96 per cent of people wearing masks are identified by improved and new facial recognition systems.

The US Department of Homeland Security (DHS) has addressed the problem with a system called West. On average, the other systems in the tests scored 77 percent with a mask and 93 percent without a mask, this system with its over 96 percent and 100 percent result is a huge improvement. Typical facial recognition technologies before the pandemic had a 50-80 percent accuracy rate in identifying masked people, meaning there was a very wide variance with a lot of room for error[46]. It must also be mentioned here that facial recognition is one of the most controversial and debated artificial intelligence technologies of our time. It has huge potential, but its excessive use can be an invasion of privacy, and its bad results, false positives, can lead to the abuse of innocent people, which has already happened several times. Concerns have of course been raised about the US system, but the DHS has also reassured the public that the new system is intended for airports (Tóth, 2021.)

CONCLUSION

Order and security have become one of the most fundamental values of our time, and today they play an increasingly important role, not only for the individual but also at all levels of society, both locally and globally. When we talk about security, we are basically assuming a state of being free from danger, free from fear of harm, and in a state of being able to carry out any activity without being disturbed. However, this approach and conceptual definition of order and security has now become much more diverse, complex, and complicated. Nowadays, security can only be understood in a complex way, which depends essentially on the areas of security risk analysis that are at the center of our analysis. It must, of course, be stressed that security is always relative, i.e. it exists only under specific environmental, social and economic conditions.

Sport, as a defining cultural phenomenon, has accompanied the history of people, of humanity, and has become a fundamental activity of human society. It can be seen as one of the most important core values of societies, and a major factor of national unity. In accordance with the Hungarian Sports Act, the community of the nation considers the cultivation of the body, sports, as a fundamental value and a desirable goal of the nation. The achievements achieved through sports and the values of sports that strengthen the individual and the community are among the values of the nation. Sport is the foundation of mental health, which is part of the value of a healthy nation, and of a citizen wishing to stay healthy.

Several of these challenges and responses are mentioned and described in detail in the study. Examples include the emergence of the phenomenon of football hooliganism and the measures taken to prevent and counter it, in particular for 2016 European Football Championship, before which the wave of terrorism sweeping across Europe made it necessary to introduce a number of new, previously unused means and measures, as well as the security and safety tasks resulting from the epidemiological management of the COVID-19 pandemic, which needed to be rethought in the context of the organization of the 2020 European Football Championship.

We considered it important to show how technological development, innovation and digitalization affect the operation and activities of law, sports law, law enforcement and sports policing, what role theoretical and practical research results, findings and their continuous representation play in increasing efficiency and in formulating quicker responses to new challenges. Innovation, info-communication, information technology, digital technologies, and the development, spread and expansion of digitalization itself at a faster pace than ever before are fundamentally changing all areas of the economy and society, and in addition to their direct value-creating role, through their multiplier effect, they catalyze the transformation of other industries and fundamentally determine the adaptability and competitiveness of businesses, and the simpler, more transparent, cheaper and more efficient operation of the state and public administration. Digital transformation, driven by innovation, information technology and the information and communications technology (ICT) sector, has already appeared (sometimes more or less strongly) in all areas of public administration, including legislation, law enforcement and sports policing, and now there is a good reason to talk about innovative legal solutions, procedures, and innovative sports policing solutions, of which the volume presents several examples.

During the writing of this publication, we have also come across several areas, the further research, detailed presentation, and analysis of which will be the focus of my future research. Such areas include, for example, the detailed analysis and presentation of event security, sports policing organization and management activities of other major international sporting events besides football, such as the FINA World Swimming Championships, the IAAF World Athletics Championships, FIA Formula 1 car races, the 2024 Summer Olympics in Paris, France, and the analysis and review of the security and specificities of E-sport events. In addition, we would like to continue the exploration and presentation of the latest legal and technological innovations for law enforcement and sports policing purposes, with a special focus on the procedures, solutions and tools based on artificial intelligence and the legal background of their use.

Otherwise, the sustainability efforts within the realm of sports, particularly football events, represent a significant step towards mitigating environmental impact and promoting responsible stewardship of resources. As the global community continues to prioritize sustainability, the integration of eco-friendly practices into football events not only enhances the overall experience for fans but also sets a precedent for future events to follow. Through collaborative efforts and ongoing commitment to sustainability, the sports industry can play a pivotal role in shaping a greener and more sustainable future for generations to come.

REFERENCES

Ahlert, G. (2006). Hosting the FIFA World Cup™ Germany 2006: Macroeconomic and regional economic impacts. *Journal of Sport Management*, *20*(1), 1–22.

Belas, J., Jan, D., Hlawiczka, R., Smrcka, L., & Khan, K. A. (2024). SMEs sustainability: The role of human resource management, corporate social responsibility and financial management. *Oeconomia Copernicana*, *15*(1), 307–342. doi:10.24136/oc.2937

Bogoch, I. I., Watts, A., Thomas-Bachli, A., Huber, C., Kraemer, M. U. G., & Khan, K. (2020). Pneumonia of Unknown Aetiology in Wuhan, China: Potential for International Spread via Commercial Air Travel. *Journal of Travel Medicine*, *27*(2), taaa008. doi:10.1093/jtm/taaa008 PMID:31943059

Boyle, P., & Haggerty, K. (2009). Spectacular security: Mega-events and the security complex. *International Political Sociology*, *3*(3), 257–274. doi:10.1111/j.1749-5687.2009.00075.x

Close, P., Askew, D., & Xin, X. (2006). *The Beijing Olympiad: The political economy of a sporting megaevent*. Routledge. doi:10.4324/9780203003015

Coaffee, J., & Wood, S. M. (2006). Security is coming home: Rethinking scale and constructing resilience in the global urban response to terrorist risk. *International Relations*, *20*(4), 503–517. doi:10.1177/0047117806069416

Drezner, J. A., Drezner, S. M., Magner, K. N., & Ayala, J. T. (2021). COVID-19 Surveillance in Youth Soccer during Small Group Training: A Safe Return to Sports Activity. *Sports Health*, *13*(1), 15–17. doi:10.1177/1941738120964458 PMID:32986534

Gonda, M. (2013). Supporters' movement. Against modern football and sport mega events: European and Polish contexts. *Przegląd Socjologiczny*, *62*(3), 85–106.

Government Decree 241/2020 (V.27.) on Outdoor Events (241/2020. (V. 27.)

Hagemann, A. (2008). *"From stadium to "Fanzone": The architecture of control.* Paper presented at the conference Security and Surveillance at Mega Sport Events, Durham University, Durham, UK.

Hall, C. M. (2006). Urban entrepreneurship, corporate interests and sports mega-events: The thin policies of competitiveness within the hard outcomes of neoliberalism. *The Sociological Review*, *54*(2), 59–70. doi:10.1111/j.1467-954X.2006.00653.x

Hyland, J. (2000). *New measures against football hooligans undermine civil liberties in Britain*. Available from https://www.wsws.org/en/articles/2000/08/act-a29. html

Imre, D. (2018). Professionalism, modesty, diligence: A festive volume in honour of 65-year- old József Boda. Budapest: Dialóg Campus Publishing House.

Kamyuka, D., Carlin, L., McPherson, G., & Misener, L. (2020). Access to physical activity and sport and the effects of isolation and cordon sanitaire during COVID-19 for people with disabilities in Scotland and Canada. *Frontiers in Sports and Active Living*, *2*, 594501. doi:10.3389/fspor.2020.594501 PMID:33426519

Lenskyj, H. (2002). *The best Olympics ever? The social impacts of Sydney 2000*. SUNY Press. doi:10.1353/book4571

Martinez-Ferran, M., de la Guía-Galipienso, F., Sanchis-Gomar, F., & Pareja-Galeano, H. (2020). Metabolic Impacts of Confinement during the COVID-19 Pandemic due to Modified Diet and Physical Activity Habits. *Nutrients*, *12*(6), 1549. doi:10.3390/nu12061549 PMID:32466598

Mátyás, Sz., Nagy-Tóth, N. A., Dávid, L. D., Gogo, A. F. C., & Bujdosó, Z. (2022). The Role of Sports Policing and Tourism Safety at the Summer Olympics. *Sustainability*, *14*(10), 1–16.

Mátyás, Sz., Tokodi, P. V., & Vince, T. M. (2021). Investigation of Certain Components of Tourism Safety in the Visegrád Countries. Internal Security, 13(1), 89-97.

McGuine, T. A., Biese, K. M., Petrovska, L., Hetzel, S. J., Reardon, C., Kliethermes, S., Bell, D. R., Brooks, A., & Watson, A. M. (2021). Mental Health, Physical Activity, and Quality of Life of US Adolescent Athletes during COVID-19-Related School Closures and Sport Cancellations: A Study of 13,000 Athletes. *Journal of Athletic Training*, *56*(1), 11–19. doi:10.4085/1062-6050-0478.20 PMID:33290516

McNiven, A., & Harris, J. (2021). Can community football clubs in Scotland develop the local partnership support they need? *Managing Sport and Leisure*, *26*(5), 346–361. doi:10.1080/23750472.2020.1838940

Miller, G. P. (2017). *The Law of Governance, Risk Management and Compliance*. Wolters Kluwer Law & Business.

Miller, H. H. (2000). Mega-events, urban boosterism and growth strategies: An analysis of the objectives and legitimations of the Cape Town 2004 Olympic bid. *International Journal of Urban and Regional Research*, *24*, 439–458.

Müller, A., Bíró, M., Bodolai, M., Hidvégi, P., Váczi, P., Dávid, L., & Szántó, Á. (2017). A 2016-os fitnesztrendek helye és szerepe a rekreációban [The 2016 fitness trends place and role of recreation]. *ACTA Academiae Paedagogicae Agriensis Nova Series: Sectio Sport*, *44*, 91–102.

Samatas, M. (2007). Security and surveillance in the Athens 2004 Olympics: Some lessons from a troubled story. *International Criminal Justice Review*, *17*(3), 220–238. doi:10.1177/1057567707306649

Schreiber, S., Faude, O., Gärtner, B., Meyer, T., & Egger, F. (2022). Risk of SARS-CoV-2 Transmission from on-Field Player Contacts in Amateur, Youth and Professional Football (Soccer). *British Journal of Sports Medicine*, *56*(3), 158–164. doi:10.1136/bjsports-2021-104441 PMID:34663570

Steene-Johannessen, J., Hansen, B. H., Dalene, K. E., Kolle, E., Northstone, K., Møller, N. C., Grøntved, A., Wedderkopp, N., Kriemler, S., Page, A. S., Puder, J. J., Reilly, J. J., Sardinha, L. B., van Sluijs, E. M. F., Andersen, L. B., van der Ploeg, H., Ahrens, W., Flexeder, C., Standl, M., ... Ekelund, U. (2020). Variations in Accelerometry Measured Physical Activity and Sedentary Time across Europe— Harmonized Analyses of 47,497 Children and Adolescents. *The International Journal of Behavioral Nutrition and Physical Activity*, *17*(1), 38. doi:10.1186/s12966-020-00930-x PMID:32183834

Stott, C. J., Hutchison, P., & Drury, J. (2001). Hooligans abroad? Inter-group dynamics, social identity and participation in collective disorder at the 1998 World Cup finals. *British Journal of Social Psychology*, *40*(3), 359–384. doi:10.1348/014466601164876 PMID:11593939

Šuput, D. (2021). FIFA and UEFA regulations against violence at football matches – the problem of overlapping competencies between the state and sports organizations. *Bezbednost*, *1/2012*, 166.

Tóth, N. A. (2019). Sports Policing in the System of Sports Administration. University of Public Service, Faculty of Public Governance and International Studies.

Tóth, N. A. (2020). Thoughts about the current issues of sports policing. *Belügyi Szemle*, *68*(Special Issue), 77–93. doi:10.38146/BSZ.SPEC.2020.3.6

Tóth, N.A. (2021). The Application of Innovative Sports Policing Tools. *Časopis Policajna Teoria a Prax, 2*, 102-114.

Tulloch, J. (2000). Terrorism, "killing events," and their audience: Fear of crime at the 2000 Olympics. In K. Schaffer & S. Sidonie (Eds.), *The Olympics at the millennium: Power, politics and the games* (pp. 224–242). Rutgers University Press.

Turek, J., Ocicka, B., Rogowski, W., & Jefmański, B. (2023). The role of Industry 4.0 technologies in driving the financial importance of sustainability risk management. *Equilibrium. Quarterly Journal of Economics and Economic Policy*, *18*(4), 1009–1044. doi:10.24136/eq.2023.032

Walters, G., & Tacon, R. (2010). Corporate social responsibility in sport: Stakeholder management in the UK football industry. *Journal of Management & Organization*, *16*(4), 566–586. doi:10.5172/jmo.2010.16.4.566

Watson, A.M., Haraldsdottir, K., Biese, K., Goodavish, L., Stevens, B., & McGuine, T. (2021). The Association of COVID-19 Incidence with Sport and Face Mask use in United States High School Athletes. *Journal of Athletic Training*.

Whitson, D., & Horne, J. (2006). Underestimated costs and overestimated benefits? Comparing the outcomes of sports mega-events in Canada and Japan. *The Sociological Review*, *54*(2), 73–89. doi:10.1111/j.1467-954X.2006.00654.x

Chapter 10

Web of Wonders:
Sustainable Digital Marketing Strategies for Unforgettable African Tourism Destinations

Bassam Samir Al-Romeedy
Faculty of Tourism and Hotels, University of Sadat City, Egypt

ABSTRACT

This chapter investigates sustainable digital marketing strategies tailored for African tourism destinations, with a focus on promoting unforgettable experiences while preserving the natural and cultural heritage of these destinations. It also highlights the importance of aligning digital marketing efforts with sustainable tourism principles, such as promoting responsible travel behavior and supporting local communities. The chapter emphasizes the need to showcase the unique natural and cultural heritage of African destinations through visually appealing content and immersive digital experiences. Furthermore, the potential of emerging technologies, such as blockchain and artificial intelligence, is explored to enhance the effectiveness and sustainability of digital marketing campaigns. The outcomes of this study aim to provide practical recommendations for African tourism destinations to enhance their digital marketing efforts while ensuring long-term sustainability and success.

INTRODUCTION

African tourism destinations boast a rich tapestry of natural and cultural heritage (Mutanga et al., 2024). To allure tourists in search of genuine and unforgettable experiences, digital marketing strategies ought to accentuate these distinctive attributes. Utilizing visually captivating content, compelling storytelling, and immersive digital experiences, destinations can forge an emotional bond with their audience. By spotlighting the awe-inspiring landscapes, wildlife interactions, and cultural traditions, digital marketing campaigns have the potential to instill a sense of wonder and curiosity in prospective travelers (El-Sherbiny & Al-Romeedy, 2023).

DOI: 10.4018/979-8-3693-3286-3.ch010

In recent years, African tourism destinations have achieved remarkable success in their marketing efforts, implementing innovative strategies to attract more visitors and enhance the overall tourism experience. One notable example is South Africa, which has effectively positioned itself as a diverse and appealing destination through its "Meet South Africa" campaign. This initiative showcases the country's abundant cultural heritage, wildlife, and adventure tourism offerings, capturing the attention of potential travelers (National Government of South Africa, 2024). Similarly, Kenya has reinvigorated its tourism sector by adopting the "Magical Kenya" branding, which highlights the nation's natural beauty, wildlife safaris, and cultural tours as key attractions (Ministry of Tourism and Wildlife, 2022). Furthermore, Ghana's "Year of Return" campaign in 2019, commemorating the 400th anniversary of the arrival of the first enslaved Africans in the United States, successfully attracted visitors from the African diaspora, providing both tourism opportunities and cultural education (United Nations, 2020).

Sustainability awareness within African tourism destinations has undergone significant development, influenced by global environmental concerns, evolving traveler preferences, and an increasing recognition of the importance of conserving natural and cultural resources for future generations. Many African destinations are now more conscious of the environmental consequences associated with tourism, including pollution, over-tourism, and disruption of habitats. This heightened awareness is driving a shift towards eco-tourism and the adoption of sustainable practices aimed at minimizing negative impacts. There is a growing realization that sustainable tourism should also contribute to the economic and social well-being of local communities. This entails initiatives such as job creation, fair wages, and utilizing tourism revenues to support community projects (Gunduz & Agayi, 2021; Binns & Nel, 2002). The preservation of cultural heritage and traditions is gaining recognition as a crucial aspect, not only for maintaining the social fabric of local communities but also as a distinctive feature that sets destinations apart. Governments and regional tourism boards are taking steps to promote sustainability by implementing more stringent regulations and guidelines. These measures include the introduction of requirements for environmental impact assessments and the adoption of sustainable management practices (Richards, 2007; Glasson & Therivel, 2013).

Harmonizing digital marketing endeavors with sustainable tourism principles is imperative in the business landscape. Marketing sustainable tourism entails advocating for responsible travel behavior, mitigating environmental impacts, and backing local communities. African tourism destinations can utilize digital platforms to spotlight their distinctive attractions, encompassing abundant biodiversity, picturesque landscapes, and dynamic cultural experiences. Digital marketing initiatives should underscore the sustainable measures and responsible tourism projects implemented by these destinations to appeal to environmentally conscious travelers (Wibawa et al., 2022; Gheorghe et al., 2023; Saayman & Saayman, 2012).

Cutting-edge technologies like blockchain and artificial intelligencc have the potential to elevate the efficacy and sustainability of digital marketing strategies. Blockchain technology offers transparency and trust by allowing tourists to authenticate sustainable assertions made by destinations. This encompasses the verification of eco-certifications, responsible tourism endeavors, and community engagement initiatives. Artificial intelligence plays a role in tailoring marketing messages, pinpointing specific audience segments, and refining campaign performance through data analysis (Al-Romeedy, 2024; Mohamed et al., 2022). The formulation of sustainable digital marketing strategies necessitates collaborative efforts and partnerships among diverse stakeholders. Tourism officials, destination managers, local communities, and digital marketing experts should collaborate to guarantee the genuineness and credibility of marketing campaigns. The involvement of local communities is vital to incorporate their viewpoints, engage them in decision-making processes, and ensure the fair distribution of tourism benefits. Collaboration also enables the exchange of best practices, knowledge, and resources, fostering the development of sustainable tourism (Farahat, 2023).

Sustainable digital marketing strategies tailored for African tourism destinations should prioritize the promotion of responsible travel behavior, the display of natural and cultural heritage, the utilization of emerging technologies, the encouragement of collaboration, and the implementation of evaluation mechanisms. By embracing these strategies, African destinations can efficiently market unforgettable experiences, draw in responsible tourists, and actively contribute to the conservation and advancement of local resources (Arasli et al., 2021; Gumede, 2019).

The significance of sustainable digital marketing as a pivotal tool for global tourism destination promotion has been established (Devkota et al., 2023). Nonetheless, the application of sustainable digital marketing strategies in African tourism destinations requires dedicated research to align with the principles of responsible tourism. This chapter is dedicated to investigating and suggesting sustainable digital marketing strategies specifically designed for African tourism destinations, emphasizing the promotion of unforgettable experiences while safeguarding the natural and cultural heritage of these locales.

Sustainable Tourism

Sustainable tourism encompasses the principles and actions within the tourism sector with the goal of reducing environmental and cultural impact. Simultaneously, it seeks to contribute to income generation, employment, and the preservation of local ecosystems. It signifies a form of responsible tourism that prioritizes ecological and cultural sensitivity (Saarinen, 2020).

At the core of sustainable tourism lies a fundamental principle: environmental conservation. This entails the reduction of tourism activities' ecological footprint through measures like waste reduction, energy and water conservation, and the advocacy for the preservation of natural resources. Sustainable tourism endeavors to safeguard biodiversity, ecosystems, and delicate environments, including coral reefs, forests, and wildlife habitats. Additionally, it advocates for sustainable transportation alternatives, such as public transit or eco-friendly vehicles, aiming to mitigate carbon emissions (Edgell, 2019; McCool & Bosak, 2016).

Another principle is the preservation of socio-cultural authenticity and the respectful treatment of local communities. Sustainable tourism strives to honor and safeguard the cultural and social heritage of host communities. This involves fostering respectful interactions with local residents, patronizing local businesses and artisans, and encouraging cultural exchange. The objective is to engage local communities in decision-making processes and guarantee that they derive equitable economic benefits from tourism activities, thereby empowering them to sustain their cultural traditions and way of life (Harris et al., 2012; Saarinen, 2020).

Economic viability stands as another pivotal principle of sustainable tourism, focusing on generating economic benefits that enhance the welfare and prosperity of local communities. The goal is to create avenues for employment, income generation, and entrepreneurship, with a particular emphasis on empowering marginalized groups. Sustainable tourism advocates for the establishment of enterprises that operate sustainably, ensuring a fair distribution of economic benefits and contributing to the support of the local economy (McCool & Bosak, 2016; Slocum et al., 2020).

Moreover, sustainable tourism underscores the importance of education and awareness. It advocates for responsible behavior among tourists by enhancing awareness of the ecological, cultural, and social consequences of their actions. Sustainable tourism encourages tourists to make well-informed choices that mitigate negative impacts and enhance positive contributions to the destinations they explore. Additionally, it endorses educational programs for both tourists and local communities, aiming to cultivate understanding, respect, and appreciation for the environment and cultural heritage (Edgell, 2019; Saarinen, 2020).

Digital Marketing in the Tourism Industry

Digital marketing strategies within the tourism sector are crucial for reaching and captivating contemporary travelers who heavily depend on digital platforms for their travel planning and reservations. These strategies empower tourism enterprises to establish global connections with prospective customers, providing them with personalized, pertinent, and timely information and services (Mohamed et al., 2022). The primary digital marketing strategies in the tourism industry encompass the following:

Website Optimization: An effectively crafted and user-friendly website is indispensable for any tourism enterprise. It should undergo search engine optimization (SEO) to enhance visibility in search engine results, be compatible with mobile devices, and furnish comprehensive and current information regarding services, destinations, and packages (Farahat, 2023).

Content Marketing: Compelling and top-notch content plays a pivotal role in drawing and retaining customers. This encompasses blog posts, videos, and photos that showcase destinations, experiences, and services. The use of storytelling can be particularly influential, establishing emotional connections with prospective travelers (Mathew & Soliman, 2021).

Social Media Marketing: Social media platforms like Instagram, Facebook, Twitter, and TikTok play a pivotal role in the promotion of tourist destinations and services. They offer a space for sharing captivating images and videos, engaging with audiences, and fostering a community around a brand. User-generated content, including reviews and testimonials, can wield significant influence (Farahat, 2023).

Email Marketing: Delivering personalized emails to prospective and previous customers proves to be an efficient method for promoting exclusive offers, dispensing travel advice, and sustaining audience interest. This direct approach helps in nurturing relationships with customers and fostering repeated bookings (Farahat et al., 2022).

Influencer Partnerships: Partnering with influencers and travel bloggers can greatly enhance visibility. These individuals possess substantial followings and can influence the travel choices of their audience. Influencer-generated content often carries a sense of authenticity and relatability that surpasses traditional advertising (Saleh, 2023).

Online Booking and Reservation Systems: User-friendly online booking systems are crucial for translating interest into sales. Incorporating a secure and efficient booking system on the website offers convenience for customers (Farahat, 2023).

Search Engine Marketing (SEM): Investing in paid advertising on search engines is beneficial for reaching potential customers actively seeking travel-related services. This encompasses pay-per-click (PPC) campaigns, display ads, and remarketing strategies (Maurer, 2021).

Mobile Marketing: In the era of growing smartphone use for travel planning, mobile marketing becomes essential. This involves mobile-optimized websites, location-based services, mobile ads, and apps crafted to enhance the travel experience (Alghizzawi et al., 2023).

Virtual and Augmented Reality (VR/AR): These technologies offer immersive experiences, enabling potential tourists to virtually explore destinations, accommodations, or experiences before making bookings (Al-Romeedy, 2024).

Data Analytics and Customer Relationship Management (CRM): By employing data analytics tools and Customer Relationship Management (CRM) software, tourism businesses can acquire insights into customer preferences and behavior, facilitating more targeted and effective marketing strategies (Mohamed et al., 2022).

Sustainable Digital Marketing in Tourism Industry

Sustainable digital marketing practices encompass a range of strategies, techniques, and approaches that prioritize environmental and social responsibility while promoting products, services, or brands through digital channels. The goal of these practices is to minimize adverse environmental effects, provide support to local communities, and promote transparency and authenticity in digital marketing endeavors (Rathore, 2017; Ottman, 2017). Given the significant impact of the tourism sector on the environment and local communities, the significance of adopting sustainable digital marketing practices cannot be overstated in today's world (Alghizzawi et al., 2023).

In the digital realm, marketing activities, while intangible compared to traditional methods, still leave an environmental footprint mainly due to energy consumption for data storage, processing, and content dissemination. Sustainable digital marketing practices have the objective of minimizing this impact. They achieve this by selecting eco-friendly hosting solutions, optimizing online content for energy efficiency, and reducing digital waste. These practices make a significant contribution to reducing the overall environmental impact of digital marketing activities. This approach is in line with the global focus on reducing carbon emissions and preserving natural resources (Thangam & Chavadi, 2023; Gigauri et al., 2023).

Sustainable digital marketing practices also have a crucial role in promoting social responsibility. They encompass ethical considerations such as representing cultures and destinations accurately, avoiding the exploitation of local communities, and ensuring that marketing messages do not perpetuate harmful stereotypes or unsustainable tourism practices. By prioritizing inclusive and respectful marketing, sustainable digital strategies contribute to the well-being and positive growth of local communities affected by tourism. This ethical engagement fosters trust and credibility among consumers who are becoming increasingly aware of and concerned about the social and environmental impacts of their travel decisions (Khan et al., 2019; Gupta et al., 2019).

From an economic standpoint, sustainable digital marketing plays a vital role in ensuring the long-term sustainability of tourism destinations. By promoting responsible travel that minimizes negative impacts and maximizes positive contributions to local economies, these marketing practices help preserve the attractiveness and viability of tourist destinations for the future. This approach helps prevent the detrimental effects of over-tourism, which can result in environmental degradation and cultural dilution, ultimately undermining the tourism potential of a destination (Beirman, 2020; Dodds & Graci, 2012).

In today's market, where consumers are placing greater emphasis on sustainability, the adoption of sustainable digital marketing practices is crucial for establishing brand value and fostering customer loyalty. By aligning marketing messages with sustainability principles and demonstrating a dedication to responsible practices, brands can set themselves apart in a competitive marketplace. This alignment not only attracts an expanding segment of environmentally conscious consumers but also cultivates stronger customer loyalty and engagement. Consumers are more inclined to develop long-term relationships with brands that reflect their values (Khandai et al., 2023; Tariq et al., 2022).

Sustainable digital marketing practices involve safeguarding marketing strategies for the future. As sustainability regulations and public expectations continue to evolve, embracing sustainable practices today ensures that businesses are prepared for forthcoming changes and obstacles. Moreover, it positions them as frontrunners in the shift towards more sustainable and responsible business models (Doe, J. K., & Hinson, 2023; Gigauri et al., 2023).

Implementing sustainable digital marketing practices offers tourism businesses the opportunity to gather and analyze data that can inform decision-making and drive ongoing improvements. By monitoring and evaluating the effectiveness of marketing campaigns, businesses can pinpoint areas for optimization and fine-tune their strategies to achieve optimal results. Digital marketing tools provide valuable insights into customer preferences, behaviors, and feedback, allowing businesses to customize their offerings and experiences to meet evolving demands in a sustainable manner. This data-driven approach enhances efficiency, minimizes waste, and ensures that marketing efforts are targeted towards the appropriate audience, resulting in more effective resource utilization and enhanced business performance (Del Vecchio et al., 2018; Farahat, 2023).

Sustainable digital marketing is of utmost importance in the tourism industry, as it aligns with the growing demand for environmentally conscious and socially responsible practices. By integrating sustainability into digital marketing strategies, businesses can effectively convey their dedication to sustainable tourism and attract travelers who prioritize environmental responsibility. This entails promoting eco-friendly accommodations, showcasing sustainable activities and attractions, and encouraging responsible travel behaviors. Sustainable digital marketing also enables tourism businesses to differentiate themselves in a competitive market by highlighting their sustainable initiatives and certifications. By emphasizing environmentally friendly practices such as waste reduction, energy efficiency, and water conservation, businesses can appeal to travelers who value sustainability and seek well-informed choices for their travel experiences (El-Sherbiny & Al-Romeedy, 2023).

Moreover, sustainable digital marketing empowers businesses to educate and actively involve travelers in sustainable tourism practices. By delivering informative content, implementing social media campaigns, and utilizing interactive platforms, businesses can enhance awareness regarding the significance of sustainable travel. They can offer tips on minimizing environmental impact, as well as promote the preservation of local culture. By equipping travelers with knowledge, businesses can foster a sense of responsibility and inspire sustainable behaviors throughout every stage of the travel experience (Hysa et al., 2021; Peceny et al., 2019).

Additionally, sustainable digital marketing plays a crucial role in ensuring the long-term sustainability of tourism destinations. By advocating for sustainable practices, businesses can contribute to the protection of natural resources, the preservation of local ecosystems, and the well-being of local communities. This approach not only safeguards unique destinations for future generations but also enhances the socio-economic advantages derived from tourism, including job opportunities, community development, and the preservation of cultural heritage (Font & McCabe, 2017; Tahiri et al., 2022).

Further, sustainable digital marketing empowers businesses to evaluate and assess the outcomes of their sustainability initiatives. By utilizing analytics and data-driven insights, businesses can measure the effectiveness of their sustainable marketing campaigns, identify areas in need of improvement, and make informed decisions. This iterative approach to improvement enables businesses to refine their sustainability strategies, allocate resources more efficiently, and foster positive transformations within the industry (Baltaci, 2021; Sivarajah et al., 2020).

In the current digital era, consumers are highly informed and interconnected. Sustainable digital marketing offers businesses the opportunity to share their sustainability narratives and initiatives with a global audience. By employing compelling content, storytelling techniques, and impactful visuals, businesses can effectively convey their commitment to sustainability and inspire travelers to adopt sustainable behaviors. This not only enhances awareness about sustainable tourism but also empowers travelers to become advocates for responsible travel practices (Kaufman et al., 2023; Ottman, 2017).

At its core, sustainable digital marketing is more than just a niche tactic; it is a holistic strategy that encompasses environmental, social, and economic considerations in the digital era. Its significance lies in its capacity to make a positive impact on the planet and society, while also ensuring the long-term effectiveness and significance of marketing endeavors in an increasingly conscientious consumer landscape.

Technology and Innovation in Sustainable Digital Marketing

Exciting possibilities arise in sustainable tourism marketing with the emergence of innovative technologies like virtual reality (VR), augmented reality (AR), and blockchain (Mohamed et al., 2022). The potential of virtual reality (VR) to transform sustainable tourism marketing is significant. By offering immersive experiences that virtually transport potential travelers to destinations, VR has the capacity to revolutionize the industry. Through VR, sustainable tourism destinations can showcase their natural beauty and cultural heritage, enabling users to explore various locations, interact with local communities, and partake in eco-friendly activities. This technology fosters a sense of presence and connection, inspiring travelers to consider visiting these destinations while minimizing the environmental impact associated with physical travel (Al-Romeedy, 2024).

As well, sustainable tourism marketing can be enhanced through the use of augmented reality (AR), which overlays digital information onto the real-world environment. AR applications offer interactive and educational experiences for tourists, providing valuable information about sustainable practices, wildlife conservation, and the cultural significance of local areas. For instance, AR can be utilized to offer virtual guided tours that emphasize sustainable features and practices at specific destinations. Additionally, AR can foster responsible behavior by providing real-time information about waste management, energy consumption, and environmental impact at tourist sites (Farahat, 2023).

Similarly, the potential of blockchain technology to enhance transparency and trust in sustainable tourism marketing is significant. Through the use of blockchain, tourism businesses can provide verifiable information about their sustainability practices, including details about their carbon footprint, waste management, and community engagement. This transparent and unchangeable ledger enables travelers to make informed decisions and support businesses that align with their sustainability values. Furthermore, blockchain can facilitate secure and decentralized transactions, enabling tourists to contribute directly to local communities or conservation efforts, bypassing intermediaries and ensuring that funds reach their intended recipients (Farahat et al., 2022).

In the realm of sustainable tourism marketing, artificial intelligence (AI) and machine learning (ML) are emerging technologies that offer considerable support. AI-powered chatbots and virtual assistants can provide individualized recommendations to travelers, suggesting sustainable accommodations, eco-friendly transportation alternatives, and responsible tourism activities. ML algorithms can analyze extensive data sets to identify patterns and trends, empowering tourism businesses to optimize their marketing strategies and effectively target the appropriate audience with sustainability-focused messages (Al-Romeedy, 2024).

Moreover, the Internet of Things (IoT) can contribute to sustainable tourism marketing by facilitating smart and interconnected destinations. IoT devices can gather real-time data on energy consumption, waste management, and environmental conditions, enabling tourism destinations to monitor and enhance their sustainability practices. This data can be utilized to educate travelers about the destination's sustainability initiatives and promote responsible behavior (Farahat, 2023).

Within the African tourism industry, there is a growing utilization of artificial intelligence (AI) to enhance sustainable digital marketing practices. By harnessing AI applications, destinations can optimize their marketing endeavors, customize traveler experiences, and effectively promote sustainable tourism practices. For instance, Egypt has implemented AI-enhanced virtual reality (VR) and augmented reality (AR) tours of historical sites. These digital tours help alleviate physical foot traffic, thereby preserving delicate archaeological sites while still providing global access to Egypt's cultural heritage. AI enhances the VR/AR experiences by offering interactive, personalized guides and educational content about sustainability initiatives aimed at preserving these sites (Farahat, 2023).

Sustainable Digital Marketing Practices

Sustainable digital marketing in the tourism sector is a dynamic area that centers around promoting travel in a manner that upholds environmental preservation, supports local communities, and aligns with sustainable development goals. It involves harnessing digital tools and platforms to foster responsible travel practices while mitigating the adverse effects of tourism (Rasoolimanesh et al., 2023).

Green Web Practices and Content Creation

An essential element of sustainable digital marketing involves adopting environmentally friendly online practices (Fazel, 2024). This entails selecting web hosting services that utilize renewable energy sources and optimizing website designs for energy efficiency, such as utilizing compressed images to enhance loading speed and reduce energy consumption. Moreover, creating content that centers around sustainability is crucial. Blogs, videos, and social media posts can serve as educational resources, providing travelers with insights into eco-friendly accommodations, responsible travel tips, sustainable tourism initiatives, and conservation endeavors. By prioritizing this type of content, tourism businesses can effectively contribute to environmental awareness and foster responsible travel behaviors (Osburg, 2017; Kilipiri et al., 2023).

Data-Driven Approaches and Personalization

Another key aspect of sustainable digital marketing is the responsible utilization of data analytics. By analyzing traveler preferences and behaviors, tourism marketers can develop targeted and impactful strategies that promote sustainable options. However, it is essential to strike a balance with ethical considerations regarding data privacy and protection. Personalization, when practiced responsibly, can significantly enhance the effectiveness of marketing campaigns by delivering tailored and meaningful messages that align with the values and interests of environmentally conscious travelers (Sivarajah et al., 2020; Hernigou & Scarlat, 2023; Inge, 2022).

Sustainable Campaigns and Storytelling

In the tourism sector, digital marketing campaigns are increasingly prioritizing storytelling, particularly narratives that highlight sustainability and responsible travel. These stories serve as potent tools for influencing traveler attitudes and behaviors towards more sustainable practices. By showcasing the positive impact of sustainable tourism on local communities and the environment, these campaigns can inspire travelers to make more conscious choices when planning their trips (Bassano et al., 2019; Abduljabbar et al., 2021).

Eco-Friendly Digital Content and Platforms

A crucial element of sustainable digital marketing entails the development and dissemination of environmentally friendly content. This involves utilizing digital platforms to educate and inform travelers about sustainable tourism practices. For example, marketing campaigns may emphasize the significance of conserving natural habitats, respecting local cultures, and selecting eco-friendly travel options. Furthermore, the choice of digital infrastructure plays a role in sustainability efforts. Opting for green web hosting services that utilize renewable energy sources, as well as optimizing websites and digital content for energy efficiency (such as faster loading times to reduce energy consumption), are integral components of this sustainable approach (Caratù et al., 2023; Buonomano et al., 2023).

Ethical Marketing and Authentic Representation

Ethical considerations play a vital role in sustainable digital marketing for tourism, particularly in how destinations and cultures are portrayed. It is essential to steer clear of commodifying local cultures and instead present them in an authentic and respectful manner. This necessitates close collaboration with local communities to ensure their culture and heritage are accurately and sensitively represented. Marketing materials should avoid stereotypes and, instead, prioritize genuine and respectful storytelling that celebrates the distinctiveness of each destination and its inhabitants (Jamal & Camargo, 2014; Avraham & Ketter, 2016).

Community-Centric Marketing Initiatives

Another crucial aspect is the emphasis on benefiting local communities. Sustainable digital marketing practices involve promoting local businesses, artisans, and cultural experiences, thereby supporting the local economy and providing tourists with an authentic experience. Showcasing these community-oriented aspects in digital content can incentivize travelers to engage in responsible tourism that directly contributes to the well-being of the local population in the destination (Coroş et al., 2017; El-Sherbiny & Al-Romeedy, 2023).

Role of Sustainable Digital Marketing in Reshaping the African Tourism Destination Branding

The intersection of digital marketing and sustainability in Africa presents a distinctive opportunity to promote sustainable tourism and business practices that align with technological advancements and environmental responsibilities (Bricker et al., 2013). Africa has witnessed a significant increase in internet penetration and mobile usage, creating a fertile ground for the implementation of digital marketing strategies. Businesses and tourism destinations are progressively leveraging digital platforms to reach a broader audience. Moreover, there is a growing awareness among African consumers and businesses regarding sustainability issues, including the impacts of climate change, loss of biodiversity, and the importance of environmental conservation (Kanyam et al., 2017; Mpofu, 2022). This evolving consciousness is leading to a rise in eco-conscious African consumers who actively seek out environmentally friendly and ethically sourced products and services (Dubihlela & Ngxukumeshe, 2016).

Sustainable digital marketing is essential for branding tourism destinations and promoting responsible and authentic experiences in line with sustainability principles. There are various avenues through which sustainable digital marketing contributes to the branding of tourism destinations (Chhabra, 2010).

Authentic Storytelling

Within the context of African tourism destination branding, sustainable digital marketing assumes a crucial role in transforming the narrative, primarily through authentic storytelling. This approach not only fosters a more nuanced and culturally sensitive representation of destinations but also aligns with the growing consumer preference for sustainable and authentic travel experiences. Sustainable digital marketing empowers destinations to authentically present their distinctive narratives and cultural heritage. In contrast to traditional marketing methods that may oversimplify or commodify cultural experiences, digital platforms provide an opportunity for more profound and nuanced storytelling. Through mediums such as online videos, blogs, and social media, tourism marketers can share the stories of local communities, showcase traditional practices, and highlight wildlife conservation initiatives. This form of storytelling deeply resonates with travelers who are seeking genuine and immersive experiences, thereby significantly improving the perception of a destination (Kumar, 2024; Kompatsiaris, 2024).

Digital marketing tools provide African tourism destinations with unparalleled visibility and access to a global audience. Through these channels, sustainable practices can be effectively showcased, appealing to the increasing number of eco-conscious travelers. Platforms such as Instagram, YouTube, and Facebook serve as dynamic platforms that vividly portray the vibrancy and diversity of African destinations, capturing the interest of potential visitors from around the world (Ketter, E., & Avraham, 2021; Farahat et al., 2022).

Trust and transparency are paramount in the realm of sustainability. Sustainable digital marketing enables destinations to openly communicate their environmental initiatives and community contributions. Digital platforms can be utilized to provide updates on how tourism revenues are utilized to support local conservation projects or community development. This transparent approach not only fosters trust but also strengthens the destination's brand as one that is responsible and dedicated to meaningful sustainability practices (Font & McCabe, 2017; Mohamed et al., 2022).

In a fiercely competitive global tourism market, differentiation is crucial. African tourism destinations can leverage sustainable digital marketing to showcase distinctive sustainable practices that set them apart from others. For instance, highlighting eco-lodges or tours that utilize renewable energy or water-saving technologies can distinguish a destination. Such marketing efforts not only promote the destination but also educate the audience about sustainable practices, thereby enhancing the value of the traveler's experience. Beyond branding purposes, sustainable digital marketing also plays a significant role in advancing broader sustainable development goals. By promoting tourism that respects environmental limits and cultural heritage, destinations can ensure long-term viability. This approach not only attracts tourists but also ensures that tourism contributes positively to the sustainable development of the local area (Farahat, 2023).

Showcasing Sustainable Practices

Sustainable digital marketing offers a powerful platform to emphasize a destination's dedication to sustainable practices. By creating captivating content, destinations can showcase their eco-friendly initiatives, including responsible waste management, conservation efforts, renewable energy adoption, and support for local communities. By effectively communicating these sustainable practices, tourism destinations can enhance their brand image and appeal to environmentally conscious travelers (Pike, 2015; Farahat et al., 2022).

Engaging with Responsible Travelers

With sustainable digital marketing, tourism destinations can effectively engage with an expanding group of travelers who prioritize sustainability and responsible travel. Through focused digital campaigns, destinations can connect with this audience by providing information about sustainable activities, eco-certifications, and responsible tourism practices. This engagement cultivates a sense of shared values and reinforces the destination's brand positioning as a responsible and appealing option for conscientious travelers (Font & McCabe, 2017; Liu et al., 2023; Kumar, 2024).

Social Media Advocacy

Sustainable digital marketing capitalizes on the influence of social media platforms to magnify the branding endeavors of tourism destinations. Social media serves as a platform for travelers to exchange their experiences, recommendations, and photos, enabling destinations to benefit from user-generated content. By actively encouraging visitors to share their sustainable experiences and tag the destination, tourism destinations can harness the expansive reach and impact of social media to elevate their brand image and allure new visitors (El-Sherbiny & Al-Romeedy, 2023).

Educational and Inspirational Content

Through sustainable digital marketing, tourism destinations have the opportunity to educate and motivate travelers regarding sustainable practices and the significance of responsible tourism. By crafting informative and visually captivating content, such as blog posts, videos, and interactive experiences, destinations can enhance awareness about sustainability issues and encourage travelers to make more conscious decisions. This educational facet of sustainable digital marketing aids in establishing the destination as a pioneer in sustainability and cultivates a favorable brand image (Kilipiri et al., 2023; Ferrer-Rosell et al., 2023).

User-generated Content and Social Proof

Sustainable digital marketing promotes the generation of user-generated content by actively engaging visitors to share their experiences and recommendations. This user-generated content acts as social proof, presenting authentic experiences and testimonials that highlight the destination's sustainability initiatives. By showcasing this content on their website, social media platforms, and other digital channels, destinations can harness the credibility and influence of travelers to bolster their brand reputation and attract a larger audience (Yamagishi et al., 2023; Neidhardt & Wörndl, 2020).

Community Engagement and Empowerment

Sustainable digital marketing opens avenues for tourism destinations to engage with and empower local communities. Destinations can utilize digital platforms to promote initiatives led by the community, support local businesses, and involve residents in sustainable tourism development. By highlighting the positive effects of tourism on the community and actively including locals in digital marketing endeavors, destinations can foster a strong sense of community pride and strengthen the destination's brand as a responsible and inclusive place to visit (Tjilen et al., 2021; Koščak & O'Rourke, 2023).

Collaboration and Partnerships

Sustainable digital marketing promotes collaboration and partnerships between tourism destinations and a range of stakeholders, such as local communities, NGOs, sustainable organizations, and influencers. By joining forces, destinations can enhance the reach of their sustainable initiatives, exchange resources, and collaboratively develop content that highlights responsible tourism practices. Through these collaborative digital marketing efforts, destinations can reinforce their brand as a frontrunner in sustainability and responsible tourism (Spadaro et al., 2023; Perkins et al., 2020).

Virtual Reality and Immersive Experiences

By harnessing virtual reality (VR) and other immersive technologies, sustainable digital marketing can create virtual experiences that showcase a destination's natural and cultural attractions. This enables prospective travelers to virtually explore the destination, immerse themselves in its distinctive features, and establish a profound connection with the place. By offering these immersive experiences, destinations can fortify their brand identity and entice travelers to visit in person, while simultaneously reducing the environmental impact typically associated with travel (Farahat, 2023).

Influencer Partnerships

Sustainable digital marketing frequently involves forming partnerships with influencers to promote tourism destinations. By collaborating with influencers who share sustainable values and possess a genuine passion for responsible travel, destinations can greatly enhance their branding endeavors. Influencers can share their experiences, provide recommendations, and create content centered around sustainability with their followers, thereby expanding the destination's visibility and appealing to a broader audience of conscious travelers (Farahat et al., 2022; Mohamed et al., 2022).

Continuous Improvement and Adaptation

Through sustainable digital marketing, destinations can collect feedback, analyze data, and assess the real-time impact of their campaigns. This data-driven approach empowers destinations to consistently enhance their marketing strategies, adjust to evolving traveler preferences, and fine-tune their sustainability initiatives. By remaining responsive and adaptable, destinations can uphold a robust brand identity and reputation as a sustainable tourism destination (Low et al., 2020; Rosário & Dias, 2023).

CONCLUSION AND RECOMMENDATIONS

African tourism destinations can reap significant advantages by implementing sustainable digital marketing strategies. It is crucial for these destinations to prioritize highlighting their distinctive cultural heritage and authenticity within their digital marketing approaches. By accentuating local traditions, cuisines, and experiences, destinations can attract travelers who are actively seeking genuine cultural immersion and authentic travel experiences.

To effectively promote African destinations, it is crucial to prioritize the creation and dissemination of authentic content that genuinely reflects the local culture and traditions. This can involve showcasing local artisans, cultural festivals, traditional music and dance performances, and indigenous practices. Utilize high-quality imagery, videos, and compelling storytelling techniques to vividly capture these authentic experiences and bring them to life for prospective travelers. Authentic content plays a pivotal role in building trust, evoking curiosity, and encouraging travelers to actively engage with the destination. Additionally, it is important to emphasize African destinations' endeavors to preserve and safeguard their cultural heritage. This can be achieved by showcasing heritage sites, museums, and historical landmarks. Collaborating with local communities and cultural organizations, destinations can create digital campaigns that raise awareness about the significance of preserving cultural heritage. By highlighting these initiatives, destinations can attract travelers who prioritize responsible and sustainable tourism practices.

Furthermore, African destinations should underscore the availability of opportunities for travelers to engage in genuine cultural immersion. This can be accomplished by promoting local homestays, cultural exchange programs, and community-based tourism initiatives. Encouraging visitors to participate in workshops, traditional ceremonies, or local celebrations enables them to gain a profound understanding of the destination's cultural heritage. By highlighting these immersive experiences in digital marketing campaigns, destinations can attract travelers who actively seek meaningful connections and transformative travel experiences.

African destinations should prioritize cultural representation with sensitivity and respect. It is imperative to ensure that digital marketing materials accurately depict local traditions and cultural practices, avoiding stereotypes or cultural appropriation. Engaging with local communities and cultural experts can help guarantee that content is culturally sensitive and authentic. By showcasing the destination's cultural heritage in a respectful manner, destinations can cultivate positive perceptions and attract travelers who highly value cultural authenticity.

Sustainable tourism practices in African destinations are predominantly centered around environmental conservation and protection. By aligning their digital marketing strategies with eco-friendly initiatives, destinations can appeal to environmentally conscious travelers and promote sustainable practices. It is crucial for African destinations to highlight their conservation efforts and initiatives in their digital marketing materials. This can involve showcasing national parks, protected areas, and wildlife reserves, while emphasizing conservation projects such as anti-poaching efforts, habitat restoration, and wildlife rehabilitation that contribute to the preservation of the destination's diverse biodiversity. By showcasing these initiatives, destinations can attract travelers who are deeply passionate about wildlife conservation and environmental sustainability.

In their digital marketing endeavors, African destinations should feature their sustainable practices and initiatives. This can involve promoting eco-friendly accommodations, energy-efficient transportation options, waste management programs, and renewable energy projects. Additionally, highlighting sustainable tourism certifications such as eco-labels or green accreditations can further emphasize the destination's dedication to environmental conservation. By showcasing these sustainable practices, destinations can attract travelers who prioritize low-impact travel and seek eco-conscious experiences.

To create a favorable perception of the destination's sustainability endeavors, it is important to infuse green messaging throughout digital marketing materials. Utilize language that highlights the destination's unwavering dedication to eco-friendliness, conservation, and responsible tourism practices. Integrate visuals that prominently display the destination's natural beauty and its initiatives to preserve the environment. By consistently communicating green messaging, destinations can establish themselves as pioneers in sustainable tourism and appeal to travelers who share these values.

For sustainable tourism, it is vital to engage with local communities. African destinations should incorporate community voices into their digital marketing content. By showcasing community-led initiatives, local entrepreneurship, and cultural events in their digital campaigns, destinations can empower local populations and also appeal to travelers who seek authentic and community-driven experiences. African destinations should highlight community-led initiatives that contribute to sustainable tourism, such as local conservation projects, community-based tourism enterprises, and initiatives that promote cultural preservation. By featuring these initiatives in digital marketing campaigns, destinations can not only raise awareness about the positive impact of community involvement but also attract travelers who value community-driven and responsible tourism practices.

Likewise, African destinations should actively promote and uplift local entrepreneurship. This can be done by highlighting local artisans, craftsmen, and entrepreneurs who offer one-of-a-kind products and services. Encourage travelers to support these local businesses by purchasing authentic souvenirs, participating in local tours or workshops, or dining at community-owned restaurants. By showcasing local entrepreneurship, destinations not only contribute to the growth of the local economy but also provide travelers with authentic and immersive experiences.

To promote sustainable tourism in African destinations, it is crucial to establish partnerships with businesses and organizations dedicated to sustainable practices. By collaborating with eco-friendly accommodations, responsible tour operators, and sustainable transportation services, destinations can enhance their credibility and appeal to conscious travelers. Prioritizing partnerships with accommodations that prioritize sustainability, such as eco-lodges, green hotels, or resorts with certified sustainable practices, allows destinations to showcase their commitment to environmentally friendly tourism options. Collaborating with responsible tour operators who prioritize sustainable practices, including minimizing environmental impact, supporting local communities, and promoting cultural preservation, enables destinations to offer sustainable and authentic experiences to travelers. Highlighting these partnerships in digital marketing campaigns can attract conscious travelers who seek immersive and responsible tourism experiences.

Digital platforms provide African destinations with a valuable tool to educate potential visitors about responsible tourism practices. By sharing informative content on sustainable travel practices, cultural etiquette, and environmental awareness, destinations can promote responsible behavior and cultivate awareness among travelers. They can emphasize waste reduction, water and energy conservation, supporting local communities, and respecting wildlife and natural habitats. Practical tips and guidelines for responsible travel, such as using eco-friendly products and opting for sustainable transportation, can be shared as well. Through education, destinations can encourage responsible behavior, minimize negative environmental impact, and raise awareness about regional environmental challenges and the importance of conservation.

Social media is a powerful tool for African destinations to engage in digital marketing. Platforms like Instagram, Twitter, and Facebook can be used to share compelling sustainability stories. By utilizing engaging visuals, user-generated content, and real-time updates, destinations can create a positive online presence that attracts socially conscious travelers and encourages them to share their experiences. Data analytics tools can also be leveraged to optimize digital marketing efforts by understanding traveler preferences, demographics, and behavior. This enables targeted marketing campaigns that effectively deliver sustainability messages to the right audience. Additionally, destinations should prioritize accessibility and inclusivity in their digital marketing strategies by providing information in multiple languages, considering diverse cultural perspectives, and ensuring that digital content is accessible to individuals with disabilities. An inclusive approach expands the appeal of the destination.

Further, African destinations should prioritize making their digital content accessible to individuals with disabilities. This involves designing websites and digital platforms in accordance with accessibility guidelines. Examples include providing alt-text for images, using descriptive headings, and ensuring compatibility with screen readers. Additionally, offering closed captions for videos and providing text alternatives for visual content can further enhance accessibility for individuals with hearing or visual impairments.

To foster sustainability awareness in Africa, a comprehensive approach that acknowledges and incorporates local cultures, economic circumstances, and environmental obstacles is essential. It is crucial to cultivate fresh perspectives that are specifically tailored to address Africa's distinct requirements and potentials, thereby driving effective sustainability initiatives. Additionally, recognizing the community-oriented nature of many African societies is vital. Sustainability endeavors should be designed to generate communal benefits and uphold communal responsibilities. Empowering communities through initiatives such as local conservation projects, community gardens for sustainable agriculture, and cooperative business models for renewable energy can foster community engagement and enable collective action towards achieving sustainability objectives.

Africa possesses a wealth of indigenous knowledge concerning environmental management and sustainable lifestyles. By acknowledging and incorporating this knowledge into contemporary sustainability strategies, culturally relevant and highly effective approaches to conservation and resource management can be developed. Initiatives that document, study, and promote indigenous practices can bridge the gap between traditional and modern techniques, fostering greater awareness and implementation of sustainability principles. Furthermore, investing in educational programs that prioritize sustainability, starting from primary schools to universities, can cultivate a generation that is well-informed and skilled in sustainable practices. Developing curricula that encompass subjects such as local biodiversity, climate change, renewable energy, and sustainable agriculture tailored to the African context is crucial in this regard.

The widespread adoption of mobile technology in Africa presents a valuable opportunity to leverage digital platforms as effective tools for promoting sustainability awareness. Mobile applications can play a pivotal role in delivering educational content, offering practical tips on sustainable practices, and facilitating the sharing of knowledge and experiences. Additionally, social media campaigns can be utilized to raise awareness and engage both urban and rural communities in discussions surrounding sustainability challenges and potential solutions. Encouraging collaborations between governments, non-governmental organizations (NGOs), and the private sector is also essential for fostering robust sustainability initiatives. These partnerships can mobilize significant resources, facilitate the exchange of knowledge, and drive innovations that are economically viable and environmentally sustainable. Furthermore, such partnerships can contribute to scaling successful local initiatives to broader regions, maximizing their impact.

REFERENCES

Abduljabbar, R., Liyanage, S., & Dia, H. (2021). The role of micro-mobility in shaping sustainable cities: A systematic literature review. *Transportation Research Part D, Transport and Environment*, *92*, 102734. doi:10.1016/j.trd.2021.102734

Al-Romeedy, B. (2024). AI and HRM in Tourism and Hospitality in Egypt: Inevitability, Impact and Future. In HRM, Artificial Intelligence and the Future of Work: Critical Insights from the Global South. Palgrave Macmillan.

Alghizzawi, M., Habes, M., Al Assuli, A., & Ezmigna, A. (2023). Digital Marketing and Sustainable Businesses: As Mobile Apps in Tourism. In *Artificial Intelligence and Transforming Digital Marketing*. Springer Nature Switzerland.

Arasli, H., Abdullahi, M., & Gunay, T. (2021). Social media as a destination marketing tool for a sustainable heritage festival in Nigeria: A moderated mediation study. *Sustainability (Basel)*, *13*(11), 6191. doi:10.3390/su13116191

Avraham, E., & Ketter, E. (2016). *Tourism marketing for developing countries: Battling stereotypes and crises in Asia, Africa and the Middle East*. Springer. doi:10.1057/9781137342157

Baltaci, A. (2021). Strategic Digital Marketing and Sustainability. In *Financial Ecosystem and Strategy in the Digital Era: Global Approaches and New Opportunities* (pp. 237–264). Springer International Publishing. doi:10.1007/978-3-030-72624-9_11

Bassano, C., Barile, S., Piciocchi, P., Spohrer, J., Iandolo, F., & Fisk, R. (2019). Storytelling about places: Tourism marketing in the digital age. *Cities (London, England)*, *87*, 10–20. doi:10.1016/j.cities.2018.12.025

Beirman, D. (2020). *Restoring tourism destinations in crisis: A strategic marketing approach*. Routledge. doi:10.4324/9781003117148

Binns, T., & Nel, E. (2002). Tourism as a local development strategy in South Africa. *The Geographical Journal*, *168*(3), 235–247. doi:10.1111/1475-4959.00051

Bricker, K. S., Black, R., & Cottrell, S. (Eds.). (2013). *Sustainable tourism & the millennium development goals: Effecting positive change*. Jones & Bartlett Publishers.

Buonomano, A., Del Papa, G., Giuzio, G., Palombo, A., & Russo, G. (2023). Future pathways for decarbonization and energy efficiency of ports: Modelling and optimization as sustainable energy hubs. *Journal of Cleaner Production*, *420*, 138389. doi:10.1016/j.jclepro.2023.138389

Caratù, M., Cherubino, P., Menicocci, S., & Martinez-Levy, A. (2023). Does sustainable communication have an impact on international social media audiences? A neuromarketing explorative study between Finland and Italy. *Italian Journal of Marketing*. doi:10.1007/s43039-023-00086-z

Chhabra, D. (2010). *Sustainable marketing of cultural and heritage tourism*. Routledge. doi:10.4324/9780203855416

Coroş, M., Gică, O., Yallop, A., & Moisescu, O. (2017). Innovative and sustainable tourism strategies: A viable alternative for Romania's economic development. *Worldwide Hospitality and Tourism Themes*, *9*(5), 504–515. doi:10.1108/WHATT-07-2017-0033

Del Vecchio, P., Mele, G., Ndou, V., & Secundo, G. (2018). Creating value from social big data: Implications for smart tourism destinations. *Information Processing & Management*, *54*(5), 847–860. doi:10.1016/j.ipm.2017.10.006

Devkota, N., Gajdka, K., Siwakoti, R., Klímová, M., & Dhakal, K. (2023). Promoting Sustainable Tourist Behavior through Promotional Marketing. *Journal of Tourism and Services*, *14*(26), 219–241. doi:10.29036/jots.v14i26.512

Dodds, R., & Graci, S. (2012). *Sustainable tourism in island destinations*. Routledge. doi:10.4324/9781849776660

Doe, J., & Hinson, R. (2023). AI-driven sustainability brand activism for family businesses: a future-proofing perspective article. *Journal of Family Business Management*. doi:10.1108/JFBM-10-2023-0217

Dubihlela, J., & Ngxukumeshe, T. (2016). Eco-friendly retail product attributes, customer attributes and the repurchase intentions of South African consumers. *The International Business & Economics Research Journal*, *15*(4), 163–174. doi:10.19030/iber.v15i4.9754

Edgell, D. (2019). *Managing sustainable tourism: A legacy for the future*. Routledge.

El-Sherbiny, M., & Al-Romeedy, B. (2023). Does loyalty moderate the impact of digital marketing on the purchase intention of customers in airlines? *Journal of the Faculty of Tourism and Hotels-University of Sadat City*, *7*(2), 1–16.

Farahat, E. (2023). *Applications of artificial intelligence as a marketing tool and their impact on the competitiveness of the Egyptian tourist destination* (Doctoral Dissertation, Minia University).

Farahat, E., Mohamed, M., & Al-Romeedy, B. (2022). Artificial intelligence applications and its impact on the competitiveness of the Egyptian tourist destination. *Research Journal of the Faculty of Tourism and Hotels Mansoura University*, *11*(2), 57–93.

Fazel, H. (2024). Digital marketing and sustainability in the era of climate change: PLS-structural equation modeling approach. *Journal of Statistics Applications & Probability*, *13*(1), 491–505. doi:10.18576/jsap/130134

Ferrer-Rosell, B., Massimo, D., & Berezina, K. (2023). *Information and Communication Technologies in Tourism 2023: Proceedings of the ENTER 2023 eTourism Conference*. Springer Nature.

Font, X., & McCabe, S. (2017). Sustainability and marketing in tourism: Its contexts, paradoxes, approaches, challenges and potential. *Journal of Sustainable Tourism*, *25*(7), 869–883. doi:10.1080/09669582.2017.1301721

Gheorghe, G., Tudorache, P., & Roşca, I. (2023). The contribution of green marketing in the development of a sustainable destination through advanced clustering methods. *Sustainability (Basel)*, *15*(18), 13691. doi:10.3390/su151813691

Gigauri, I., Palazzo, M., & Ferri, M. (2023). *Handbook of Research on Achieving Sustainable Development Goals with Sustainable Marketing*. IGI Global. doi:10.4018/978-1-6684-8681-8

Glasson, J., & Therivel, R. (2013). *Introduction to environmental impact assessment*. Routledge.

Gumede, T. (2019). Heritage tourism as a strategy for the local economic development in KwaZulu-Natal, South Africa. *African Journal of Hospitality, Tourism and Leisure*, *8*(3), 1–15.

Gunduz, E., & Agayi, C. (2021). Assessment of the State and Impact of Tourism Activities in Kenya. *Kent Akademisi*, *14*(1), 174–185. doi:10.35674/kent.822245

Gupta, A., Dash, S., & Mishra, A. (2019). All that glitters is not green: Creating trustworthy ecofriendly services at green hotels. *Tourism Management*, *70*, 155–169. doi:10.1016/j.tourman.2018.08.015

Harris, R., Williams, P., & Griffin, T. (2012). *Sustainable tourism*. Routledge. doi:10.4324/9780080498362

Hernigou, P., & Scarlat, M. (2023). Two minutes of orthopaedics with ChatGPT: It is just the beginning; it's going to be hot, hot, hot! *International Orthopaedics*, *47*(8), 1887–1893. doi:10.1007/s00264-023-05887-7 PMID:37428231

Hysa, B., Karasek, A., & Zdonek, I. (2021). Social media usage by different generations as a tool for sustainable tourism marketing in society 5.0 idea. *Sustainability (Basel)*, *13*(3), 1018. doi:10.3390/su13031018

Inge, C. (2022). *Marketing metrics: leverage analytics and data to optimize marketing strategies*. Kogan Page Publishers.

Jamal, T., & Camargo, B. (2014). Sustainable tourism, justice and an ethic of care: Toward the just destination. *Journal of Sustainable Tourism*, *22*(1), 11–30. doi:10.1080/09669582.2013.786084

Kanyam, D. A., Kostandini, G., & Ferreira, S. (2017). The mobile phone revolution: Have mobile phones and the internet reduced corruption in Sub-Saharan Africa? *World Development*, *99*, 271–284. doi:10.1016/j.worlddev.2017.05.022

Kaufman, I., Horton, C., & Soltanifar, M. (2023). *Digital Marketing: Integrating Strategy, Sustainability, and Purpose*. Taylor & Francis. doi:10.4324/9781351019187

Ketter, E., & Avraham, E. (2021). # StayHome today so we can# TravelTomorrow: Tourism destinations' digital marketing strategies during the Covid-19 pandemic. *Journal of Travel & Tourism Marketing*, *38*(8), 819–832. doi:10.1080/10548408.2021.1921670

Khan, A., Wang, M., Ehsan, S., Nurunnabi, M., & Hashmi, M. (2019). Linking sustainability-oriented marketing to social media and web atmospheric cues. *Sustainability (Basel)*, *11*(9), 2663. doi:10.3390/su11092663

Khandai, S., Mathew, J., Yadav, R., Kataria, S., & Kohli, H. (2023). Ensuring brand loyalty for firms practising sustainable marketing: A roadmap. *Society and Business Review*, *18*(2), 219–243. doi:10.1108/SBR-10-2021-0189

Kilipiri, E., Papaioannou, E., & Kotzaivazoglou, I. (2023). Social Media and Influencer Marketing for Promoting Sustainable Tourism Destinations: The Instagram Case. *Sustainability (Basel)*, *15*(8), 6374. doi:10.3390/su15086374

Kompatsiaris, P. (2024). *Curation in the Age of Platform Capitalism: The Value of Selection, Narration, and Expertise in New Media Cultures*. Taylor & Francis. doi:10.4324/9781003182801

Koščak, M., & O'Rourke, T. (2023). *Ethical and responsible tourism: managing sustainability in local tourism destinations*. Taylor & Francis.

Kumar, S. (2024). The Role of Digital Marketing on Customer Engagement in the Hospitality Industry. In *Leveraging ChatGPT and Artificial Intelligence for Effective Customer Engagement*. IGI Global. doi:10.4018/979-8-3693-0815-8.ch010

Liu, C., Horng, J., Chou, S., Yu, T., Lee, M., & Lapuz, M. (2023). Discovery sustainable servicescape on behavioural intention practices and nationality: The moderating role of parasocial interaction. *Journal of Retailing and Consumer Services*, *71*, 103213. doi:10.1016/j.jretconser.2022.103213

Low, S., Ullah, F., Shirowzhan, S., Sepasgozar, S., & Lin Lee, C. (2020). Smart digital marketing capabilities for sustainable property development: A case of Malaysia. *Sustainability (Basel)*, *12*(13), 5402. doi:10.3390/su12135402

Mathew, V., & Soliman, M. (2021). Does digital content marketing affect tourism consumer behavior? An extension of t echnology acceptance model. *Journal of Consumer Behaviour*, *20*(1), 61–75. doi:10.1002/cb.1854

Maurer, C. (2021). Digital marketing in tourism. In *Handbook of e-Tourism*. Springer International Publishing. doi:10.1007/978-3-030-05324-6_82-1

McCool, S., & Bosak, K. (2016). *Reframing sustainable tourism*. Springer. doi:10.1007/978-94-017-7209-9

Ministry of Tourism and Wildlife. (2022). *New Tourism Strategy for Kenya 2021-2025*. Republic of Kenya.

Mohamed, M., Farahat, E., & Al-Romeedy, B. (2022). The Mediating Role of Marketing Effectiveness in the Relationship between Artificial Intelligence and Destination Competitiveness. *Minia Journal of Tourism and Hospitality Research*, *14*(1), 113–129.

Mpofu, F. Y. (2022). Green Taxes in Africa: Opportunities and challenges for environmental protection, sustainability, and the attainment of sustainable development goals. *Sustainability (Basel)*, *14*(16), 10239. doi:10.3390/su141610239

Mutanga, C., Kolawole, O., Gondo, R., & Mbaiwa, J. (2024). A review and SWOC analysis of natural heritage tourism in sub-Saharan Africa. *Journal of Heritage Tourism*, *19*(1), 49–67. doi:10.1080/1743873X.2023.2255689

National Government of South Africa. (2024). *South African Tourism Annual Report 2021/2022*. South African Tourism Board.

Neidhardt, J., & Wörndl, W. (2020). *Information and Communication Technologies in Tourism 2020*. Springer International Publishing. doi:10.1007/978-3-030-36737-4

Osburg, T. (2017). *Sustainability in a digital world needs trust*. Springer International Publishing. doi:10.1007/978-3-319-54603-2

Ottman, J. (2017). *The new rules of green marketing: Strategies, tools, and inspiration for sustainable branding*. Routledge. doi:10.4324/9781351278683

Peceny, U. S., Urbančič, J., Mokorel, S., Kuralt, V., & Ilijaš, T. (2019). Tourism 4.0: Challenges in marketing a paradigm shift. In Consumer behavior and marketing. IntechOpen.

Perkins, R., Khoo-Lattimore, C., & Arcodia, C. (2020). Understanding the contribution of stakeholder collaboration towards regional destination branding: A systematic narrative literature review. *Journal of Hospitality and Tourism Management*, *43*, 250–258. doi:10.1016/j.jhtm.2020.04.008

Pike, S. (2015). *Destination marketing: essentials*. Routledge. doi:10.4324/9781315691701

Rasoolimanesh, S., Ramakrishna, S., Hall, C., Esfandiar, K., & Seyfi, S. (2023). A systematic scoping review of sustainable tourism indicators in relation to the sustainable development goals. *Journal of Sustainable Tourism*, *31*(7), 1497–1517. doi:10.1080/09669582.2020.1775621

Rathore, B. (2017). Aligning Profitability and Environmental Responsibility: A Study on Sustainable Marketing Strategies. *Eduzone: International Peer Reviewed/Refereed Multidisciplinary Journal, 6*(2), 7-15.

Richards, G. (2007). *Cultural tourism: Global and local perspectives*. Psychology Press.

Rosário, A., & Dias, J. (2023). How has data-driven marketing evolved: Challenges and opportunities with emerging technologies. *International Journal of Information Management Data Insights*, *3*(2), 100203. doi:10.1016/j.jjimei.2023.100203

Saarinen, J. (2020). *Tourism and sustainable development goals: Research on sustainable tourism geographies*. Routledge.

Saayman, M., & Saayman, A. (2012). Shopping tourism or tourists shopping? A case study of South Africa's African tourism market. *Tourism Economics*, *18*(6), 1313–1329. doi:10.5367/te.2012.0169

Saleh, M. (2023). Advantaging Tourism Through Influencers: Applying Transaction Cost Theory to Recognize Top Hero, Hub, and Hygiene Content Tactics for Tourism Marketing. *Journal of Travel Research*, 00472875231214727. Advance online publication. doi:10.1177/00472875231214727

Sivarajah, U., Irani, Z., Gupta, S., & Mahroof, K. (2020). Role of big data and social media analytics for business to business sustainability: A participatory web context. *Industrial Marketing Management*, *86*, 163–179. doi:10.1016/j.indmarman.2019.04.005

Slocum, S. L., Aidoo, A., & McMahon, K. (2020). *The business of sustainable tourism development and management*. Routledge. doi:10.4324/9781351031066

Spadaro, I., Pirlone, F., Bruno, F., Saba, G., Poggio, B., & Bruzzone, S. (2023). Stakeholder Participation in Planning of a Sustainable and Competitive Tourism Destination: The Genoa Integrated Action Plan. *Sustainability (Basel)*, *15*(6), 5005. doi:10.3390/su15065005

Tahiri, A., Kovaçi, I., & Trajkovska Petkoska, A. (2022). Sustainable tourism as a potential for promotion of regional heritage, local food, traditions, and diversity—Case of Kosovo. *Sustainability (Basel)*, *14*(19), 12326. doi:10.3390/su141912326

Tariq, E., Alshurideh, M., Akour, I., Al-Hawary, S., & Kurdi, B. (2022). The role of digital marketing, CSR policy and green marketing in brand development. *International Journal of Data and Network Science*, *6*(3), 995–1004. doi:10.5267/j.ijdns.2022.1.012

Thangam, D., & Chavadi, C. (2023). Impact of Digital Marketing Practices on Energy Consumption, Climate Change, and Sustainability. *Climate and Energy*, *39*(7), 11–19. doi:10.1002/gas.22329

Tjilen, A., Ririhena, S., & Manuhutu, F. (2021). Model Ecotourism-Based Tourism Development Strategy and Empowerment of Local Communities in Merauke Regency. *Journal of Economics, Trade and Marketing Management*, *3*(01), 48–59. doi:10.22158/jetmm.v3n1p48

United Nations. (2020). Beyond the Year of Return: Africa and the diaspora must forge closer ties. Author.

Wang, Z., Li, M., Lu, J., & Cheng, X. (2022). Business Innovation based on artificial intelligence and Blockchain technology. *Information Processing & Management*, *59*(1), 102759. doi:10.1016/j.ipm.2021.102759

Wibawa, I., Astawa, I., Sukmawati, N., Arjana, I., & Sutarma, I. (2022). Digital Marketing and Sustainable Tourism for Tourist Villages in Bangli Regency. *International Journal of Glocal Tourism*, *3*(2), 88–99. doi:10.58982/injogt.v3i2.209

Yamagishi, K., Canayong, D., Domingo, M., Maneja, K., Montolo, A., & Siton, A. (2023). User-generated content on Gen Z tourist visit intention: a stimulus-organism-response approach. *Journal of Hospitality and Tourism Insights*. doi:10.1108/JHTI-02-2023-0091

Chapter 11
Smart Data Analysis and Prediction of Responsible Customer Behaviour in Tourism:
An Exploratory Review of the Literature

Naoual Bouhtati
Sidi Mohamed Ben Abdellah University, Fez, Morocco

Lhoussaine Alla
https://orcid.org/0000-0002-7238-1792
Sidi Mohamed Ben Abdellah University, Fez, Morocco

Ikram Ed-Daakouri
Universiét Ibn Zohr, Agadir, Morocco

ABSTRACT

This research seeks to analyze how the use of smart data analytics solutions by tourism stakeholders can promote the prediction of responsible customer behavior. After a theoretical and conceptual framework, the theoretical study is based on an exploratory literature review. It revealed the adoption of a range of smart data analytics solutions by businesses and tourist destinations in order to better predict attitudes and concrete responsible actions reflecting the eco-responsible behavior of tourists. Despite the inherent limitations of the approach, the results show multiple good practices that can be adopted by industry players to better understand, anticipate, and even guide the responsible behavior of their customers and prospects.

DOI: 10.4018/979-8-3693-3286-3.ch011

INTRODUCTION

In the current context of the abundance of big data on the customer or even on the prospect, companies are obliged to ensure the proper management for judicious exploitation to make relevant marketing decisions and be able to adapt continuously with the evolution of the needs of the customer, its attitudes and behaviours to update its value proposition, satisfy and streamline its customers, stand out from competitors, improve its performance, etc. Tourism is one of the sectors most affected by the digital revolution and digital transformation (Pencarelli, 2020 ; Lazić et al., 2023).

The rise of digital technologies such as data analytics, artificial intelligence (AI) and digital marketing has profoundly transformed tourism behaviours, marketing strategies and business models in the tourism industry. Tourists now have access to an unprecedented amount of information about tourist destinations, hotels, attractions, restaurants, and experiences.

In this sense, various studies have shown that the use of intelligent customer data management solutions is now essential to predict consumer behaviours, particularly in terms of responsible consumption, through dynamic interaction between responsible attitudes and behaviours.

The prediction of tourist behaviour is very complex given the specificities of the tourist activity (complexity and integration of the tourist product vs volatility of the activity). A second difficulty is added in terms of the unpredictability of tourist behavior marked by contradictions inherent to its ethical and responsible commitment and the maximization of pleasure and entertainment).

Also, research on the issue finds justification in many gaps that it tries to address, in this case, contribute to improving the understanding of the effectiveness of AI Solutions in predicting responsible customer behaviours (Puntoni et al., 2021) and reconsider the role of ethics in the use of AI Solutions to predict responsible customer behaviour (Martínez et al., 2022).

To what extent does the adoption of AI solutions by tourism businesses help improve the effectiveness of predicting customer needs? This is the central question to which our research attempts to provide answers.

To address this issue, we plan to conduct an exploratory literature review by examining results from a selection of empirical studies that have addressed the relationship between AI methods and the effectiveness of predicting responsible guest behaviours in tourism establishments.

Certainly, due to the innovative nature of the theme, the exploratory approach of the literature review is the most recommended. However, this methodological choice is not perfect and remains open to criticism. Indeed, the approach promotes the identification of emerging trends in the use of smart data and the analysis of responsible customer behavior in the tourism sector (Snyder, 2019; Munn et al., 2018), helps to better map research areas and thus help guide future research in this area (Tricco et al., 2018; Peters et al., 2015), identify gaps and guide researchers towards promising new avenues of research on the analysis of intelligent data and the prediction of responsible customer behaviour in tourism (Levac et al., 2010; Munn et al., 2018). While other authors highlight the lack of methodological rigor of the approach in terms of reliability and validity of the conclusions drawn (Tricco et al., 2018; Peters et al., 2015), or limited generalization deficit due to potential lack of depth and detail on specific aspects of smart data analysis and prediction of responsible customer behavior in tourism (Munn et al., 2018; Levac et al., 2010), or significant risk of bias (Tricco et al., 2018; Peters et al., 2015).

Our chapter is structured in three parts, highlighting successively the theoretical and conceptual framework of our research, the methodological approach adopted, the results of the study and finally, the discussion of the results and the contributions of the research.

1. THEORETICAL AND CONCEPTUAL FRAMING

1.1. Responsible Consumption and Responsible Customer Behaviour

The customer's behaviours are the model used by the customer to conduct research or purchase, by choosing the most appropriate products to satisfy his needs and desires, the uses he makes of them, and the reasons and factors that determine such attitudes and actions. According to Al-Muezzin (2008), this includes all visible and invisible acts and actions by people to obtain a particular good and service from a certain place and at a specific time.

According to Nwulu and Ateke (2014), consumer behaviour encompasses activities in which the individual participates when they research, evaluate, select, purchase, consume and dispose of products to satisfy their needs and desires, including the decision-making processes that precede and determine them. We can thus approach consumer behavior as the process by which the customer makes his purchase decision to satisfy his needs and desires, according to his tastes and aversions, under the constraints of purchasing power, availability of products, moral attitudes, and beliefs, etc.

Responsible consumption can be approached as the integration in consumption choices of the impacts that the production, marketing, and consumption of products can have on the environment, public health and more broadly society as a whole (Giesler & Veresiu, 2014).

It follows that, to be responsible, consumers should include an ethical dimension in their consumer decisions by considering, beyond their particular and immediate interest, the consequences of their choices in a long-term perspective and the consideration of the costs of negative externalities of their consumption (Séré de Lanauze et al., 2021). Such behaviour corresponds to «all voluntary acts in the sphere of consumption carried out following the realization of negative consequences of consumption on the outside world» (Özçağlar-Toulouse, 2009).

In the particular context of tourism, responsible consumer behaviour takes the form of a virtuous process of adopting responsible attitudes and concrete committed actions reflecting its effective involvement in reducing its environmental footprint, from the simple "awareness of the fragility of the environment and the limits to the use of natural resources" (Krause, 1993) to a real "responsibility and ethics to contribute to the fight against climate change, ensuring sustainable resources and supporting the well-being of the current and future population (Rousseau et al., 2007, Giesler & Veresiu, 2014; Giuliano & Eloy, 2021).

On board, responsible tourism behaviour should translate into committed attitudes and intentions towards the environment, the local population, and all stakeholders (Table 1).

Logically, these virtuous attitudes on the part of the tourist should be expressed practically in the form of concrete actions of the same nature, broadly summarized as follows:

Overall, and as concluded by Xiang & Fesenmaier (2017), these dimensions of responsible customer behaviours in smart tourism are essential to creating a balance between traveller satisfaction and preserving the environment and local cultures.

Suppose studies have validated the gap between attitude and consumer behaviours («attitude-behaviour gap»), by dismantling that the majority of consumers with very positive responsible attitudes, do not consistently engage in significant responsible behaviour (Feldmann & Hamm, 2015). Other research, on the other hand, has shown that very strong attitudes can also outweigh a context unfavourable to the achievement of behaviours (Zepeda & Deal, 2009).

These multiple contradictions justify the concerns of companies and tourism destination stakeholders regarding the relevance of the prediction of responsible behavior of their customers.

Table 1. Typology of responsible tourist attitudes

Attitude	Definition	References
Ecological awareness	Recognition of the importance of environmental preservation and adoption of travel practices that minimize ecological impact.	El Archi & Benbba (2022)
Consumer ethics	The choice of ethical and sustainable products and services reflects a preference for practices that support community conservation and well-being.	Lignon-Darmaillac (2016)
Community engagement	The intention is to engage with local communities respectfully and contribute to the local economy through conscious purchasing.	François-Lecompte et al. (2013)
Cultural respect	Appreciation and respect for local cultures and traditions, avoiding activities that could compromise or devalue them.	Lahcene & Naili (2019).
Social responsibility	Awareness of the social implications of tourism and the promotion of practices that respect human rights and promote social equity.	Rakotomalala-Ramandimbiarison (2019)

Source: Authors

Table 2. Typology of responsible tourist actions

Behaviours	Definition	Implications for Tourism	References
Choice of sustainable tourism destinations and services	The choice of destinations and alternative modes of transport refers to the decision of individuals to favour environmentally friendly destinations and to opt for sustainable tourism services.	This responsible behaviour of customers helps predict their preference for environmentally friendly destinations and modes of transport, which can be useful in developing sustainable tourism policies and offerings.	Xiang & Fesenmaier (2017) Bernard & Parker (2021) Durif et al. (2017)
Rationing of consumption of resources	Resource rationing refers to the practice of limiting one's consumption of non-renewable natural resources to preserve the environment and promote a sustainable lifestyle. This behaviour of the customer implies an awareness of resource limitations and a willingness to use them in an eco-responsible way.	By managing resources wisely, tourism destinations can reduce their ecological footprint and ensure long-term sustainability. This benefits both travellers and local communities.	Sigala (2020) Sheth et al. (2011) Hmioui et al. (2022)
Social and cultural commitment within companies and territories	Social and cultural engagement within companies and territories refers to the involvement of consumers in initiatives aimed at promoting social well-being, cultural diversity and the development of local communities. This reflects the willingness of consumers to support socially and culturally engaged businesses and territories.	This commitment promotes the preservation of cultural traditions, local economic development and the creation of authentic links between travellers and locals. It also contributes to a more rewarding and meaningful travel experience.	Fall Diallo et al. (2015) Trentmann (2008) Scarano et al. (2023)

Source: Authors

1.2. Prediction of Responsible Consumer Behaviour

To try a predictive approach, we first seek to understand the theoretical foundations of responsible consumer behaviours by mobilizing explanatory theories, including:

- **Alphabet Theory:** This theory, combines the VBN (Values-Beliefs-Norms) theory of Stern et al. (1999) and the ABC (Attitude-Behaviour-Context) theory of Guagnano et al. (1995), by adding four new variables, including knowledge, the search for information, habits, and demographics (Zepeda & Deal, 2009; Feldmann & Hamm, 2015).

- **Theory of Planned Behaviours:** This theory, extending the Theory of Reasoned Action, seeks to explain certain behaviours from predictor variables, including attitude towards behaviours, the subjective norm, the perception of control, and the intention to behave or not (Ajzen, 1991). This theory has helped explain the adoption of pro-environmental and sustainable behaviours (Bissonnette & Contento, 2001; Sparks & Shepherd, 1992; Vermeir & Verbeke, 2008).

Admittedly, most researchers opt for this last theory of predicting consumer behaviours, especially the one who wants to be or claims to be responsible, to shed more light on the elements that contribute to creating the breach, between attitude and behaviour.

In the tourism sector, the prediction of responsible customer behaviour makes it possible to understand the place of ethical and sustainability values among the determinants of their purchasing behaviour, in their scale of needs, among the factors of their satisfaction and loyalty, to better adapt the offer strategy of companies and destinations (Khelifi, 2023; Ackermann & Sellitto, 2022). It also aims to better involve tourists in sustainability and CSR programmes, as an interested and active stakeholder.

1.3. Smart Data Analytics and Prediction of Responsible Customer Behaviour

Both businesses and tourist destinations are thus invited to develop their predictive systems of tourist behaviour, especially in their responsible and sustainable dimension. This is followed by massive use of new solutions offered by digitalisation and emerging technologies, commonly known as Smart Data analytics and prediction of responsible customer behaviour. The literature review highlights the following nine features:

To better understand the application potential of Smart Data Analytics in the prediction of responsible customer behaviours especially in the tourism field, we plan to conduct a literature review on the subject. The objective is to try a better understanding of the practices and to draw a sketch of the model of analysis, highlighting the possible causal relationship between the use of such solutions by companies and destinations and the effectiveness of predicting such attitudes and behaviours on the part of customers and visitors.

2. METHODOLOGY

2.1. Methodological Approach

To explore the impact of applying solutions from emerging technologies to better predict the responsible behavior of customers in tourism businesses and destinations, given the scarcity of empirical studies devoted to the problem, we opted for the exploratory literature review.

We believe that the passage through an exploratory literature review is necessary for this research question not sufficiently explored and in continuous reconfiguration. This work also seeks to identify the main practices for drawing a sketch of an analysis model, to be tested later by a quantitative study.

Table 3. Key smart data analytics for predicting responsible customer behaviors

Smart Data Analytics	Definition	References
Spatial Big Data (SBD)	Big data spatial (SBD) refers to all big data that have a spatial component of geolocation, which can come from social networks, online sensors, GPS, etc.	Park et al. (2020) Tingting Shi, 2019
Scientific mapping techniques	These are methods of visualization and analysis of geographical data in an interactive way, used in particular in tourism to map sites of tourist interest, travel routes, etc.	Blanco-Moreno et al. (2023)
Artificial Intelligence (AI)	Artificial intelligence is the ability of a computer system to mimic human intelligence, including learning, problem-solving, and making autonomous decisions. In the field of tourism, AI is used to personalize tourist offers, recommend destinations and optimize the customer experience.	Rahmadian et al. (2022)
Online search and booking platforms	These are websites or mobile apps that allow travellers to find and book accommodation, activities, transportation, etc. online.	Aleksandra Minkwitz (2018)
Big Data Analytics	It refers to all the techniques and tools used to analyze and use large amounts of data to draw insights and make strategic decisions. In tourism, Big Data Analytics is used to analyze tourist behaviour, anticipate market trends, etc.	ALLA et al. (2022) Lhoussaine et al. (2022) Xiang & Fesenmaier (2017)
Machine Learning	It is a branch of artificial intelligence that allows computers to learn from data and improve their performance without being explicitly programmed. In tourism, it is used to predict the behaviours of tourists, customize tourist offers, etc.	De Oliveira et al. (2023)
Data mining	This is the process of extracting meaningful and useful information from large databases. In tourism, this tool is used to identify trends, discover patterns of behaviours, etc.	Park et al. (2020) Jivrajani et al. (2023) Geetha et al. (2021)
Social Media analytics based on location (LBSNs)	These are social media analytics techniques focused on the geographic location of users. In tourism, they are used to analyze tourists' interactions with social networks, target geolocated promotions, etc.	
Prediction algorithms (SeqLearn)	They are computer models that predict future events based on historical data. In tourism, these algorithms are used to anticipate market trends, forecast tourist flows, etc.	Guo et al. (2021)

Source: Authors

The "scoping review" is a more recent approach. It aims to make a quick inventory of knowledge on a subject (scientific research and/ or other type of literature). It is particularly interesting to have an interdisciplinary overview of a research theme, some topics being studied more and more by plural and multi-disciplinary approaches (this is the case of business research, Snyder 2019). The exploratory review aims to examine the scope, variety, and nature of research activity on a topic, clarify key concepts (how are they understood and defined?) and identify gaps in research (Munn et al. 2018). To do this, this type of literature review answers a broad exploratory research question by searching, selecting, and synthesizing existing research. Thus, unlike a systematic literature review, it does not answer a specific research question but can be a precursor to such a review, assessing its feasibility. The exploratory literature review therefore remains primarily an identification research work, essentially descriptive, which will include broader inclusion criteria than a systematic review (Tricco et al., 2018).

Although less rigorous as a literature review approach, compared to the systematic approach, for our exploratory review, we adopted the following framework (Figure 1):

Figure 1. Methodological outline of the literature review
Tricco et al. (2018)

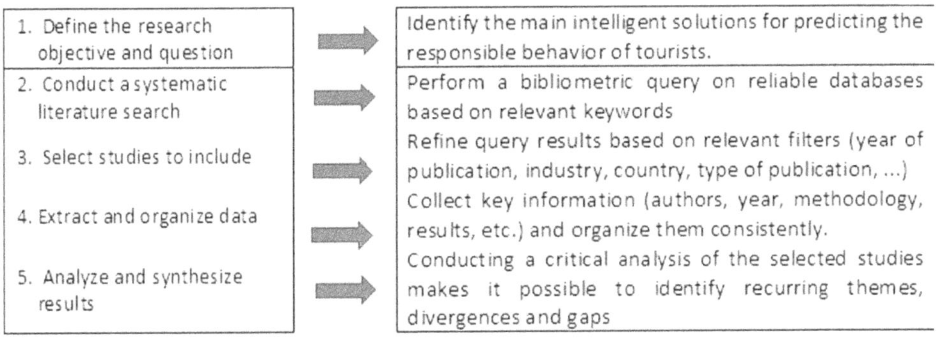

Our query was based on relevant keywords (smart SC in EPC, PC logistics digitization, emerging SC technologies in EPC, etc.) in reputable databases such as PubMed, IEEE Xplore, ScienceDirect, SCOPUS, SPRINGER, J STOR and CAIRN.

Following the protocol below (Figure 1), our initial query generated 127 results. After cleaning by filters (publication from 2016, AI Solutions, visitor behavior, Open Access publication, newspaper articles), our sample of articles was reduced to 49 articles.

2.2. Profile of the References Studied

Our sample of scientific articles (Annex) is marked by the more recent nature of the publications, a predominance of qualitative and exploratory studies.

The series of studies on the application of Big Data and related technologies in the tourism and hospitality sector, covering a range of publication years from 2016 to 2023 (Figure 2).

Figure 2. Breakdown of articles by year of publication
Authors

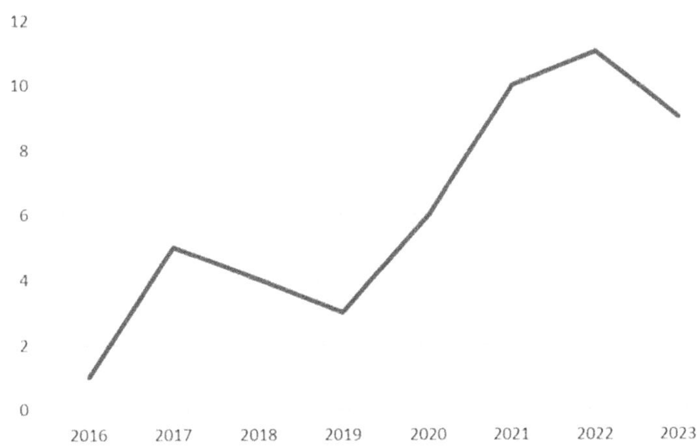

These studies explore various dimensions of integration of smart solutions for predicting responsible customer behavior in the tourism field, using different methodological research approaches (Figure 3). The predominance of studies in the form of literature reviews and qualitative studies is explained by the still embryonic level of research on the issue. Indeed, the first aims to synthesize existing knowledge on the use of Big Data in tourism and to identify current trends, research gaps and future opportunities. The seconds are more oriented towards a better understanding of the phenomenon. while quantitative studies remain very rare and seek to analyze data and identify behaviours or trends in an often-exploratory way.

Figure 3. Distribution of references by type of methodological approach
Authors

	Occurrences
Literature review	22
Qualitative studies	15
Quantitative studies	12
Total	49

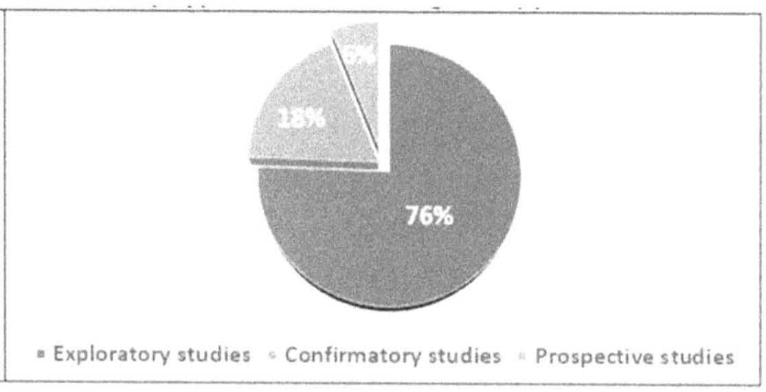

When analyzed according to the depth of the research (Figure 4), we note a predominance of exploratory studies (aiming to discover new insights and understand the extent of the application of Big Data in the sector), followed by confirmatory studies (seeking to validate existing hypotheses or theories in the context of Big Data in tourism) and finally prospective studies (seeking to project into the future, trying to predict trends and future directions for research and practice in the field).

Overall, our sample of articles studied presents so many attributes of relevance and reliability sufficient to constitute such a solid text corpus based on our thematic qualitative analysis to identify the main Smart data Analytics solutions mobilized by tourism businesses or by tourist destinations to develop their capacity to predict the behaviour of tourists, both in terms of attitudes and intentions and concrete responsible actions, reflecting an eco-responsible commitment of the customer.

Figure 4. Breakdown of references by the level of research depth
Authors

	Occurrences
Exploratory studies	37
Confirmatory studies	9
Prospective studies	3
Total	49

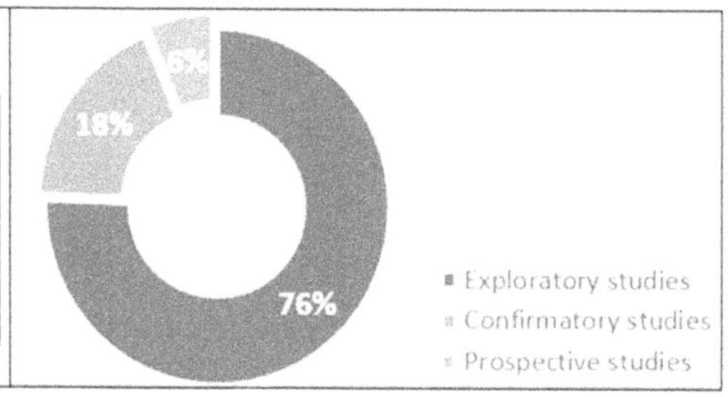

3. SEARCH RESULTS AND DISCUSSION

3.1. Key Data Analytics Prediction Solutions Identified

Based on the analysis of the text corpus relating to the basket of articles studied, we can identify a range of Smart Data Analytics Solutions adopted by tourism companies to predict the responsible behaviours of their customers, which we can group into ten main tools (Table 4):

Table 4. Synthesis of smart tourism data analysis

• Spatial big data (SBD) • Bibliometric analysis with SciMAT • Scientific mapping techniques • Artificial Intelligence (AI) • Online search and booking platforms	• Big Data Analytics • Machine Learning • Data mining • Social Media analytics based on location (LBSNs) • Prediction algorithms (SeqLearn)

Source: Authors

In summary, the main tools of Smart Data Analytics for predicting the responsible behaviours of customers of tourist establishments include advanced data analysis, AI, Data Mining and Predictive Modeling. These techniques allow tourism companies to better understand the needs and expectations of their customers, and to offer more personalized and responsible services and experiences.

By comparing these results with those obtained by comparable studies, we observe a certain convergence, which reinforces the reliability of our study. Indeed, studies have highlighted the importance of big data analysis (Chandra & Verma, 2023), analysis of content generated by users and marketers on social networks (Goh et al., 2013; Liu et al., 2023), analysis of social networks (Sümer & Parilti, 2023), the use of AI, blockchain, IoT, AR and VR (Rane et al., 2023), big data analytics and neuromarketing (Azman et al., 2023), and the use of embedded AI voice recognition in mobile applications (Arachchi & Samarasinghe, 2023)to predict sustainable consumption behaviors.

By combining these different approaches and taking advantage of technological advances in Smart Data Analytics, it is possible to develop effective solutions to predict the responsible behaviours of guests of tourist establishments and adapt marketing strategies accordingly.

3.2. Responsible Intentions of Identified Clients

Based on the results of our literature review and following the above classification (Table 2), we were able to identify the main responsible attitudes of the tourist likely to promote his responsible behaviour (Table 3).

The results regarding the responsible attitude of visitors remain broadly in line with those obtained by similar works. First, Goh et al. (2013) emphasized the interest of tourists in understanding the relative impact of user-generated content and advertisers on consumer behaviour in social media brand communities. Then, Chandra & Verma (2023) focus on the environmental and social concerns of the visitors. While, Sümer & Parilti (Eds.) (2023) allude to the importance of privacy and data security. Similarly, Azman et al. (2023) values the ethical and transparent dimension in the tourism industry. Finally, Arachchi & Samarasinghe (2023) highlight the role of generational differences in the purchase intent of vistors.

Table 5. Main responsible attitudes adopted by tourists

1. Attitudes of ecological awareness	
→ Environmental sensitivity → Intent to reduce carbon footprint → Preference for green travel → Interest in ecological tourist destinations	→ Intent to visit ecological tourist destinations → Respect for biodiversity → Concern for environmental protection and pollution reduction, …
2. Consumer Ethics Attitudes	
→ Importance of reducing resource consumption (water, electricity, gas, food, etc.)	→ Concern to save the use of natural and spatial resources of the destination, …
3. Community Engagement Attitudes	
→ Interest in local traditions → Appreciate and encourage the best community-based eco-responsible practices → Concern about reducing tourism load and pressure	→ Tolerance of local ethnic, historical and cultural specificities → Integrating the ethical dimension into the process of choosing tourist destinations, …
4. Attitudes of cultural respect	
→ Awareness of the importance of local culture → Respect for local culture and traditions	→ Interest in understanding local culture and traditions, …
5. Social Responsibility Attitudes	
→ Preference for sustainable destinations → Commitment to sustainable tourism → Interest in adopting innovative technologies → Integrating sustainability as a parameter for the consumption of tourism services	→ Intent to adopt emerging technologies → Commitment to sustainable tourism → Believe in the reliability of shared information on sustainable tourist destinations → …

Source: Authors

In principle, the anchoring of these different attitudes and responsible intentions in the tourist behaviours, throughout their tourist journey, is to influence effective behaviours during their stay towards more eco-responsible commitment.

3.3. Predicted Responsible Customer Behaviour

The main types of responsible attitudes adopted by Responsible Actions clients:

In reality, multiple factors prevent or at least hinder, the translation of the good intentions of the consumer into real responsible actions, including the feeling of scepticism about companies and their implications, commercial overexploitation of the DD concept, such as greenwashing (Monnot & Reniou, 2013). At the limit, some authors evoke the *myth* of the responsible consumer (Carrigan & Attalla, 2001; Devinney & al., 2010).

Other research has revealed the different types of barriers to the transformation of CR intentions into behaviours (Sempels, 2009). Thus, a range of psychological mechanisms can be put in place by consumers to overcome the dissonance that results from the mismatch between ethical values and real behaviours, including rationalization and justification techniques inspired by neutralization theory (Sykes & Matza, 1957).

Table 6. Main responsible actions and practices undertaken by tourists

1. Choice of sustainable tourism destinations and services	
→ Preferences for sustainable destinations → Choice of sustainable modes of transport → Use of collective and collaborative transportation → Choice of sustainable accommodation	→ Preferences for less crowded destinations and establishments → Purchases of local products → Choice of social and solidarity accommodation → Preferences for unconventional circuits, ...
2. Resource Consumption Rationing	
→ Rationalize the use of infrastructure, and equipment during the tourist stay → Avoid All-inclusive formulas	→ Optimization of water, electricity, gas, and food consumption, ... during the stay, ... → ...
3. Social and cultural engagement within businesses and territories	
→ Participation in local activities → Integration of green ICTs → Commitment to sustainable tourism → Share more reliable and relevant information about rural tourist destinations online	→ Sharing sustainable tourism experiences online → Online recommendation of sustainable tourism services → Using big data to promote sustainability, ... → ...

Source: Authors

3.3. Synthesis and Analysis Model Sketch

In summary, the model suggests a new approach to predicting responsible customer behaviour through the use of Smart Data Analytics. What is likely to help companies and managers of tourist destinations to better understand such behaviors (attitudes, intentions and actions), thanks to the effectiveness in the analysis of behavioral data, purchasing and navigation to identify trends, motivations and factors influencing tourist decisions. But also, through machine learning algorithms to detect complex behavioral patterns and customer segments.

Similarly, these Smart Data Analytics Solutions (SDAS) promote the anticipation and prediction of customer behavior, via advanced predictive models to anticipate the choices and actions of tourists according to multiple variables, and the simulation of behaviour scenarios according to different parameters (price, offers, communication, etc.).

As a synthesis of this theoretical study on the impact of the adoption of SDAS to predict the responsible behaviours of the customer in the tourism field, we propose a framework of analysis model (Figure 5).

This model sketch aims to measure the causal link between the adoption and anchoring of SDAS on the improvement of the prediction of the responsible behaviours of the customer before, during and after his tourist stay. Multiple committed attitudes behind these responsible behaviours include a preference for eco-tourism practices, adoption of innovative technologies, and a desire for green travel and sustainable destinations. However, these good intentions must be translated into real concrete actions, such as the optimization of advertising campaigns to promote sustainable tourism, the selection of ecological accommodation, and participation in local and ecotourism activities (El Jaouhari & Lhoussaine, 2021) for the creation of global and shared value within tourism businesses and territories (Alla et al., 2018).

Figure 5. Analysis model sketch (Authors)

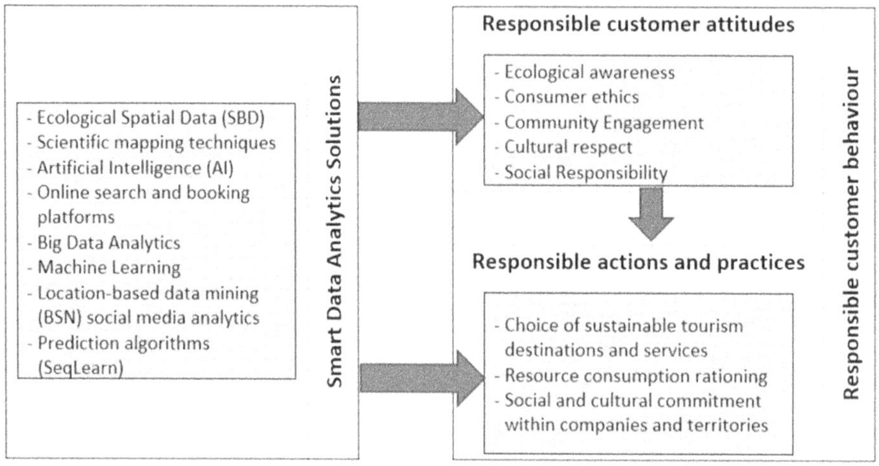

4.3. Conclusion and Managerial Contributions of Research

Through the multiple solutions offered by AI and data engineering, companies come to understand globally the reasons for responsible behaviours of the customer as well as the factors that determine such a direction in terms of attitudes and behaviours, particularly in the specific context of tourism.

Such customer intelligence is likely to allow the company in the sector to better understand the buying motivations of its customers to better satisfy and retain them, by constantly adapting its business value proposition and engaging them more with its brand. Indeed, greater engagement usually means stronger relationships between consumers and the brand, greater satisfaction and loyalty, and more purchases (Brodie et al., 2013; Gummerus et al., 2012; Laroche et al., 2013; Vivek et al., 2012).

Thus, the Smart approach to responsible customer behaviours, and therefore the ability of the company to understand, anticipate and «manage», through solutions offered by emerging technologies is to allows it to better control the multiple dimensions of such behaviours, in cognitive, emotional and behaviours terms, and thus promote customer engagement with its brand (Creevey et al., 2019) and thus promote more success for marketing actions (Da & Wang, 2021).

Despite its limitations, compared to other approaches to literature review, including the systematic approach, the exploratory way is a research method in its own right and whose purpose will be the results of literature research and analysis. The literature review is a particularly relevant method for taking stock of knowledge on subjects that are rapidly evolving and subject to a strong production of studies.

In summary, these scientific articles highlight different independent and dependent variables seeking to understand the impact of using SDAS to better predict the responsible behavior of the customer in the tourism field.

Independent variables include big spatial data (MDS), scientific mapping techniques, artificial intelligence (AI), online search and booking platforms, big data analytics, machine learning, data mining, social media analysis based on location (BSN), prediction algorithms (SeqLearn) etc. Dependent sub-variables measuring the responsible behavior of the tourist respectively count ecological awareness, consumer ethics, community engagement, cultural respect and social responsibility for responsible attitudes, choice of sustainable tourism destinations and services, choice of sustainable products, rationing the consumption of resources and social and cultural commitment within companies and territories for responsible actions.

However, our study did not reveal the mediating, moderating or control variables; Also, it did not contribute to highlighting the effect of the interaction between AI and humans on the effectiveness of the prediction of responsible behaviours of customers (Zhang et al., 2021).

Far from being a mechanical process, this evolution in customer behaviour comes up against several difficulties, notably the lack of criteria for measuring the effectiveness of the prediction of responsible customer behaviour following the adoption of Smart Data Analytics solutions (Lee & Lim (2013), and in particular about the effectiveness of predictive algorithms in maximising their impact on responsible customer behaviour (Kotler et al., 2022). In this sense, some authors argue that the transparency of predictive algorithms can enhance customer trust and engagement, leading to better prediction of responsible behaviour (Hoffman et al., 2018).

We, therefore, plan to launch perspectives for this study, in particular, the development and purification of our analysis model through more rigorous and analytical literature reviews (systematic and prospective), supported by more in-depth qualitative analyses (via textual and thematic analysis software), before empirically testing it with companies, tourist destination managers and tourists (quantitative analysis via regression analysis software).

REFERENCES

Ackermann, A. E., & Sellitto, M. A. (2022). Méthodes de prévision de la demande: Une revue de la littérature. *Innovar (Universidad Nacional de Colombia)*, *32*(85), 83–99. doi:10.15446/innovar.v32n85.100979

Agrawal, R., Wankhede, V. A., Kumar, A., Luthra, S., & Huisingh, D. (2022). Big data analytics and sustainable tourism: A comprehensive review and network based analysis for potential future research. *International Journal of Information Management Data Insights*, *2*(2), 100122. doi:10.1016/j.jjimei.2022.100122

Aguirre Montero, A., & López-Sánchez, J. A. (2021). Intersection of data science and smart destinations: A systematic review. *Frontiers in Psychology*, *12*, 712610. doi:10.3389/fpsyg.2021.712610 PMID:34393952

Ajzen, I. (1991). The theory of planned behaviours. *Organizational Behavior and Human Decision Processes*, *50*(2), 179–211. doi:10.1016/0749-5978(91)90020-T

Al-Muezzin, M. (2008). *Principles of Marketing* (Vol. 1). Dar Althaqafa.

Alla, L., Hmioui, A., & Taghzouti, A. (2018). Impact de la création de valeur pour le client sur la performance des entreprises touristiques marocaines: Etude exploratoire des pratiques des hotels de Fès. *Les Cahiers du CEDIMES*, 75.

AllaL.KamalM.BouhtatiN. (2022). Big data et efficacité marketing des entreprises touristiques : Une revue de littérature. *Alternatives Managériales Economiques, 4*(0). doi:10.48374/IMIST.PRSM/ame-v1i0.36928

Álvarez-García, J., González-Vázquez, E., Del Río-Rama, M. D. L. C., & Durán-Sánchez, A. (2019). Quality in customer service and its relationship with satisfaction: An innovation and competitiveness tool in sport and health centers. *International Journal of Environmental Research and Public Health*, *16*(20), 3942. doi:10.3390/ijerph16203942 PMID:31623254

Arachchi, H. D. M., & Samarasinghe, G. D. (2023). Impact of embedded AI mobile smart speech recognition on consumer attitudes towards AI and purchase intention across Generations X and Y. *European Journal of Management Studies.*

Azman, N., Albattat, A., & Valeri, M. (2023). The Importance of Big Data Analysis: Developing Neuromarketing in Tourism Industry. In Tourism Innovation in the Digital Era: Big Data, AI and Technological Transformation (pp. 1-24). Emerald Publishing Limited.

Bachouche, H., & Sabri, O. (2022). The impact of consumers' status-participant or nonparticipant-on the evaluation of empowerment strategies. *Recherche et Applications en Marketing*, *37*(3), 149–175. doi:10.1177/20515707221118811

Baker, M. A., Davis, E. A., & Weaver, P. A. (2014). Eco-friendly attitudes, barriers to participation, and differences in behaviours at green hotels. *Cornell Hospitality Quarterly*, *55*(1), 89–99. doi:10.1177/1938965513504483

Bank, A. D. (2021). *Big data for better tourism policy, management, and sustainable recovery from COVID-19.* Asian Development Bank.

Benbba, B., Ismaili, F. E. A., & El Archi, Y. (2022). Smart tourism destination et développement durable: Quels apports à l'expérience touristique? *Alternatives Managériales Economiques, 4*, 1-18. https://revues.imist.ma/index.php/AME/article/view/36926

Bernard, M., & Parker, L. (2021). *The effect of conscious consumerism on purchasing behaviours: The example of greenwashing in the cosmetics industry*. Academic Press.

Bissonnette, M. M., & Contento, I. R. (2001). Adolescents' perspectives and food choice behaviours in terms of the environmental impacts of food production practices: Application of a psychosocial model. *Journal of Nutrition Education*, *33*(2), 72–82. doi:10.1016/S1499-4046(06)60170-X PMID:12031187

Blanco-Moreno, S., González-Fernández, A. M., & Muñoz-Gallego, P. A. (2023). Big data in tourism marketing: past research and future opportunities. *Spanish Journal of Marketing-ESIC.*

Brodie, R. J., Ilic, A., Juric, B., & Hollebeek, L. (2013). Consumer engagement in a virtual brand community: An exploratory analysis. *Journal of Business Research*, *66*(1), 105–114. doi:10.1016/j.jbusres.2011.07.029

Caia, Y., Lib, G., Wena, L., & Liua, C. (2023). *Intellectual Landscape and Emerging Trends of Big Data Research in Tourism and Hospitality: A Scientometric Analysis*. Academic Press.

Carneiro, T., Picoto, W. N., & Pinto, I. (2023). Big Data Analytics and Firm Performance in the Hotel Sector. *Tourism and Hospitality*, *4*(2), 244–256. doi:10.3390/tourhosp4020015

Carrigan, M., & Attalla, A. (2001). The myth of the ethical consumer–do ethics matter in purchase behaviour? *Journal of Consumer Marketing*, *18*(7), 560–578. doi:10.1108/07363760110410263

Chandra, S., & Verma, S. (2023). Big data and sustainable consumption: A review and research agenda. *Vision (Basel)*, *27*(1), 11–23. doi:10.1177/09722629211022520 PMID:36810315

Cobanoglu, C., Terrah, A., Hsu, M. J., Corte, V. D., & Gaudio, G. D. (2022). A systematic review of big data: Research approaches and future prospects. *Journal of Smart Tourism*, *2*(1), 21–31.

Cousin, S., Chareyron, G., & Jacquot, S. (2017). Big data and tourism. In L. L. Lowry (Ed.), *The SAGE international encyclopedia of travel and tourism*. Sage publications.

Creevey, D., Kidney, E., & Mehta, G. (2019). From dreaming to believing: A review of consumer engagement behaviours with brands' social media content across the holiday travel process. *Journal of Travel & Tourism Marketing*, *36*(6), 679–691. doi:10.1080/10548408.2019.1624242

Cunha, C. R., Morais, E. P., & Martins, C. A. (2018). Can we use big data analytics to leverage tourism in rural tourism destinations? In *Proceedings of the 32nd International Business Information Management Association Conference*. International Business Information Management Association (IBIMA).

Dai, Y., & Wang, T. (2021). Prediction of customer engagement behaviour response to marketing posts based on machine learning. *Connection Science*, *33*(4), 891–910. doi:10.1080/09540091.2021.1912710

De Oliveira, P. S. G., Wada, E. K., Lopes, A. S., & da Silva, L. F. (2023). *Machine learning and sentiment analysis to assess the evolution of the COVID-19 pandemic and the impacts on tourism*. Anais Brasileiros de Estudos Turísticos.

Della Corte, V., & Del Gaudio, G. (2022). A Systematic Review of Big Data: Research Approaches and Future Prospects. *Journal of Smart Tourism*.

Devinney, T. M., Auger, P., & Eckhardt, G. M. (2010). *The myth of the ethical consumer hardback with DVD*. Cambridge University Press.

Dimitrios. (2021). The Use of Big Data in Tourism: Current Trends and Directions for Future Research. Diverses institutions incluant l'University of Thessaly, Arab Open University - Kuwait, Panteion University of Social and Political Sciences, et le Metropolitan College.

Durif, F., François-Lecompte, A., & Prim-Allaz, I. (2017). Pratiques de tourisme durable: proposition d'une typologie croisée avec les comportements de consommation. *Téoros. Revue de recherche en tourisme, 36*(1).

Durmaz, Y., & Başer, M. Y. (2023). *A Systematic Literature Review on Artificial İntelligence Applications in Tourism Marketing*. Academic Press.

El Archi, Y., & Benbba, B. (2022). Ecotourisme et développement durable: Étude empirique sur la perception des consommateurs. *Revue Internationale des Sciences de Gestion*, *5*(2).

El Archi, Y., & Benbba, B. (2023). The Applications of Technology Acceptance Models in Tourism and Hospitality Research: A Systematic Literature Review. *Journal of Environmental Management & Tourism*, *14*(2), 379–391. doi:10.14505/jemt.v14.2(66).08

El Archi, Y., Benbba, B., Nizamatdinova, Z., Issakov, Y., Vargáné, G. I., & Dávid, L. D. (2023). Systematic Literature Review Analysing Smart Tourism Destinations in Context of Sustainable Development: Current Applications and Future Directions. *Sustainability (Basel)*, *15*(6), 5086. doi:10.3390/su15065086

El Jaouhari, S., & Lhoussaine, A. L. L. A. (2021). Approche participative en tourisme et gouvernance territoriale, quel apport pour le développement territorial? *Alternatives Managériales Economiques*, *3*(2), 257–277.

Fall Diallo, M., Diop-Sall, F., Leroux, E., & Valette-Florence, P. (2015). Comportement responsable des touristes: Le rôle de l'engagement social. *Recherche et Applications en Marketing*, *30*(3), 88–108. doi:10.1177/0767370115571048

Feldmann, C., & Hamm, U. (2015). Consumers' perceptions and preferences for local food: A review. *Food Quality and Preference*, *40*, 152–164. doi:10.1016/j.foodqual.2014.09.014

François-Lecompte, A., Prim-Allaz, I., & Durif, F. (2013). Le tourisme durable: Les sept clés du succès 1. *Gestion*, *38*(3), 84–96. doi:10.3917/riges.383.0084

García, F. C., Valverde, M. I., Mascuñano, P. J., & Gimeno, V. M. (2018). Quality Implications of the use of Big Data in Tourism Statistics: Three exploratory examples. *Boletin de estadistica e Investigacion Operativa, 34*, 138-148.

Giesler, M., & Veresiu, E. (2014). Creating the responsible consumer: Moralistic governance regimes and consumer subjectivity. *The Journal of Consumer Research*, *41*(3), 840–857. doi:10.1086/677842

Giuliano, R., & Eloy, F. (2021). Le comportement responsable des consommateurs face à l'agriculture belge. *Question (s) de management*, (2), 143-156.

Goh, K. Y., Heng, C. S., & Lin, Z. (2013). Social media brand community and consumer behavior: Quantifying the relative impact of user-and marketer-generated content. *Information Systems Research*, *24*(1), 88–107. doi:10.1287/isre.1120.0469

Guagnano, G. A., Stern, P. C., & Dietz, T. (1995). Influences on attitude-behaviours relationships: A natural experiment with curbside recycling. *Environment and Behavior*, *27*(5), 699–718. doi:10.1177/0013916595275005

Gummerus, J., Liljander, V., Weman, E., & Pihlström, M. (2012). Customer engagement in a Facebook brand community. *Management Research Review*, *35*(9), 857–877. doi:10.1108/01409171211256578

Guo, L., Zhang, B., & Zhao, X. (2021). A Consumer Behaviours Prediction Model Based on Multivariate Real-Time Sequence Analysis. Hindawi Mathematical Problems in Engineering.

Hartmann, K., & Lederer, M. (2021). The current state of Big Data research in tourism: Results of a systematic literature analysis. *Zeitschrift für Tourismuswissenschaft*, *13*(2), 209–226. doi:10.1515/tw-2021-0015

Hmioui, A., Alla, L., & Bentalha, B. (2022). *Perception of sustainable tourism by international clients: Case of the visit to a hotel establishment*. Academic Press.

Hoffman, R. R., Klein, G., & Mueller, S. T. (2018, September). Explaining explanation for "explainable AI". *Proceedings of the Human Factors and Ergonomics Society Annual Meeting*, *62*(1), 197–201. doi:10.1177/1541931218621047

Kalıpçı, M. B., & Şimşek, E. K. (2023). Exploring The Intersection of Big Data and Hospitality: A Bibliometric Analysis. *Journal of Business Research-Turk*, *15*(3), 2096–2110.

Khelifi, K. (2023, October). *Conceptualisation des déterminants du choix d'une destination touristique par la valeur perçue. 9èmes Rendez-vous Champlain.*

Kotler, P., Kartajaya, H., Setiawan, I., & Vandercammen, M. (2022). Chapitre 9. Le marketing prédictif. *Marketing*, 169-183.

Krause, D. (1993). Environmental consciousness: An empirical study. *Environment and Behavior*, *25*(1), 126–142. doi:10.1177/0013916593251007

Kumar, K. S., Kaur, J., & Ramani, S. (2017). Consumer Behvior Analysis using Big Data Analytic. *International Journal of Control Theory and Applications, 10*(30).

Lahcene, S. A., & Naili, N. (2019). *Les attentes des consommateurs touristiques en matière de tourisme durable* (Doctoral dissertation, Université Mouloud Mammeri).

Laroche, M., Habibi, M. R., & Richard, M. O. (2013). To be or not to be in social media: How brand loyalty is affected by social media? *International Journal of Information Management*, *33*(1), 76–82. doi:10.1016/j.ijinfomgt.2012.07.003

Lazić, M., Bradić-Martinović, A., & Banović, J. (2023). Digital skills in tourism and hospitality as a precondition for the sector resilient growth: The case of Serbia. *Менаџмент у хотелијерству и туризму, 11*(1), 25-40.

Lee, H. S., & Lim, J. H. (2013). Statistical package for the social sciences. *JypHyunJae Publication.*

Lee, S., Sun, K. A., Wu, L., & Xiao, Q. (2018). A moderating role of green practices on the relationship between service quality and customer satisfaction: Chinese hotel context. *Journal of China Tourism Research*, *14*(1), 42–60. doi:10.1080/19388160.2017.1419897

Levac, D., Colquhoun, H., & O'brien, K. K. (2010). Scoping studies: Advancing the methodology. *Implementation Science : IS*, *5*(1), 1–9. doi:10.1186/1748-5908-5-69 PMID:20854677

Lhoussaine, A. L. L. A., Kamal, M., & Bouhtati, N. (2022). Big data et efficacité marketing des entreprises touristiques: Une revue de littérature. *Alternatives Managériales Economiques*, *1*, 39–58.

Li, J., & Cao, B. (2022). Study on Tourism Consumer Behaviours and Countermeasures Based on Big Data. Hindawi Computational Intelligence and Neuroscience.

Li, J., & Cao, B. (2022). Study on tourism consumer behaviours and countermeasures based on big data. *Computational Intelligence and Neuroscience.*

Lignon-Darmaillac, S. (2016). *Œnotourisme et éthique de la consommation.* Academic Press.

Liu, C. H., Horng, J. S., Chou, S. F., Yu, T. Y., Huang, Y. C., & Lin, J. Y. (2023). Integrating big data and marketing concepts into tourism, hospitality operations and strategy development. *Quality & Quantity*, *57*(2), 1905–1922. doi:10.1007/s11135-022-01426-5 PMID:35729961

Liu, Y., Naveed, R. T., Kanwal, S., Tahir Khan, M., Dalain, A. F., & Lan, W. (2023). Psychology in action: Social media communication, CSR, and consumer behavior management in banking. *PLoS One*, *18*(8), e0289281. doi:10.1371/journal.pone.0289281 PMID:37590276

Lyu, J., Khan, A., Bibi, S., Chan, J. H., & Qi, X. (2022). Big data in action: An overview of big data studies in tourism and hospitality literature. *Journal of Hospitality and Tourism Management*, *51*, 346–360. doi:10.1016/j.jhtm.2022.03.014

Majeed, S., Zhou, Z., Lu, C., & Ramkissoon, H. (2020). Online tourism information and tourist behaviours: A structural equation modeling analysis based on a self-administered survey. *Frontiers in Psychology*, *11*, 599. doi:10.3389/fpsyg.2020.00599 PMID:32373008

Makridis, I. (2018). *Big data analytics and knowledge discovery through location-based social networks*. LBSN.

Mariani, M., Baggio, R., Fuchs, M., & Höepken, W. (2018). Business intelligence and big data in hospitality and tourism: A systematic literature review. *International Journal of Contemporary Hospitality Management*, *30*(12), 3514–3554. doi:10.1108/IJCHM-07-2017-0461

Mariani, M. M., & Baggio, R. (2022). Big data and analytics in hospitality and tourism: a systematic literature review. *International Journal of Contemporary Hospitality Management, 34*(1), 231-278. doi:10.1108/IJCHM-03-2021-0301

Martínez, P., Herrero, Á., & García de los Salmones, M. D. M. (2022). An examination of the determining factors of users' intentions to share corporate CSR content on Facebook. *Current Issues in Tourism*, *25*(13), 2159–2176. doi:10.1080/13683500.2021.1946019

Miah, S. J., Vu, H. Q., Gammack, J., & McGrath, M. (2017). A big data analytics method for tourist behaviour analysis. *Information & Management*, *54*(6), 771–785. doi:10.1016/j.im.2016.11.011

Minkwitz, A. (2018). TripAdvisor as a source of data in the planning process of tourism development on a local scale. *Turyzm (Lódz)*, *28*(2), 49–55. doi:10.2478/tour-2018-0014

Monnot, E., & Reniou, F. (2013). Ras le bol d'entendre parler d'écologie! comprendre la contestation des discours écologiques par les consommateurs. *Décisions Marketing*, 93-109.

Morrongiello, C. (2014). *De l'empowerment à l'engagement du client sur les plateformes en ligne: ou comment favoriser l'activité des clients sur Internet* (Doctoral dissertation, Université de Grenoble).

Mousavian, S., Miah, S. J., & Zhong, Y. (2023). A design concept of big data analytics model for managers in hospitality industries. *Personal and Ubiquitous Computing*, *27*(4), 1–11. doi:10.1007/s00779-023-01714-3 PMID:36818420

Munn, Z., Peters, M. D., Stern, C., Tufanaru, C., McArthur, A., & Aromataris, E. (2018). Systematic review or scoping review? Guidance for authors when choosing between a systematic or scoping review approach. *BMC Medical Research Methodology*, *18*(1), 1–7. doi:10.1186/s12874-018-0611-x PMID:30453902

Munn, Z., Stern, C., Aromataris, E., Lockwood, C., & Jordan, Z. (2018). What kind of systematic review should I conduct? A proposed typology and guidance for systematic reviewers in the medical and health sciences. *BMC Medical Research Methodology*, *18*(1), 1–9. doi:10.1186/s12874-017-0468-4 PMID:29316881

Naif, M. R. K. (2023). *Big Data Analytics for Predicting Consumer Behaviour*. Academic Press.

Noby, O., Rady, A., & Abd Eljalil, S. G. (2021). Big Data Analytics' Utilization in Egyptian Hotels' Decision-Making: What are The Challenges of Applying BDA? *Journal of Association of Arab Universities for Tourism and Hospitality*, *21*(5), 156–170. doi:10.21608/jaauth.2021.106301.1265

Nwulu, C. S., & Ateke, B. W. (2014). The impact of marketing information credibility on consumer purchase behaviour in the pharmaceutical industry in Port Harcourt Nigeria. *West African Journal of Business and Management Science*, *3*(3), 41–49.

Özçağlar-Toulouse, N. (2009). Quel sens les consommateurs responsables donnent-ils à leur consommation? Une approche par les récits de vie. *Recherche et Applications en Marketing*, *24*(3), 3–23. doi:10.1177/076737010902400306

Palomo, J., Figueroa-Domecq, C., Flecha-Barrio, D., & Segovia-Pérez, M. (2017). The Use of New Data Analysis Techniques in Tourism: A Bibliometric Analysis in Data Mining, Big Data and Structural Equations Models. *Ereview of Tourism Research*, 8.

Park, S., Xu, Y., Jiang, L., Chen, Z., & Huang, S. (2020). Spatial structures of tourism destinations: A trajectory data mining approach leveraging mobile big data. *Annals of Tourism Research*, *84*, 102973. doi:10.1016/j.annals.2020.102973

Pencarelli, T. (2020). The digital revolution in the travel and tourism industry. *Information Technology & Tourism*, *22*(3), 455–476. doi:10.1007/s40558-019-00160-3

Peters, M. D., Godfrey, C. M., Khalil, H., McInerney, P., Parker, D., & Soares, C. B. (2015). Guidance for conducting systematic scoping reviews. *JBI Evidence Implementation*, *13*(3), 141–146. PMID:26134548

Puntoni, S., Reczek, R. W., Giesler, M., & Botti, S. (2021). Consumers and artificial intelligence: An experiential perspective. *Journal of Marketing*, *85*(1), 131–151. doi:10.1177/0022242920953847

Quynh, T. D., & Dung, H. T. T. (2021). Prediction of Customer Behaviours using Machine Learning: A Case Study. *Proceedings of the 2nd International Conference on Human-centered Artificial Intelligence (Computing4Human 2021).*

Rahmadian, E., Feitosa, D., & Zwitter, A. (2022). A systematic literature review on the use of big data for sustainable tourism. *Current Issues in Tourism*, *25*(11), 1711–1730. doi:10.1080/13683500.2021.1974358

Rakotomalala-Ramandimbiarison, L. M. (2019). Les déterminants dans l'adhésion à la Charte du Tourisme Durable. *La Revue des Sciences de Gestion*, (5-6), 99–112. doi:10.3917/rsg.299.0099

Rane, N., Choudhary, S., & Rane, J. (2023). Sustainable tourism development using leading-edge Artificial Intelligence (AI), Blockchain, Internet of Things (IoT), Augmented Reality (AR) and Virtual Reality (VR) technologies. *Blockchain, Internet of Things (IoT), Augmented Reality (AR) and Virtual Reality (VR) technologies.*

Rodrigues, V., Eusébio, C., & Breda, Z. (2023). Vitor Rodrigues, Celeste Eusébio, Zélia Breda. (2022), "Enhancing sustainable development through tourism digitalisation: A systematic literature review. *Information Technology & Tourism*, *25*(1), 13–45. doi:10.1007/s40558-022-00241-w

Scarano, M. C., Regany, F., & Özçağlar-Toulouse, N. (2023). L'engagement du consommateur envers la marque: Perspective socio-culturelle et voies de recherche. *Recherche et Applications en Marketing*, *38*(4), 71–97. doi:10.1177/07673701231180409

Sempels, C., & Lille, E. S. C. (2009). Vers une meilleure compréhension des écarts entre discours et comportements dans le cadre d'une consommation responsable: apports de la littérature. *International Marketing Trends Congress.*

Séré de Lanauze, G., Lallement, J., & De Ferran, F. (2021). Représentations sociales du consommateur responsable: les autres consommateurs ont-ils vraiment envie de le suivre? *Revue management et avenir*, (5), 141-160.

Shamim, S., Yang, Y., Zia, N. U., & Shah, M. H. (2021). Big data management capabilities in the hospitality sector: Service innovation and customer generated online quality ratings. *Computers in Human Behavior*, *121*, 106777. doi:10.1016/j.chb.2021.106777

Sheth, J. N., Sethia, N. K., & Srinivas, S. (2011). Mindful consumption: A customer-centric approach to sustainability. *Journal of the Academy of Marketing Science*, *39*(1), 21–39. doi:10.1007/s11747-010-0216-3

Shi, T. (2019). Spatial Data Mining and Big Data Analysis of Tourist Travel Behaviours. *Ingenierie Des Systemes D'information, 24*(2).

Sigala, M. (2020). Tourism and COVID-19: Impacts and implications for advancing and resetting industry and research. *Journal of Business Research*, *117*, 312–321. doi:10.1016/j.jbusres.2020.06.015 PMID:32546875

Small, D. A., & Cryder, C. (2016). Prosocial consumer behaviours. *Current Opinion in Psychology*, *10*, 107–111. doi:10.1016/j.copsyc.2016.01.001

Snyder, H. (2019). Literature review as a research methodology: An overview and guidelines. *Journal of Business Research*, *104*, 333–339. doi:10.1016/j.jbusres.2019.07.039

Song, J. (2017). *A Systematic Literature Review: Existing Hospitality & Tourism Research on Big Data.* Academic Press.

Sparks, P., & Shepherd, R. (1992). Self-identity and the theory of planned behaviours: Assesing the role of identification with" green consumerism. *Social Psychology Quarterly*, *55*(4), 388–399. doi:10.2307/2786955

Stern, P. C., Dietz, T., Abel, T., Guagnano, G. A., & Kalof, L. (1999). A value-belief-norm theory of support for social movements: The case of environmentalism. *Human Ecology Review*, 81–97.

Stroumpoulis, A., Kopanaki, E., & Varelas, S. (2022). Role of artificial intelligence and big data analytics in smart tourism: A resource-based view approach. *WIT Transactions on Ecology and the Environment*, *256*, 99–108. doi:10.2495/ST220091

Stylos, N., & Zwiegelaar, J. (2019). *Big data as a game changer: how does it shape business intelligence within a tourism and hospitality industry context?* Springer Singapore.

Sümer, S. I., & Parilti, N. (Eds.). (2023). *Social Media Analytics in Predicting Consumer Behavior*. CRC Press. doi:10.1201/9781003200154

Sykes, G. M., & Matza, D. (1957). Techniques of Neutralization: A Theory of Delinquency'. *American Sociological Review*, *22*(6), 664. doi:10.2307/2089195

Torre, C., Guazzo, G. M., Çekani, V., & Bacco, V. (2022). The relationship between big data and decision making. A systematic literature review. *Journal of Service Science and Management*, *15*(2), 89–107. doi:10.4236/jssm.2022.152007

Trentmann, F. (2008). Le consommateur en tant que citoyen: Synergies et tensions entre bien-être et engagement civique. *Economics and Politics*, 7–20.

Tricco, A. C., Lillie, E., Zarin, W., O'Brien, K. K., Colquhoun, H., Levac, D., Moher, D., Peters, M. D. J., Horsley, T., Weeks, L., Hempel, S., Akl, E. A., Chang, C., McGowan, J., Stewart, L., Hartling, L., Aldcroft, A., Wilson, M. G., Garritty, C., ... Straus, S. E. (2018). PRISMA extension for scoping reviews (PRISMA-ScR): Checklist and explanation. *Annals of Internal Medicine*, *169*(7), 467–473. doi:10.7326/M18-0850 PMID:30178033

Vassakis, K., Petrakis, E., Kopanakis, I., & Makridis, J. (2021). Location intelligence for tourism destinations: A big data comparative analysis through location-based social networks. *International Journal of Tourism Policy*, *11*(3), 247–264. doi:10.1504/IJTP.2021.118664

Vassakis, K., Petrakis, E., Kopanakis, I., Makridis, J., & Mastorakis, G. (2019). Location-based social network data for tourism destinations. *Big data and innovation in tourism, travel, and hospitality: Managerial approaches, techniques, and applications*, 105-114.

Vermeir, I., & Verbeke, W. (2008). Sustainable food consumption among young adults in Belgium: Theory of planned behaviour and the role of confidence and values. *Ecological Economics*, *64*(3), 542–553. doi:10.1016/j.ecolecon.2007.03.007

Vivek, S. D., Beatty, S. E., & Morgan, R. M. (2012). Customer engagement: Exploring customer relationships beyond purchase. *Journal of Marketing Theory and Practice*, *20*(2), 122–146. doi:10.2753/MTP1069-6679200201

White, K., Habib, R., & Hardisty, D. J. (2019). How to SHIFT consumer behaviours to be more sustainable: A literature review and guiding framework. *Journal of Marketing*, *83*(3), 22–49. doi:10.1177/0022242919825649

Wong, I. A., Wan, Y. K. P., Huang, G. I., & Qi, S. (2022). Green event directed pro-environmental behaviours: An application of goal systems theory. In *Events and Sustainability* (pp. 210–231). Routledge. doi:10.4324/9781003314295-12

Xiang, Z., & Fesenmaier, D. R. (2017). Big data analytics, tourism design and smart tourism. *Analytics in smart tourism design: concepts and methods*, 299-307.

Xin, L. X., & Law, R. (2020). *Network analysis of big data research in tourism*. Academic Press.

Yallop, A., & Seraphin, H. (2020). Big data and analytics in tourism and hospitality: Opportunities and risks. *Journal of Tourism Futures*, *6*(3), 257–262. doi:10.1108/JTF-10-2019-0108

Ye, B. H., Ye, H., & Law, R. (2020). Systematic review of smart tourism research. *Sustainability (Basel)*, *12*(8), 3401. doi:10.3390/su12083401

Zajadacz, A., & Minkwitz, A. (2020). Using social media data to plan for tourism. *Quaestiones Geographicae*, *39*(3), 125–138. doi:10.2478/quageo-2020-0027

Zepeda, L., & Deal, D. (2009). Organic and local food consumer behaviour: Alphabet theory. *International Journal of Consumer Studies*, *33*(6), 697–705. doi:10.1111/j.1470-6431.2009.00814.x

Zhang, Y., Xu, S., Zhang, L., & Yang, M. (2021). Big data and human resource management research: An integrative review and new directions for future research. *Journal of Business Research*, *133*, 34–50. doi:10.1016/j.jbusres.2021.04.019

APPENDIX

Table 7. Description of the references studied

References	Year of Publication	Approach to Research			Further Research		
		Literature Review	Qualitative Research	Quantitative Research	Exploratory Study	Confirmatory Study	Prospective Study
Alla et al. (2022)	2022	X	X		X		
Miah et al. (2017)	2017			X	X		
Li & Cao (2022)	2022	X		X		X	
El Archi et al. (2023)	2023	X	X		X		X
El Archi & Benbba (2023)	2023	X			X		X
Blanco-Moreno et al. (2023)	2023	X		X	X		X
Palomo et al. (2017)	2017	X			X		
Dimitrios et al. (2021)	2021	X			X		
Shi (2019)	2019			X	X		
Majeed et al. (2020)	2020			X	X		
Li Jingjing et al. (2018)	2018	X	X		X		X
Yallop & Seraphin (2020)	2020	X		X	X		
Rahmadian et al. (2022)	2022	X		X	X		X
Mariani et al. (2022)	2022	X			X		
Miah et al. (2023)	2023	X		X		X	
Liu et al. (2022)	2022	X	X		X		X
Agrawal et al. (2022)	2022	X	X	X	X		X
Li Xin & Law (2020)	2020	X		X	X		X
Aguirre Montero & López-Sánchez (2021)	2021	X		X	X		X
Noby et al. (2021)	2021	X		X	X		X
Shamim et al. (2021)	2021			X	X		X
Cunha et al. (2018)	2018	X		X	X		X
Stroumpoulis et al. (2022)	2022	X		X	X		X
Song (2017)	2017	X			X		
Mariani et al. (2018)	2018	X		X	X		
Hartmann & Lederer (2021)	2021	X	X		X		
Ye et al. (2020)	2020	X		X	X		X
Cobanoglua et al. (2022)	2022	X		X	X		X
Rodrigues et al. (2022)	2022	X		X	X		X
Durmaz & Başer (2023)	2023	X		X	X		

Table 7 continued

References	Year of Publication	Approach to Research			Further Research		
		Literature Review	Qualitative Research	Quantitative Research	Exploratory Study	Confirmatory Study	Prospective Study
Torre et al. (2022)	2022	X		X	X		X
Carneiro et al. (2023)	2023	X		X			X
Stylos & Zwiegelaar (2019)	2019	X	X		X		
Makridis (2018)	2018		X	X	X		
Minkwitz (2018)	2018		X	X	X		
Caia et al. (2023)	2023		X	X	X		
Bank (2021)	2021		X	X	X		
Vassakis et al. (2019)	2019		X	X	X		
Zajadacz & Minkwitz (2020)	2020		X	X	X		
Vassakis et al. (2021)	2021		X	X	X		
Cousin et al. (2017)	2017		X	X	X		
García et al. (2018)	2018		X	X	X		
Kalıpçı & Şimşek (2023)	2023		X	X	X		
Lyu et al. (2022)	2022		X	X	X		
Kumar et al. (2017)	2017		X	X	X		
Geetha et al. (2021)	2021			X	X		
Guo et al. (2021)	2021			X	X		
Naif (2023)	2023			X	X		
Quynh & Dung (2021)	2021			X	X		

Chapter 12
Exploring Sustainability in Tourism Marketing Through Digital and Social Networks:
A Literature Review

Faissal El Mattichi
https://orcid.org/0009-0007-1717-8402
Sidi Mohamed Ben Abdellah University, Fez, Morocco

Abderrazzak Elabbadi
Sidi Mohamed Ben Abdellah University, Fez, Morocco

Aziz Hmioui
https://orcid.org/0000-0002-9032-4978
Sidi Mohamed Ben Abdellah University, Fez, Morocco

Aziz Barhmi
Mohammed V University, Morocco

ABSTRACT

Digital technology has become an essential tool for the tourism sector (comparing destinations, choosing hotels, booking flights, etc.). As a result, the majority of tourism bookings are now made using online travel platforms. This revolution in the use of digital technologies and artificial intelligence by travel companies and tourists has been amplified by the spread of the COVID-19 pandemic, prompting managers to switch to e-marketing strategy. Internet use has certainly contributed to accelerated growth in the tourism sector, but the question is how do tourists and companies use social media to make their choices? Using the PRISMA method, the authors attempt to explore the state of the art on the use of social networks and artificial intelligence by professionals and users in the tourism sector. The results show a positive impact of the use of digital technology on the evolution of the tourism sector.

DOI: 10.4018/979-8-3693-3286-3.ch012

INTRODUCTION

Over the last ten years, the tourism sector has experienced a steady expansion. This development is essentially due to its interconnectivity with other economic sectors such as transport, culture, accommodation and other economic axes, which has given the tourism sector an important place in the global economy. Today, the sector is seen as a lever for socio-economic and environmental development, actively contributing to the economic dynamism of countries. According to the World Travel and Tourism Council (WTTC), by 2022, the tourism sector's contribution to global GDP will have risen to 7.6%, with an increase of over 22% compared to 2021. Again, the sector has contributed to the creation of 22 million new jobs, up 7.9% on 2021, and is expected to generate profits of up to 500 billion US dollars by 2029 (Alnajim and Fakieh, 2023). However, the tourism sector was faced with a difficult situation following the spread of the Covid-19 pandemic, forcing it to change strategy and adapt to the situation. This change found its impetus in exploiting digital developments by implementing a marketing strategy based on digital and information and communication technologies (ICT's).

The significant development of the digital world has had a positive impact on all sectors of economic activity, including tourism. Internet connectivity, for example, has grown at an annual rate of 62.5%. By 2021, Internet users had reached nearly 4.9 billion, while networking platforms had reached nearly 3.8 billion (Balatska et al., 2023). Indeed, the rapid progress of the Internet has influenced the travel and hospitality market, providing tourists with highly effective digital tools. This has given rise to a new relationship between consumers of tourism products and the tourism industry. Today, the Internet is undoubtedly the most important source of information, especially for the tourism industry. Social networks and various digital applications have revolutionized the tourism market, giving consumers the power to demand the best product at the right price. This feat has been reinforced by the abundance of internet and digital applications, their ease of use and low cost (Myat et al., 2023).

The digitalization of the tourism industry has contributed to the transparency and clarity of tourism services presented online, notably by eliminating barriers to accessing information, enabling price consultation and comparison, diversifying marketing channels, and increasing the competitiveness of the products on offer. This new trend has been amplified by the assured access to digital tourism applications via smartphones and other electronic devices (Carlisle et al., 2023). In addition, (Gutierriz et al., 2023) consider that the digitalization of tourism products has helped to make tourism services one of the best-selling products online, prompting professionals to give greater importance to e-marketing by remaining constantly attentive to changing customer needs.

Thanks to the Internet, the tourism sector has taken on a new face, enabling tourists to organize and plan their trip without actually testing the product on offer. In fact, with the digitalization of tourism services, consumers can now plan all stages of their trip online (buying plane tickets, booking hotels and excursions, renting cars, etc.). Consumers can also ask other tourists for their opinions on a tourist product. Basically, the Internet has contributed to the emergence of a new commerce based on technology and digital (Alnajim and Fakieh, 2023). Sin et al. (2020), reveal that 40% of travelers change their travel plans following recommendations available on social networks and the Internet. And that 70% of tourists use social networks for obtaining product information, rather than consulting the supplier directly. This points to the power of social networks and specialized applications in guiding travel decisions. Online reviews and comments on one product or another have more credibility and importance than you might imagine, especially when it concerns official sites (Sakas et al., 2022).

The Internet is becoming a relentless factor in the tourism industry, changing the marketing strategies followed by professionals in the field. Specialized websites and social media are indispensable sources of information thanks to comments, images and videos, and the majority of bookings are made on these platforms. A whopping 240k photos are shared via Facebook every minute, and 575 tweets are posted on Twitter, as well as 65k photos shared on Instagram. Google searches exceed 6 million and e-commerce traffic approaches 6 million transactions. All this clearly shows the importance and power of the digital world for the tourism market. Users of digital applications are more comfortable in making their tourism choices, thanks to the easy management of these applications and the feedback they generate. This change in behavior has contributed to the emergence of a digital consumer quite different from the classic consumer (Sakas et al., 2022).

The widespread use of smartphones and other digital devices has boosted the popularity of phone apps among tourists and professionals alike. Data put forward by Statista (2020) show that users downloaded 116 billion mobile apps in 2019, and this figure will reach 184 billion by the year 2024. These applications are useful not only for tourists, making it easier for them to make the right choices, but also for tourism businesses, enabling them to tailor tourism offers to needs on the basis of information generated through these applications (Hamouda, 2022; Quiroz-Fabra et al., 2022). Indeed, thanks to ICT, the tourism sector has been able to develop digital applications that offer new digitized services and enhanced experiences by opting for interactive platforms that provide information on the tourism product and its features, and also promote sites to visit with a clear idea of the product (Quiroz-Fabra et al., 2022).

The pre-digital era in the tourism sector was marked by the decision to choose a tourism product without sufficient information on the nature of the product chosen and its characteristics. However, with the advent of digitalization in the tourism industry, it is now possible to evaluate the proposed service in advance, and the consumer can even live a virtual experience and test the product before making a purchasing decision (Saseanu et al., 2020). Another form of digital marketing that is very common in the tourism industry is the word-of-mouth or "E-WOM" technique. This is a technique that "*allows someone to have a reason to talk about a certain product from a company and facilitates the engagement of a conversation, which is also closely related to "customer-to-customer" marketing*" (Amalia and Hanika, 2021). Thus, it enables tourists to share their opinions and views on a tourism product via the Web, and other tourists take these opinions into consideration when making decisions. Thanks to the evolution of technology, E-WOM is becoming an effective marketing strategy, especially for platforms with a certain credibility and notoriety. According to the research firm Statista, 49% of people surveyed said they trusted reviews posted by influencers and celebrities. This shows the impact of e-WOM in the tourism sector (F. S. Yilmaz and Aytekin, 2018).

The use of digital technology in tourism can also be seen in 'virtual reality', which is defined as "*the use of a computer-generated environment [...] that can be navigated and interacted with, resulting in a real-time simulation of one or more of the user's five senses*" (Saseanu et al., 2020). This technique is very useful in this sector, as it offers tourists the possibility of getting closer to the reality of the tourist areas visited and at 360° (Rauscher et al., 2020). Technological innovation in the tourism industry has also favored its exploitation in sustainable development and green tourism, by inculcating the spirit of environmental protection through social networks, or by using virtual reality and artificial intelligence to simulate fragile tourist areas for tourists (Ramon-Rodriguez et al., 2021; Nascimento and Loureiro, 2023). The challenge for professionals and governments is to train teams capable of mastering these technological tools and using them for sustainable tourism (Rodrigues et al., 2023).

By adopting a systematic literature review, we will look at the current articles (academic journals, conference proceedings and books published between 2018 and 2024) that address the state of the art on the use of digital technologies in the tourism sector with the aim of ensuring its sustainability. In addition, the present study seeks to clarify the degree to which digital technologies are being integrated by the tourism sector in pursuit of its sustainability.

This review paper has been designed to provide answers to the following questions: (1) identify the principal topics and subjects of study in the literature about the adoption of digital technologies in sustainable tourism (2) what are the most important years of publication in the set of articles covered? (3) which scientific journal(s) has/have published the most articles on the importance of social media in achieving sustainable tourism? What are the most frequent words in the corpus?

The PRISMA methodology comprises four stages: identification, screening, eligibility and inclusion, all of which help us to retrieve relevant articles. Thus, this paper will take the form of a literature review of the works of authors and scholars who have discussed the context of digitalization in tourism marketing. Based on the PRISMA (preferred reporting standard for systematic review and meta-analysis) methodology, we will analyze the documents that have studied this theme with precision and rigor. Our work will be divided into four sections. The first three deal with the introduction, methodology and results. The last will be devoted to results and discussion.

LITERATURE REVIEW

Nowadays, the tourism industry is a strategic economy for several countries and plays a major role in the international economy. However, the emergence of the Covid-19 pandemic has exposed it to unprecedented difficulties. Yet, despite these difficulties, the tourism industry has shown remarkable resilience and fluidity in the face of a humanitarian crisis, notably in exploiting the technological revolution and integrating it into the sector, which has fostered the emergence of innovative communication and marketing tools. Continuous technological development has revolutionized the tourism industry, both in terms of the way it operates and the creation of new management systems (Farhangi and Alipour, 2021).

Information and communication technologies (ICT) have been instrumental in bringing out new concepts, strategic orientations, more modern practices and, consequently, new approaches. Advances in the use of artificial intelligence to carry out searches, the increasing use of conversational software to interact with tourists, the rise of video marketing to promote tourism products and services, and the combination of artificial intelligence and machine learning to assist customers in programming and determining what they want to do, are all examples of the integration of digital technologies in the tourism industry. The combination of the above innovations and those linked to the Industry 4.0 concept will help transform tourism and marketing, as it will enhance communication between users and machines through relationships and communications on social networking sites using a smartphone and other electronic devices, while increasing the desire to make good choices and decisions (Maquera et al., 2022). Ultimately, the birth of various technologies, including the Internet, interactive screens, computers and smartphones, has greatly influenced the way people communicate with each other. Digital innovations in communication and technological media have changed the notion of knowledge and culture, and redefined the way tourism is perceived and experienced. Activities relating to ICT and the tourism sector aim to develop a distributed infrastructure that supports users and businesses and adds value, with a particular focus on virtual and augmented reality experiences. In short, ICT is considered essential for the good or bad functioning of the tourism sector and will have a big impact on virtual and sustainable tourism, as well as on good tourism and hotel management, as it will become increasingly easy to implement (Lima et al., 2024). Activities in the field of ICT in tourism aim to develop a design that is adaptable to sustainable tourism standards (Zafar et al., 2023). Ultimately, there is a close correlation between the use of information and internet technologies and the development of travel technology. Today's tourism trade and market are characterized by the use of the Internet as a means of communication and marketing between companies and their customers.

Using digital and social networks to enable tourists to research and gather basic information from the comfort of their own homes can be useful. Travel organizations rely on online booking platforms to process information, carry out transactions and offer immediate customer support. Thus, the internet has become a tool that supports marketing strategies following the expansion of data generated by internet users through social networks (i.e. Facebook and Twitter) and the spread of new technologies frequently used in the tourism field and attached to "Web 4.0" (i.e. AI, AR/VR, Machine Learning) and which have transited the sector to the next stage (Alnajim and Fakieh, 2023). This technological mutation in the tourism sector requires professionals in the field to train teams with subtle skills capable of making good use of these highly technical tools. Thanks to Web 4.0, we're witnessing the rise of new professions in the field. The list of jobs related to technological redesign is not limited to the use of social media and digital applications, but extends to skills linked to robotics, Machine Learning, artificial intelligence, virtual reality and many others. The intense competition in the tourism market calls for sound processing of the information and data generated from customers and consumers, with the aim of adapting the tourism offering to eminent needs. Thanks to the exploitation of new technologies, this task is becoming a ritual for tourism businesses (Balatska et al., 2022).

Today's technologies have the potential to create experiences identical to the real world, such as augmented reality, virtual reality and mixed reality. Virtual reality creates a non-real world in 3-D mode, while augmented reality and mixed reality mesh the real and the virtual, making the tourist experience clearer and more fun, especially for places under threat (Monaco and Sacchi, 2023).

The technological revolution has dramatically affected the way people travel, the way they book and the way they use other users' suggestions to make decisions (Gutierriz et al., 2023). Local companies have been heavily impacted by these upheavals and must now share their profits with the international companies that manage the applications. Disruption is inevitable in today's digital society. Visitors can efficiently find out about hotels and restaurants and read customer reviews on a digital platform, enabling them to make a better choice of vacation destination (Farhangi and Alipour, 2021). The relentless development of technological innovation and the use of social networks to find out about tourist destinations represents a feat that the tourism sector must fully grasp today.

The tourism industry has frequently capitalized on the vocation of keeping in touch with its customers. In recent years, tourists have become accustomed to being questioned via web surveys and automated telephone calls. They have also begun to practice electronic marketing, which can be seen as a much more sophisticated form of digital means of keeping in touch. A large number of hotels and tourist establishments offer free Internet access, with or without additional charges, local communications and fax transmission by the page (Myat et al., 2023). These personalized offers and services allow tourists to see technology as a two-way means of communication. Technology is used to offer individualized service while improving revenues in hotels, airplanes, transportation centers...etc. The other facet of integrating digital technologies into the tourism industry is to support sustainable tourism through the protection of natural resources and the development of sustainable tourism (Saseanu et al., 2020). Countries around the world have taken steps to mitigate the effects of the tourism industry on the environment (Popova et al., 2023). These include initiatives recognized for their environmental responsibility, promotional labels and the enhancement of nature reserves, while integrating artificial intelligence and the Internet of Things into this strategy (Scrioşteanu et al., 2022).

All these technological innovations fit into the digital marketing box, which is defined as "the set of activities and processes facilitated by technological tools with the aim of establishing communication with consumers" (Gutierriz et al., 2023). This new concept can be resumed in the following digital tools:

Important Digital Tools for Tourism Industry

Social networks are of vital importance to digital marketing. In fact, social networks foster a direct, personalized relationship between the consumer and the tourist service delivery personnel, helping to strengthen this relationship and build loyalty. The latter can be defined as "a group of Internet-based applications, where Internet users create content and make it available to the public, rather than having it created by marketers or suppliers" (Sin et al., 2020).

Tourism professionals are creating interactive content on social networks such as Facebook, Instagram and Twitter, then moving on to Customer To Customer strategies through other networks such as LinkedIn in order to forge more tailored and specific relationships. Social networks have the ability to promote tourism products and services more effectively than traditional media (TV, radio, etc.), and at the lowest possible cost. Studies have shown the positive impact of social media use in the tourism industry, helping tourists to compare the price and nature of the products and services on offer, to enrich their experience through user feedback, and to make the right decision. As far as professionals are concerned, the use of social media is characterized by reduced marketing costs, simplified adaptation to market needs, improved services, customer loyalty and, most importantly, the ability to build relationships in real time, 7/24. The use of social platforms represents a strategic tool in the hands of managers to gather data and information on consumer patterns, market turbulence and new tourism products, as well as to come up with the right ideas. In short, social media are an essential component of any tourism company's marketing strategy, enabling them to keep a close eye on the market and stay abreast of any changes (Alnajim and Fakieh, 2023; Amalia and Hanika, 2021).

Another role played by social media is its contribution to renewing the images of areas suffering from political conflict by playing on users' cognitive and actual perceptions (i.e., the case of Saudi Arabia, Iran, Latin American countries). Social networks such as Facebook, Twitter and Instagram, and specialized applications such as Trivago and TripAdvisor, have broken down the barriers to accessing information and contributed to the creation of social relationships between suppliers of products and services on the one hand, and tourists on the other (Carlisle et al., 2023).

Social networks are also highly effective tools for reinforcing sustainable and green tourism. The sustainability of tourist sites and respect for the environment are focal points in the minds of states and governments (Spadaro et al., 2023). Tourism professionals have fully embraced this vocation, which aims to minimize the negative impact on tourist sites and the environment in general. Tools such as TripAdvisor, Booking and Trivago offer personalized tourism products and services that respect the environment. In addition, through social networks, travel companies encourage and incite non-destructive behavior in fragile and endangered destinations (Sakas et al., 2022).

Electronic Word of Mouth (E-WOM)

One of the most effective techniques in e-marketing today is the “word of mouth” or E-WOM technique. The latter is defined as “viral marketing which is capable of leading the consumers to write or to upload photos or videos about the products and services from a company online on social media” (Farhangi and Alipour, 2021). So, this technique is very useful in getting Internet users to discuss a product or service and recommend it to others on the basis of positive reviews. It’s a marketing method based on “customer to costumer”, where a tourist talks about his or her experience of using a product or service to other potential consumers. The “word-of-mouth” technique is divided into five elements. The first element concerns the sender of the information, who has power and influence over those around him or her. The second element concerns the subject or object discussed, which can be a product, a service or a brand. The third element is the medium used, such as social networks. The fourth element concerns those who discuss the subject and finally the evaluation survey will follow to give a positive or negative opinion of the product/service. For Carlisle et al. (2023), E-WOM is “as any positive or negative statement made by potential, actual or former customers about a product or company, that is made available to a multitude of people and institutions via the internet”. The effect occurs when a customer shares his or her opinion of a product he or she has already consumed with other Internet users.

Live Streaming Cluster

The digital revolution in the Internet, manifested by 5G and WI-FI, has favored the emergence of a new marketing technique: live streaming. Live streaming has been defined as the recording and broadcasting of videos in real time by individuals (broadcasters), enabling multimodal interactions (video, image and text) between broadcasters and viewers (Balatska et al., 2022). This technique has been well introduced in the tourism industry (such as Douyin, Wechat, Taobao, Weibo and Xiaohongshu) to promote products and services live through the internet medium.

Internet of Things

The Internet of Things (IoT) represents “a network of physical devices, vehicles, appliances and other physical objects equipped with sensors, software and network connectivity that enable them to collect and share data”. The importance of this technology in the tourism industry lies, on the one hand, in streamlining the decisions and strategies taken by professionals, since these tools present precise, processed data on consumers. On the other hand, they provide real-time information to tourists, enabling them to improve their trips. Other technological innovations are used in vocation, such as artificial intelligence and big data (Maquera et al., 2022).

Virtual Reality and Augmented Reality

VR usually removes visitors from their surroundings and immerses them in 3D worlds, augmented reality and mixed reality have the potential to make travel simpler, more comfortable, educational and safer, by enriching these experiences rather than substituting for them (Carlisle et al., 2023; El Archi and al., 2023). These new technologies have revolutionized the tourism sector, providing tailored information and clear tourist guidance. Professionals in the field can use virtual reality to give customers an idea of what's on offer (try-before-you-buy), and augmented reality to guide tourists to local tourist sites, available means of transport and online restaurant reservations and much more (Rauscher et al., 2020).

METHODOLOGY

The main objective of this paper is to conduct a qualitative method based on a literature review in harmony with PRISMA flow diagram (Preferred Reporting Items for Systematic Reviews and MetaAnalysis) guideline standards (Liberati et al., 2009; D'Eusanio et al., 2019; Hameduddin and amp; Engbers, 2022; Parmentola et al., 2022; El Ouarrak et al., 2024). Through the analysis of scientific databases, we will try to answer the problematic linked to the "literature review on the importance of digital marketing for tourism and sustainability in recent years". In addition, we will examine the positive or negative impact of technological tools on the tourism industry. The PRISMA analysis includes systematic reviews and meta-analyses of journal articles, book chapters and conference papers for the last 7 years (2018-2024), enabling us to identify topical results for the issue in question. The databases included in this study are global in scope, in principle Scopus Database and Web of Science.

The path followed when launching the search introduced the combination of specific keywords such as "digital marketing", "tourism sector" and "sustainability" while opting for the coordinating conjunction "AND" and the coordinating conjunction "OR" to target the "abstract", "title" and "keywords" at the same time. All documents that do not comply with the selected keywords and study topics are excluded from the study. The search process followed for each database is detailed as follows:

For the Scopus database: TITLE-ABS-KEY (digital AND marketing AND tourism AND sector OR sustainability).

For the WoS database: ALL= (digital marketing AND tourism sector).

Yet another criterion is taken into consideration, namely the requirement for records to be written in English.

The PRISMA flow diagram carried out in our study has enabled us to draw the shape shown in Figure I. The identification phase enabled us to identify 581 documents which will be processed to eliminate duplicates on the basis of author, article title and publication date. This phase excluded 191 documents. The screening phase studied documents with reference to abstracts. In this phase, 144 articles were identified for unanimous review, based on the following criteria:

- articles dealing with literature reviews;
- articles attacking the importance of digital tools for the tourism sector;
- articles analyzing the importance of digital marketing for sustainability.

The final number of articles included in this study was set at 30 documents.

The articles included meet the determined inclusion criteria, while noting that during the eligibility phase, we were unable to include a number of articles because they are not accessible, which did not allow us to verify whether or not they comply with the objectives of our systematic review of the literature.

Figure 1. The PRISMA analysis flowchart for our study

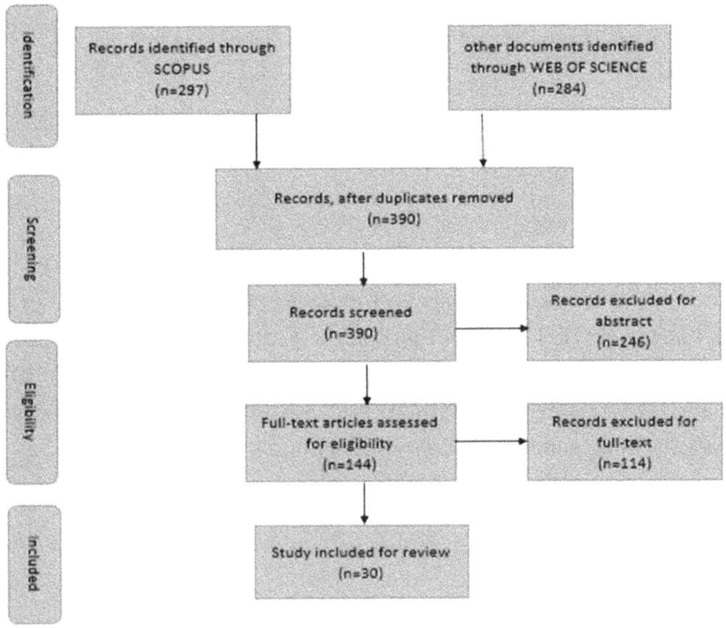

RESULTS

The study of the 30 articles included in our corpus has enabled us to identify the first results in the form of the general characteristics of the articles treated, and then to move on to the outputs identified by each article to understand the impact and importance of digital marketing for the tourism sector and sustainability.

As we showed at the outset, this work was based on two scientific databases: Scopus and Web of science. Thus, the results show that the number of articles included in our corpus for the Scopus database is (20) articles, while it is (10) for the WOS database. so we focused more on Scopus than Wos because the former database is the largest in terms of peer-reviewed documents. See Table 1:

Table 1. List of scientific databases included in our research

Databases	Frequency	Percentage
Scopus	20	66.7%
Web of science	10	33.3%
Total	30	100,0

For the type of articles introduced in our corpus, the results show that journal articles are very important in the composition of our corpus with (26) cases, while for conference articles, there are three cases (3). Finally, the section of a book in our corpus is only one (01). Table 2 clarifies this situation:

Table 2. Type of references and number of sources included in corpus

Type of references	Number of corresponding sources
Journal articles Conference papers Book chapter	26 03 01
Total	30

In terms of publications per year and taking into account the diagram below which represents the results from the year 2018 onwards, as shown on the fold-out screen, the number of articles published per year is very high for the year 2022 (8 articles) and the year 2023 (11 articles) and generally increases from one year to the next. These results demonstrate the importance of the field of "introducing digital tools into tourism marketing to achieve sustainability". As a result, this area of research has become of vital importance to researchers, academics, and professionals alike.

Figure 2. Number of publications by year

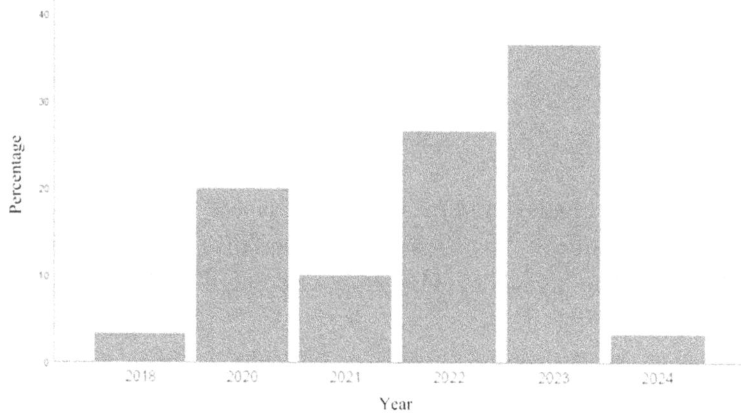

Figure 3. Most frequent words in the corpus

The word cloud included in our corpus enabled us to identify Figure 3. The most important words in this study are "marketing", "tourism", "social", "media", "digital" and "sustainable".

From our corpus, the list of journals with a significant number of publications on our theme is presented in the table below. The "Sustainability (Switzerland)" journal tops the list in terms of number of publications, with six. The other journals have one publication each.

The variety of journal disciplines indicates that the research topic studied is a multidisciplinary field of research.

Table 3. Type of references and number of sources included in corpus

Publication	Number	Percentage
Sustainability (Switzerland)	6	20%
Enlightening Tourism. A Pathmaking Journal	1	3.3%
Data	1	3.3%
Information	1	3.3%
Tourism and International Interdisciplinary Journal	1	3.3%
Journal of Tourism Analysis	1	3.3%
Innovative Marketing	1	3.3%
Journal of Eastern European and Central Asian Research	1	3.3%
Financial and Credit Activity-Problems of Theory and Practice	1	3.3%
Tourism and Hospitality Research	1	3.3%
Economic Research-Ekonomska Istrazivanja	1	3.3%
Information Technology and Tourism	1	3.3%
International Journal of Technology and Human Interaction	1	3.3%
Journal of Theoretical and Applied Electronic Commerce Research	1	3.3%
Benchmarking-an International Journal	1	3.3%
IOP Conference Series: Earth and Environmental Science	1	3.3%
Contemporary Studies in Economic and Financial Analysis	1	3.3%
Human Behavior and Emerging Technologies	1	3.3%
Administrative Sciences	1	3.3%
E3S Web of Conferences	1	3.3%
Journal of Tourism Futures	1	3.3%
Advances in Hospitality and Tourism Research	1	3.3%
Springer Proceedings in Business and Economics	1	3.3%
Journal Komunikasi: Malaysian Journal of Communication	1	3.3%
Academica Turistica	1	3.3%
Total	30	100%

DISCUSSION

For Alnajim and Fakieh (2023), they study the importance of digital marketing through social media for Saudi SMEs. Technological development, in particular through machine learning, will enable us to study the most important intentions that can affect tourists' choices during the consumption process. The main results show the importance for Saudi SMEs of launching a marketing campaign through social media, characterized by simplicity of use and design, and which satisfies the desired need. However, the marketing strategy must be adapted to the characteristics of the target audience. Amalia and Hanika (2021), consider that a digital marketing strategy based on the Instagram tool is highly effective in developing the tourism sector. Again, the E-WOM technique is of great importance in guiding tourists' intention to visit Indonesia. Indeed, according to Balatska et al. (2022), the spread of the COVID-19 pandemic has played a crucial role in shifting tourism and hospitality professionals towards a marketing strategy focused primarily on digital technology. Digital marketing techniques such as search engine optimization (SEO), social media marketing (SMM), search engine marketing (SEM), influencer marketing, content marketing and advertising on digital assets, as well as many others, have proven their positive impact on the tourist experience. The next step is to train marketing managers to master these new digital tools, particularly in the fields of artificial intelligence and virtual (augmented) reality. Count to Carlisle et al. (2023), the study attempted to address the incompatibility of skills needs in the use of digital tools in the European tourism sector between the present and the future. Thus, the use of social media, AI, virtual reality, Big-data and robotics in future tourism marketing strategy will be confronted with a skills shortage that will affect the tourism industry. Ramon-Rodriguez et al. (2021) study the competitiveness and sustainability of the Spanish tourism sector in a turbulent and competitive environment, which has been affected by the COVID-19 pandemic. In this context, professionals need to harness digital technologies to put in place indicators that can gauge the health of the tourism sector and its performance.

Farhangi and Alipour (2021), attacked the angle of the importance of social media to improve the image of tourist destinations and more precisely to defend the notoriety of destinations with a bad political image. Through social networks, managers must implement a digital marketing strategy to reach their targets in social networks by defending the image, perception and prerequisites on sites that suffer from a political blockage. This strategy must be the concern of all players in the tourism industry, especially as these tools are poorly exploited by professionals to promote their territories. Lima et al. (2024) Analyze the impact of the adoption of digital marketing by small and medium-sized tourism enterprises (SMEs) in Cusco, Peru, to guide customers' tourism choices. They found that digital marketing is positively affecting these SMEs, with the aim of guaranteeing their sustainability and adaptability to the needs of the tourism market. By conducting a systematic literature review, Gutierriz et al. (2023) studies digital transformations for the tourism sector. The results show a trend for the tourism industry to focus increasingly on technological innovation to ensure the sector's development. Zafar et al. (2023) considers that the integration of ICT in the tourism sector at the level of the BRICS countries can mitigate the negative impact of the tourism sector on sustainability and environmental protection by reducing carbon emissions and limiting dependence on traditional energies. Rodrigues et al. (2023) attempt to analyze the importance of technological innovation for sustainable and intelligent tourism. The state of the art on this subject shows a discrepancy between the technological advances made in tourism and the real impact of these advances on sustainability and environmental protection. Foris et al. (2020) are trying to show that through tourism booking sites, good practices in sustainable tourism and environmental protection can be reinforced among tourists and professionals alike. On the one hand, tourists have become more sensitive to the issue of sustainability, and on the other, hoteliers are obliged to adapt to this new situation. (Popova et al., 2023) recognize the importance of the digital ecosystem for sustainable tourism. In this way, the digital ecosystem creates harmony and digital links between all tourism stakeholders to achieve sustainable tourism objectives. Spadaro et al. (2023) consider the COVID-19 pandemic to be a period that prompted all stakeholders to review the traditional management of the tourism sector, thinking in terms of sustainable tourism and exploiting technological innovation in the sector. Thus, through the implementation of a municipal planning tool, all stakeholders can contribute to the promotion of sustainable tourism. Scrioşteanu et al. (2022) attempt to assess the factors influencing sustainable tourism in the European context. Social innovation, digitization, crisis management and human capital development are all factors affecting the sustainability of the tourism sector in the current era. Monaco and Sacchi (2023) discusses the impact of Web 3.0 and the Metaverse in particular on the tourism sector, demonstrating its usefulness in offering customers virtual experiences. The adoption of these technologies raises questions about social acceptance, affordability, and environmental sustainability.

Referring to Garrido et al. (2022), they seek to study the importance of digital technologies for a special segment of tourism, namely caravanning. Indeed, the motivations for using digital applications in this segment are generally linked to their ease of use, perceived value, personalized nature and, ultimately, the satisfaction they can create for the user. Hamouda (2022), attempts to analyze another aspect of the use of digital tools in the tourism sector. This concerns the use of digital applications for tourism marketing, while studying the implications that drive consumers to use a given application. The results show that the non-complexity and simple use of the digital application is decisive in this behavior, with the aim of reducing the time needed to manipulate it. It is therefore recommended to work on the content of the application and the information it generates for the user, in order to make it attractive.

Maquera et al. (2022), look at the importance of information and communication technologies for the tourism industry, particularly rural tourism. The results show a strong transition towards digital marketing for the post COVID period for tourism in general, but this transition is reluctant for rural tourism. As a result, there is a lack of information on the activities provided by rural tourism, which calls for a redoubling of effort. The results of Myat et al. (2023) study, shows that ICTs play a vital role in the tourism sector, particularly when it comes to high-quality Internet connections for hotels in Myanmar. Thanks to the Internet, hotels can offer a quality service that meets the needs of their customers. The Internet is also useful for hotel managers, enabling them to analyze the data collected from customers, which in turn has an impact on the services offered, by adapting the services to the needs expressed. Purwaningwulan and Ramdan (2022), study the use of digital tools for tourism promotion in Indonesia. The results show that the use of social media, particularly Instagram, to promote tourist areas can provide useful information to the public, which will contribute to the country's tourism development. The study also discussed the virtual reality techniques used by professionals to provide tourists with near-reality experiences using only smartphones. Thanks to applications that can be downloaded from the Google App store or Play store, tourists can take a virtual tour of 64 tourist sites. Quiroz-Fabra et al. (2022), analyze the use of technological and digital tools to enhance knowledge of cultural and heritage tourism. The results highlighted the importance of e-learning and gamification techniques as useful tools for providing information and knowledge in an interactive and dynamic way about tourist queries for tourist sites. Rauscher et al. (2020) and Nascimento and Loureiro (2023), analyze digital tools in the tourism industry, focusing on virtual reality. In this way, VR is used to provide tourists with a tourism experience that comes close to reality for tourist destinations. As a result, and if the virtual world can be judged to be different from reality, this technique can be useful for areas that tourists don't have the means to visit, or places that are by nature difficult to visit. At this level, professionals need to present virtual experiences with content that adapts to customers' needs, to encourage them to experiment with these new technologies. Sakas et al. (2022), try to explore the usefulness of digital tools for advertising and marketing in the hotel and tourism industry. Indeed, social media have proven their worth in studying user behavior on digital platforms. Analyzing the exchange of information on social media about tourism products and services provides managers with 360° visibility, enabling them to adapt their marketing strategies to expressed needs. So, managers need to be attentive to the changing needs of tourists, and also to digital marketing techniques, which today are becoming a competitive advantage in the hands of those who have the skill to use them well. Tourists have become concerned by the sharing of information (reviews, comments, tweets, images and videos...etc.) that circulate through social networks (Facebook, Instagram, YouTube...etc.) and which effectively contribute to shaping travel decisions and tourism choices. This new era based on digital marketing has created fierce competition in this field, pushing professionals to adapt to such a situation. Saseanu et al. (2020), study the effectiveness of digitization in driving forward the tourism sector in general, and the green and sustainable aspect of this sector in particular, for European countries. The results show a positive impact of the use of technology in tourism, promoting green and environmentally-friendly tourism and encouraging behavior through social networks that encourages tourists to respect the environment of the tourist sites they visit. For Sin et al. (2020), they tried to dissect the context of digital technology use in the hotel and tourism industries, highlighting the importance of social media such as Facebook, Instagram and YouTube in the marketing strategy of tourism companies. The results of this study make it possible to analyze the perceived value of customers as a decisive factor in the decision-making process. Thus, sharing tourism experiences on social media will help to impact Internet users and their desires. The Tarazona-Montoya et al. (2020), deals with the context in which digital marketing is used by tourism professionals in Colombia. Indeed, thanks to the expansion of information on hotels and tourist residences, consumers have become more demanding in their needs, prompting professionals to adapt on a daily basis to this new situation marked by the development of ICTs. In addition, managers at hotel units are expected to perfect their knowledge of the usefulness of each digital tool, exploiting them to the full in order to derive results that will shape the marketing roadmap. Social networks such as Instagram and Face book, as well as specialist applications such as Trivago, Tripadvisor and Atrápalo, play a key role in promoting tourism in the country. E. Yilmaz (2020) studies the effectiveness of the digital tool Tripadvisor in shaping tourist decisions. Hotel and resort managers are obliged to monitor the comments and opinions of all Internet users, who may turn into potential tourists. This tool has proven its power to guide the choice of tourism products and services, and the ratings awarded to hotel residences have a positive impact on the decision-making process. This is a win-win situation for customers, enabling them to make the right decision, and a win-win situation for professionals, giving them the opportunity to correct their marketing strategy and introduce corrective action on the products and services on offer. E. S. Yilmaz and Aytekin (2018), have somewhat confirmed the results obtained by E. Yilmaz (2020), regarding the importance of the digital application specialized in the tourism field "Tripadvisor". Zarrouk and El Aidouni (2023), focused on the Moroccan experience of using digital technologies in the tourism industry. Thanks to the integration of these digital tools into its marketing strategy, Morocco has been able to make great strides in promoting its tourism offering. The use of digital platforms and specialized applications has also had a positive impact on friendly behavior towards the environment.

CONCLUSION

The interest of our book chapter is to strengthen the literature on such an important topic as "the importance of digital marketing for the tourism sector and sustainability", through the analysis of peer-reviewed journal articles using the PRISMA method. This goal was achieved through the discussion section, which presented the results obtained for all the articles included in our study. Thus, our research addressed a very important aspect of the use of digital tools for a tourism sector better adapted to the current and sustainable context. Technologies such as artificial intelligence, virtual reality, E-wom and the use of social networks to market and adapt the tourism sector to the digital age. This reality requires researchers and managers to strengthen the skills of users of these technologies, so that they become a useful tool for the tourism industry and its sustainability.

In conclusion, it must be said that the use of digital marketing in the tourism sector has transformed the industry by introducing new tools that have contributed to a radical change in marketing techniques and opened our eyes to sustainable and intelligent tourism. On the one hand, social networks enable professionals to promote their products and services by implementing a marketing strategy that's well adapted to turbulent needs. Consumers and tourists, on the other hand, can exploit these social networks to shape their decision to consume a tourist service or make a decision that satisfies their search. This has become easy thanks to the information available on these networks and the evaluations given by other consumers. Another useful aspect of digital technologies is their use by professionals to provide a virtual experience that can replace the real thing. Thus, through the use of artificial intelligence and virtual reality techniques, tourists can have an idea before the final decision-making process. Again, these tools strengthen green and sustainable tourism by equipping the sector with techniques that reinforce environmental protection and sustainable development.

The digital age represents a major challenge for industry professionals, forcing them to be attentive to changing consumer habits and trends on the Web. As a result, tourism companies need to stay in constant touch with the tourism market, something that has become possible thanks to information and communication technologies. The second challenge lies in the need for tourism businesses to train their management teams to master digital tools and techniques, so that they can take full advantage of the opportunities they offer.

Finally, the theoretical implications of our book chapter will contribute to the existing literature by analyzing previously published journal articles, conference papers and book chapters on the topic in question, thus helping to increase the knowledge generated on the subject in question. For managerial implications, I consider that by conducting this study we aim to provide tourism researchers, professionals, and scholars with an overview of the progress made in the adoption and integration of digital technologies and social media in the industry tourism, while striving to ensure its sustainability. This study also focused on the new facets of the use of artificial intelligence, virtual reality, augmented reality, and many other techniques to adapt the tourist offer to the expressed demand, while analyzing the opportunities and the challenges offered by these techniques and how they can affect the future of tourism and hospitality.

REFERENCES

Alnajim, R., & Fakieh, B. (2023). A Tourist-Based Framework for Developing Digital Marketing for Small and Medium-Sized Enterprises in the Tourism Sector in Saudi Arabia. *Data*, *8*(12), 179. Advance online publication. doi:10.3390/data8120179

Amalia, J., & Hanika, I. (2021). Tourism in Digital Era : The Influence of Digital Marketing on the Intention to Visit Museum. *Jurnal Komunikasi-Malaysian Journal of Communication*, *37*(3), 274–287. doi:10.17576/JKMJC-2021-3703-16

Balatska, N., Radkevych, L., Robul, Y., Vdovichena, O., & Strenkovska, A. (2022). Digital technologies and digital marketing: New opportunities for tourism and hospitality business. *Financial and Credit Activity-Problems of Theory and Practice*, *6*(47), 424–432. doi:10.55643/fcaptp.6.47.2022.3924

Carlisle, S., Ivanov, S., & Dijkmans, C. (2023). The digital skills divide : Evidence from the European tourism industry. *Journal of Tourism Futures, 9*(2), 240-266. Advance online publication. doi:10.1108/JTF-07-2020-0114

D'Eusanio, M., & Zamagni, A., & Petti, L. (2019). Social sustainability and supply chain management : Methods and tools. *Journal of Cleaner Production, 235*, 178-189. Advance online publication. doi:10.1016/j.jclepro.2019.06.323

El Archi, Y., Benbba, B., Kabil, M., & Dávid, L. D. (2023). Digital Technologies for Sustainable Tourism Destinations: State of the Art and Research Agenda. *Administrative Sciences*, *13*(8), 184. doi:10.3390/admsci13080184

El Ouarrak, Y., Hmioui, A., Rajouani, B., Alla, L., & Benjelloun, A. (2024). A Systematic Review of the Relationship Between Supply Chain Risk Management and Sustainability. In A. Iglesias, J. Shin, B. Patel, and A. Joshi (Éds.), *Proceedings of World Conference on Information Systems for Business Management* (p. 41-52). Springer Nature. 10.1007/978-981-99-8349-0_5

Farhangi, S., & Alipour, H. (2021). Social media as a catalyst for the enhancement of destination image : Evidence from a mediterranean destination with political conflict. Sustainability (Switzerland), 13(13). *Sustainability (Basel)*, *13*(13), 7276. Advance online publication. doi:10.3390/su13137276

Foris, D., Crihalmean, N., & Foris, T. (2020). Exploring the environmental practices in hospitality through booking websites and online tourist reviews. *Sustainability (Switzerland), 12*(24), 1-18. Advance online publication. doi:10.3390/su122410282

Garrido, R. H., Grávalos-Gastaminza, A., Escobar-Rodríguez, T., & Pérez-Calañas, C. (2022). Determining factors influencing use of caravanning tourism apps. *Journal of Tourism Analysis, 29*(1), 160-184. Advance online publication. doi:10.53596/jta.v29i1.399

Gutierriz, I., Ferreira, J. J., & Fernandes, P. O. (2023). Digital transformation and the new combinations in tourism : A systematic literature review. *Tourism and Hospitality Research*. Advance online publication. doi:10.1177/14673584231198414

Hameduddin, T., & Engbers, T. (2022). Leadership and public service motivation : A systematic synthesis. *International Public Management Journal, 25*(1), 86-119. Advance online publication. doi:10.1080/10967494.2021.1884150

Hamouda, M. (2022). Mobile Applications in Tourism : Examining the Determinants of Intention to Use. *International Journal of Technology and Human Interaction*, *18*(1), 1–13. Advance online publication. doi:10.4018/IJTHI.293198

Liberati, A., Moher, D., Tetzlaff, J., & Altman, D. G. (2009). Preferred reporting items for systematic reviews and meta-analyses: The PRISMA statement. *PLoS Medicine*, *6*(7), e1000097. doi:10.1371/journal.pmed.1000100 PMID:19621072

Lima, N. T. Y., Haro-Zea, K. L., & Saavedra, R. A. D. (2024). Digital marketing and customer orientation as predictors of sustainability in tourism SMEs. *Innovative Marketing, 20*(1), 160-171. Advance online publication. doi:10.21511/im.20(1).2024.14

Maquera, G., da Costa, B., Mendoza, O., Salinas, R., & Haddad, A. (2022). Intelligent Digital Platform for Community-Based Rural Tourism-A Novel Concept Development in Peru. *Sustainability (Basel)*, *14*(13), 7907. Advance online publication. doi:10.3390/su14137907

Monaco, S., & Sacchi, G. (2023). Travelling the Metaverse : Potential Benefits and Main Challenges for Tourism Sectors and Research Applications. *Sustainability (Basel)*, *15*(4), 3348. Advance online publication. doi:10.3390/su15043348

Myat, A. A., Sharkasi, N., & Rajasekera, J. (2023). Myanmar's tourism : Sustainability of ICT to support hotel sector for online booking and digital marketing. *Benchmarking, 30*(7), 2486-2508. Advance online publication. doi:10.1108/BIJ-07-2017-0200

Nascimento, J., & Loureiro, S. M. C. (2023). The Impact of Augmented and Virtual Reality for Sustainable Tourism. In Springer Proc. Bus. Econ. (p. 148-156). Springer Nature. doi:10.1007/978-3-031-25390-4_14

Parmentola, A., Petrillo, A., & Tutore, I., & De Felice, F. (. (2022). Is blockchain able to enhance environmental sustainability? A systematic review and research agenda from the perspective of Sustainable Development Goals (SDGs). *Business Strategy and the Environment, 31*(1), 194-217. Advance online publication. doi:10.1002/bse.2882

Popova, P., Petrova, M., Popov, V., Marinova, K., & Sushchenko, O. (2023). Potential of the digital ecosystem for the sustainable development of the tourist destination. *IOP Conference Series. Earth and Environmental Science*, *1126*(1), 012021. Advance online publication. doi:10.1088/1755-1315/1126/1/012021

Purwaningwulan, M., & Ramdan, T. (2022). Digital promotion of local tourist destinations in the new normal era and its effect on the economy in Indonesia. *Journal of Eastern European and Central Asian Research*, *9*(1), 29–40. doi:10.15549/jeecar.v9i2.873

Quiroz-Fabra, J., Valencia-Arias, A., Londoño-Celis, W., & García-Pineda, V. (2022). Technological Tools for Knowledge Apprehension and Promotion in the Cultural and Heritage Tourism Sector : Systematic Literature Review. *Human Behavior and Emerging Technologies*, *2022*, 1–12. Advance online publication. doi:10.1155/2022/2851044

Ramon-Rodriguez, A. B., Such-Devesa, M. J., Perles-Ribes, J. F., & Moreno-Izquierdo, L. (2021). Competitiveness, intelligence and sustainability in consolidated destinations : What we know and pending research challenges. *Tourism, 69*(2), 281-299. Advance online publication. doi:10.37741/t.69.2.8

Rauscher, M., Humpe, A., & Brehm, L. (2020). Virtual reality in tourism : Is it « Real » enough? *Academica Turistica, 13*(2), 127-138. Advance online publication. doi:10.26493/2335-4194.13.127-138

Rodrigues, V., Eusébio, C., & Breda, Z. (2023). Enhancing sustainable development through tourism digitalisation : A systematic literature review. *Information Technology and Tourism, 25*(1), 13-45. Advance online publication. doi:10.1007/s40558-022-00241-w

Sakas, D. P., Reklitis, D. P., Terzi, M. C., & Vassilakis, C. (2022). Multichannel Digital Marketing Optimizations through Big Data Analytics in the Tourism and Hospitality Industry. *Journal of Theoretical and Applied Electronic Commerce Research, 17*(4), 1383-1408. Advance online publication. doi:10.3390/jtaer17040070

Saseanu, A., Ghita, S., Albastroiu, I., & Stoian, C. (2020). Aspects of Digitalization and Related Impact on Green Tourism in European Countries. *Information (Basel), 11*(11), 507. Advance online publication. doi:10.3390/info11110507

Scrioşteanu, A., Barbu, C. M., & Criveanu, M. M. (2022). Sustainable tourism in the current European context. In Contemp. Stud. Econ. Financ. Anal. (Vol. 108B, p. 17-36). Emerald Group Holdings Ltd.; Scopus. doi:10.1108/S1569-37592022000108B031

Sin, K.-Y., Mohamad, A. A., & Lo, M.-C. (2020). A critical review of literature in the rising tide of social media towards promoting tourism. *Enlightening Tourism, 10*(2), 270-305. Advance online publication. doi:10.33776/et.v10i2.4887

Spadaro, I., Pirlone, F., Bruno, F., Saba, G., Poggio, B., & Bruzzone, S. (2023). Stakeholder Participation in Planning of a Sustainable and Competitive Tourism Destination : The Genoa Integrated Action Plan. *Sustainability (Basel), 15*(6), 5005. Advance online publication. doi:10.3390/su15065005

Tarazona-Montoya, R., Peris-Ortiz, M., & Devece, C. (2020). The value of cluster association for digital marketing in tourism regional development. *Sustainability (Switzerland), 12*(23), 1-18. Advance online publication. doi:10.3390/su12239887

Yilmaz, E. (2020). The effects on consumer behavior of hotel related comments on the tripadvisor website: an istanbul case. *Advances in hospitality and tourism research-ahtr, 8*(1), 1-29. doi:10.30519/ahtr.536303

Yilmaz, E. S., & Aytekin, M. (2018). The determinants of tourist preferences of the mediterranean region in Turkey. *Administrative Sciences, 8*(4), 81. Advance online publication. doi:10.3390/admsci8040081

Zafar, S. Z., Zhilin, Q., Mabrouk, F., Ramirez-Asis, E., Alzoubi, H. M., Hishan, S. S., & Michel, M. (2023). Empirical linkages between ICT, tourism, and trade towards sustainable environment : Evidence from BRICS countries. *Ekonomska Istrazivanja, 36*(2), 2127417. Advance online publication. doi:10.1080/1331677X.2022.2127417

Zarrouk, I., & El Aidouni, A. (2023). The Digital Marketing Practiced by the Moroccan Tourism Sector : National-scale Communication Actions and Environmental Management. EDP Sciences. doi:10.1051/e3sconf/202341201035

Chapter 13
Assessing Loss Due to COVID-19 in the Homestay Sector and Comprehending Resilience Strategies

Sakhhi Chhabra
https://orcid.org/0000-0002-0705-0030
Xavier School of Management, India

Ramkrishan Padhy
Indian Institute of Management, Odisha, India

ABSTRACT

In this research, the authors estimate the revenue loss due to COVID-19 on homestays in 13 districts of Uttarakhand, India. They use novel techniques and unique variables to calculate revenue and expenditure. They also attempted to uncover resilience strategies adopted by homestay owners to recover their respective businesses. They found six themes from the eleven semi-structured interviews with these homestay hosts. They defined these themes as untact hospitality, upgrading homestay services with digital technologies, digitally transforming booking policies, sustainability with legal registration, supporting intrapreneurship, and social entrepreneurship model. This study contributes to advancing the crisis management literature within the tourism sector by evaluating the impact of the pandemic on homestays of Uttarakhand and thereafter elaborating on resilience strategies adopted. Practitioners and policymakers can use the study outcomes to re-evaluate the existing guidelines concerning homestay operations according to the new normal solitary concept.

DOI: 10.4018/979-8-3693-3286-3.ch013

1. INTRODUCTION

The unprecedented disaster due to COVID-19 led to severe losses to the travel & tourism sector worldwide (Worldometer, 2020). A significant drop was witnessed in travel demand across the globe due to the spread of anxiety and uncertainty (Wassler and Talarico, 2021). The impact of loss due to the pandemic was also seen in India since tourism contributes significantly to the economy in terms of revenue and employment (UNWTO, 2020). A 68% fall was observed in India's foreign tourist arrival rate from February to March 2020. Hence, foreign exchange earnings fell by 66.32% (Ministry of Tourism, 2020).

Uttarakhand, a state in northern India crossed by the Himalayas, accounts for over 50% of the state's GDP from tourism suffered huge losses due to the pandemic. The state attracted many tourists owing to its natural beauty, pilgrimage values, ayurveda and natural health resorts, and varied geography. However, due to COVID-19, many hospitality units operated at the local level underwent crisis (Eichelberger and Peters, 2021), especially homestays, which have significantly increased in the state over time. Even though they are famed as the best way to experience the Himalayan hospitality of Uttarakhand, they endured damage because they are mainly family-operated and highly vulnerable to such externalities (Dahles and Susilowati, 2015).

There have been many studies that have investigated the actual and potential effects of the pandemic on various tourism sectors like cruise lines, hotels, dining, and tourist attractions (Gossling, Scott, and Hall, 2020; Pappas, 2021; Ugur ˘ and Akbıyık, 2020). However, there has been little attempt to investigate the impact of COVID-19 on revenue loss of homestays, specifically in the Uttarakhand state of India. There have been studies on response, adaptation, and coping strategies to crises and disasters (e.g., Dahles and Susilowati, 2015; Prayag, 2018; Sharma et al., 2021) in tourism and hospitality businesses. More recently, a growing body of literature has developed on understanding business resilience, especially during the pandemic. Nevertheless, there is no work owing to resilience strategies followed by homestay owners. Studies have captured the behaviors of guests and travelers, while the perspectives of homestay owners and hosts have not been accounted for (Farmaki et al. 2020). Thus, in this study, we elaborate on the case of homestays in Uttarakhand and estimate the loss of revenue and employment in the first phase of COVID-19. Further, we explore resilience strategies executed by homestay owners to recover from the COVID-19 pandemic.

This study was conducted in two phases. During the first phase conducted in June 2020, we estimate the impact of revenue loss for homestays of Uttarakhand. The challenge was to assess the loss realistically as most of the sector was unorganized, and insufficient reliable data was available. The primary data was collected from Uttarakhand Tourism Development Board (UTDB) officials and homestay owners. The variables used for calculating the revenue and expense for 13 state districts were assumed upon discussion with the homestay owners. Both peak season and off-peak season were included. In the second phase of the study conducted post-pandemic in June 2022, eleven semi-structured interviews were taken with homestay owners to understand recovery strategies. Interviews were analyzed thematically, and data revealed resilience strategies employed by homestay owners to mitigate the impact of the COVID-19 pandemic.

The present study showcases innovative methods of assessing revenue loss by using innovative variables and extrapolating the data. It would be useful for researchers and policymakers to estimate losses in emerging economies that lack organized tourism-related data. This study can also help micro, small, and medium enterprises in understanding how to contribute to the understanding of how micro, small, and medium-sized enterprises (MSME) can adapt to unexpected losses due to externalities. It could help formulate mitigation strategies to overcome the loss. Insights from this inquiry will be valuable to homestay owners and their stakeholders, especially in building their adaptive capacities during and after prolonged crises. These results could help homestay owners, guesthouse owners, small lodges, restaurant owners, and other MSMEs in different regions formulate a forward plan and guidelines to overcome the loss during the pandemic.

The chapter comprises six sections. The first section presents the introduction. The subsequent section offers a literature review on homestays and studies assessing the impact of COVID-19 and resilience strategies adopted in the hospitality sector. The methodology section elucidates the process of estimating revenue loss, followed by in-depth interviews with homestay owners. The fourth section presents the calculation of revenue loss due to COVID-19 on homestays in Uttarakhand. The fifth section presents homestay owners' resilience strategies, and the last section presents implications and conclusions.

2. LITERATURE REVIEW

2.1 Homestays and COVID-19 Pandemic

Homestay is a relatively recent novelty in the community-based tourism (CBT) accommodation market. The concept escalated in Europe in the late 1970s when people were looking for means to spend vacations outside their homes (Yong and Hassan, 2004). For this study, we consider "Homestay" as an accommodation where tourists or guests pay to stay in private homes, with recurrent prospects to interact with the host family and experience their living style (Di Domenico and Lynch, 2007; Lynch, 2005). Homestay is considered a prospective alternative to generate revenue for local communities, especially in less developed countries (Kayat, 2010), and indirectly contributes to the nation's tourism industry. While for guests, Peer to Peer (P2P) accommodation has emerged as a disruptor in the traditional hospitality sector (Sigala, 2017) since it offers an opportunity to experience local cuisine and culture (Leh and Hamzah, 2012).

Even though, over the years, CBT has become a research topic, post-COVID-19 homestays continue to struggle with various challenges. And mainstream academic research has focused on large-scale hospitality sectors, deserting homestays despite its apparent susceptibility to the effects of the pandemic (Hassanli et al., 2016; Lynch and MacWhannell, 2000). The initial study regarding the impact of COVID-19 on different accommodation types by Bresciani et al. (2021) concluded that the assurance of physical distance reduced tourists' concerns about shared accommodation. In their qualitative study, Farmaki et al. (2020) found that the pandemic had varying levels of impact on different hosts by investigating the impacts of the COVID-19 pandemic on hosting practices. Stone et al. (2021), in their study, found intensified anxiety about the future of homestays, restaurants, eco-lodges, etc., due to the pandemic in Botswana's nature-based tourist industry. This anxiety resulted in occupation losses for locals and decreased the pace of resilience to restart normal operations. Furthermore, Hussain (2021) emphasized that the pandemic has raised concerns about the tourism sector's revival, even though community-based tourism operations have been justifying efforts to maintain business amid the pandemic by locals.

Astoundingly, earlier investigations have focused more on guests' protective behaviors, while hosts' perspectives and protective behaviors have been ignored in the literature (Ghaderi, Butler, and B'eal, 2022). There are no specific studies related to homestays, especially in emerging markets of India, which focus on the resilience strategies adopted due to the COVID-19 pandemic (Doan et al., 2022). This study fills this research void.

2.2 Studies Assessing the Impact of COVID-19

There have been many studies in the literature that have attempted to measure the impact of COVID-19 on the tourism and hospitality industry. In a review article by Davahli et al. (2020), it was found that the impact of the pandemic on the hospitality industry has been studied using discrete models and analyses, which include seasonal autoregressive integrated moving average model, scenario analysis, and trend analysis. Centeno and Marquez (2020) forecast a total loss of around $3.37 billion from COVID-19 until July 2020 for the Philippines' tourism and hospitality industry by developing seasonal autoregressive integrated moving average models. Günay et al. (2020) applied a scenario analysis technique to predict Turkey's tourism and hospitality industry's total loss of revenues in the best and worst-case scenarios due to COVID-19. For India's economy, Mehta and Jha (2020) estimated an earnings loss of about $28 billion in 2020 due to COVID-19. Different studies have used diverse methodologies and numerous variables for measurements. The impact of COVID-19 on the global tourism and hospitality industry was also examined using trend analysis (Priyadarshini, 2020). Pramana et al. (2021) clustered the provinces of Indonesia based on the room occupancy rate to understand and investigate the impact of the COVID-19 pandemic. In this study, we narrow the scope of assessing the impact of COVID-19 on homestays of the Uttarakhand region of India by estimating revenue loss using the number of rooms, approximate tariff per day, occupancy rate, peak season, off-season, salary to helpers, electricity bill, water bill, and maintenance bill.

2.3 Business Resilience

To ensure the continuity and success of tourism businesses, developing the capability to adapt in a swiftly changing environment is essential (Hinson and Slade, 2011). Many recent studies have comprehended business resilience (Sarker and Clegg, 2021), especially during the COVID-19 pandemic. Resilience can be recognized as a proactive and reactive strategy (Mayson and Barrett, 2017) that organizations and individuals apply. Business resilience in the context of tourism and hospitality has been described as – response, adaptation, and coping strategies to crises and disasters (Sharma et al., 2021; Prayag, 2018; Dahles and Susilowati, 2015).

The majority of research on resilience in the travel context has fundamentally concentrated on external events such as natural disasters (Prayag et al., 2020), destinations' adaptation to climate change (Kajan and Saarinen, 2013; Njoroge, 2015), and response, adaptation, and coping strategies to crises and disasters (Dahles and Susilowati, 2015) and organizational resilience (Prayag, 2018). Within the hotel industry, the latest studies have examined the responses and adaptive procedures of big hotels, particularly in terms of their human resource practices (Su et al., 2021), marketing and promotions (Hoang et al., 2021), general managers' responses (Giousmpasoglou et al., 2021), or corporate social responsibility (Marco-Lajara et al., 2022). Nikos et al. (2021) studied the resilience of the travel and hospitality industry in England. King et al. (2021) studied the resilience framework, examining the socio-ecological systems such as social structure and livelihood capital in Indonesia during the crisis of COVID-19. McCartney (2020) studied the impact of a pandemic on Macao, from tourism lockdown to recovery.

However, there is no specific scholarly reconnaissance related to the resurrection of homestay businesses due to the COVID-19 pandemic. This leaves a significant knowledge gap within the tourism resiliency sector. Drawing on the concepts of business resilience, this research aims to explore the approaches embraced by homestay owners to recover from the crisis. By analyzing data collected from Uttarakhand homestay owners, the paper addresses the subsequent research questions:

- What is the estimated loss of revenue and employment in homestays due to COVID-19?
- How do homestay businesses modify their processes to overcome loss due to the COVID-19 pandemic?

To answer these research questions, the research objectives are:

1. Compute loss of revenue and employment in homestays of Uttarakhand due to COVID-19 Pandemic
2. Comprehending resilience strategies adopted by homestay owners to overcome the loss due to the pandemic.

3. METHODOLOGY

Homestays within the Indian tourism sector have gained importance over time due to government support and increased awareness among tourists and homeowners. We calculate revenue loss to assess the impact of the COVID-19 pandemic on the homestays of the Uttarakhand region. The scope of the study is confined to registered homestays as recognized by the Uttarakhand Tourism Development Board (UTDB), which is the state's tourism authority. We divide this section into two parts. First, we discuss the process of estimating loss incurred for homestays. In the second part, we discuss the in-depth interview procedures used to understand the resilience strategies adopted by homestay owners to overcome the loss due to the COVID-19 pandemic.

3.1 Process of Estimation of Revenue Loss for Homestays

The first data point required to estimate the impact of revenue loss was the number of homestays in Uttarakhand. We used the UTDB registered data to find the number of rooms registered in each homestay of the 13 districts registered under the "*Deen Dayal Upadhyaya Griha Awas homestay regulations*" scheme. We collected the data regarding the contact details of homestay owners.

The sampling strategy used was stratified sampling, whereby we divided the homestays into smaller strata region-wise (Malhotra and Dash, 2010) and thereafter collected data through simple random sampling. The data from sample homestays consisted of only active homestays, which responded to either of three calls made to each homestay over 45 days. When the call was not connected, they were considered inactive homestays. We surveyed them telephonically to collect primary data regarding the following information: number of rooms, approximate tariff per day, occupancy rate, peak season, off-season, salary to helpers, electricity bill, water bill, and maintenance bill. It was assumed that all homestays registered till March 2018 were considered operational and active. For the homestays registered after March 2018, a suitable correction factor was applied for each district. It was calculated separately for each district. The calculation for the correction factor was as follows:

Factor of activity (Correction Factor) = 1- { ((Proportion of calls Not Connected)*(Number of homestays registered after Mar 18)) / Total Number of Homestays }

To calculate total revenue, the variable used was the sample average tariff (per room) provided by the sample homestay owners. We also calculate the weighted average of the tariffs in peak and off-peak seasons. This gave us the approximate monthly revenue, which was later adjusted to calculate the adjusted revenue according to the correction factor. The numbers provided by Homestay owners were further verified with the secondary data of Average Tariff (Per room) provided by UTDB. It was calculated as the weighted average of the maximum tariff set during registration for homestays in each district.

For the purpose of this study, 'peak season' has been defined in terms of the months in which the tourists' visits are highest for each district. This variable has been suitably assumed upon discussion with homestay owners, further validating the best seasons to visit the tourist places. Then, we calculated the sample average occupancy (per month) for each district, segregated based on peak and off-season as specified by the homestay owners sampled.

Average expense was calculated from the data collected through primary research samples, including utilities (electricity and water bills) and employee compensation. This was multiplied by the number of homestays sampled in each district and adjusted according to the correction factor. Finally, with both revenue and expense, we estimated the loss incurred during the COVID-19 period.

To estimate the employment loss, we found the average monthly employee salary in each sample homestay of 13 districts. We estimated the total number of people employed by dividing the average expense per month for all homestays by the average salary of employees. We calculated the same for peak and off-peak seasons since employment changes as demand increases in summer.

Figure 1. Steps in estimating revenue loss

3.2 Depth Interviews to Understand Resilience Strategies Adopted by Homestay Owners

We conducted semi-structured interviews with homestay owners in the Uttarakhand region to understand the resilience strategies adopted to overcome loss caused by COVID-19. Qualitative research methodology was suitable since homestay owners used different strategies (Creswell and Tashakkori, 2007). Purposive sampling was used to select samples since it is a non-probability sampling method, and respondents are selected based on their ability to answer the research objectives (Malhotra and Dash, 2010). In this case, we recruited participants based on the primary researcher's judgment to utilize the time spent on the interview (Brinkmann and Kvale, 2015), keeping the study's objective in mind.

We designed the interview protocol to collect information about the impact of COVID-19 on the business and resilience strategies adopted (refer to Appendix 1). Based on these questions, we urged the respondents to share their specified views and probed them based on the individuals' responses. Refer to Appendix 2 for the respondent profile. We also discussed some of the questions at length, such as which strategies or tactics benefitted you most from recovering from the pandemic. Participants were laddered on specific answers and asked to detail their opinions (Chhabra, 2022).

All respondents were pre-informed about the purpose of the interview. They were briefed about the type of questions that would be asked. Eleven interviews were conducted telephonically between July 2021 and August 2021. Each interview lasted approximately 30-40 minutes. We recorded the interviews and transcribed and analyzed the transcripts. Thematic analysis was done to identify themes, sub-themes, and codes (Bryman, 2012). First, we read and re-read the interview transcripts to familiarize ourselves with the data. This helped us generate an initial list of ideas (Braun and Clark, 2006). In the next stage, we listed codes and segregated them into their potential themes, arranged all data extracts within the identified themes, and named them based on the essence of each theme.

4. ESTIMATION OF REVENUE LOSS FOR HOMESTAYS

To estimate the impact of revenue loss for homestays, we found the number of registered homestays in Uttarakhand on the UTDB website. The total number of homestays was 1954 in each of the 13 districts registered (Figure 2). Out of which, 83% of the total homestays were registered after March 2018 under the "*Deen Dayal Upadhyaya Griha Awas homestay regulations*" Scheme. Figure 3 gives the district-wise allocation of active and inactive homestays in Uttarakhand as registered on the UTDB website.

Out of the 1954 total homestays, 59.4% of the total registered homestays were found to be active, which counted as 1161 and 793 were inactive. The number of rooms registered was 7126, and rooms qualified from active homestays were 4400. From the call statistics, it was found that out of the 641 sampled homestays that were contacted telephonically, only 23.4% of homestay owners responded and cooperated in providing the data, i.e., 12.9% or 150 of the active homestays were taken as the sample (Figure 4). District-wise and overall call statistics are given in Figure 5.

Further data collected from homestay owners was used to understand other variables necessary to calculate revenue loss due to COVID-19. The Average Peak Season Tariff was calculated as Rs1428/-, and the off-season tariff was Rs 967/-. Average Peak Season Occupancy was calculated as 64.5%, and Average off-season Occupancy was 25%. The average number of rooms was informed as 3.8 per homestay, and the average duration of peak season was 5.6 months. The average monthly revenue was computed as Rs. 65789/- and the average monthly maintenance cost was Rs.13772/-. The average number of employees in a homestay was 1.7, and the average salary to employees was calculated as Rs7994/- across all 13 districts, as shown in Figure 6.

Figure 2. Registered homestays in Uttarakhand

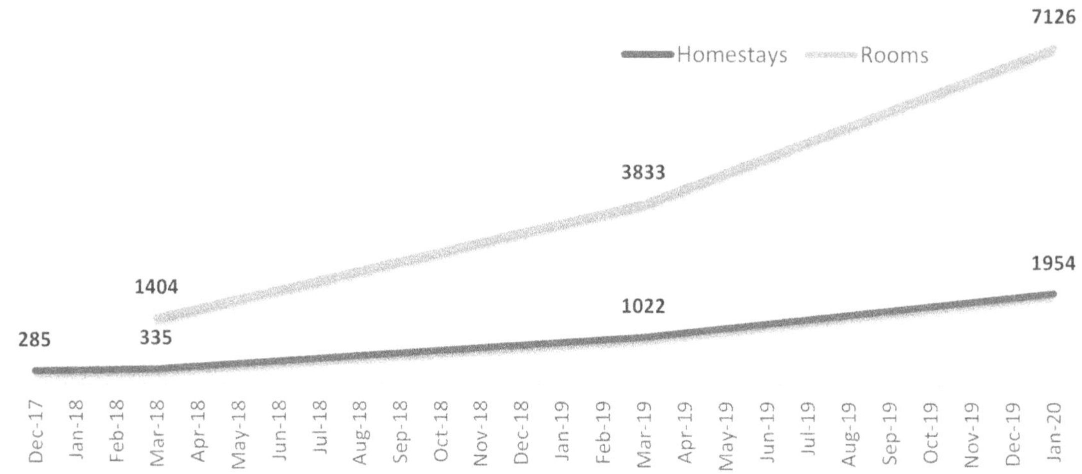

Figure 3. District-wise homestays listed on the UTDB website

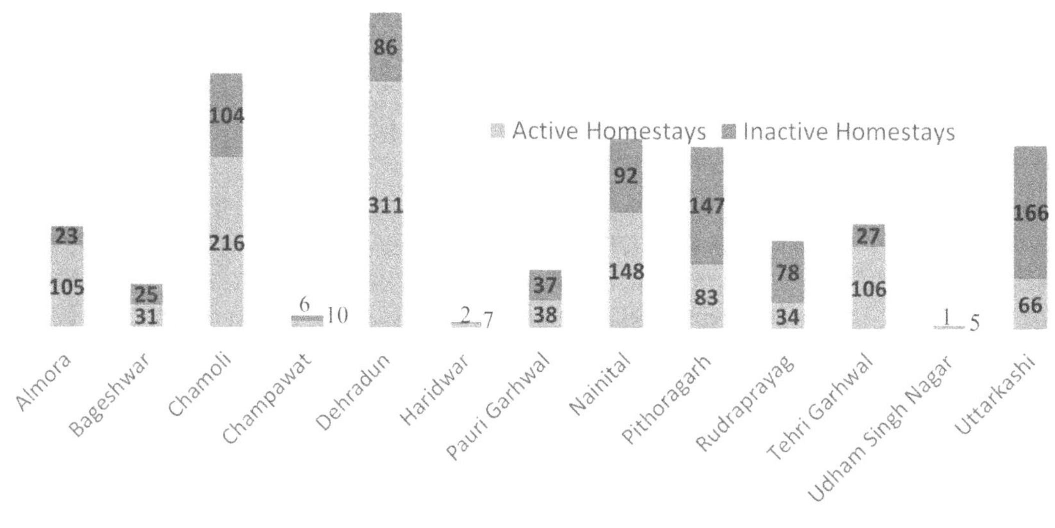

Figure 4. Sampled homestays

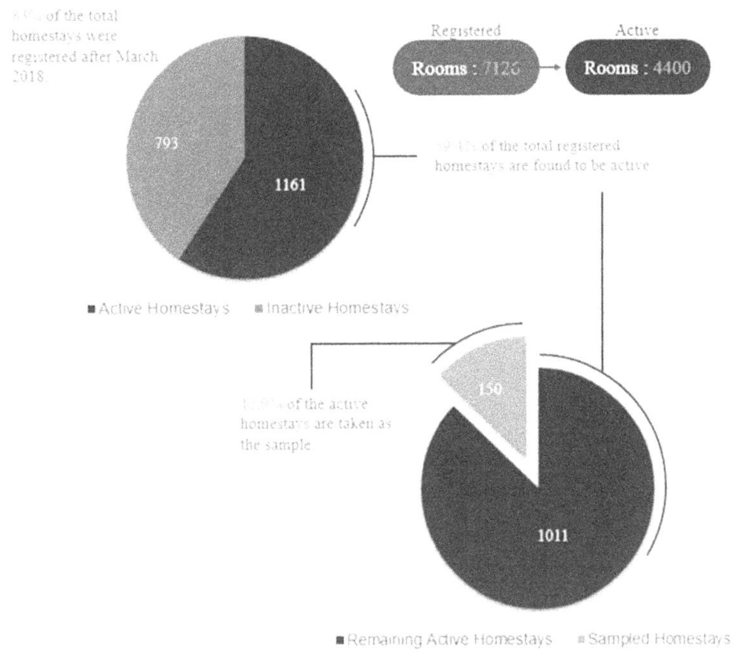

Figure 5. District-wise and overall call statistics

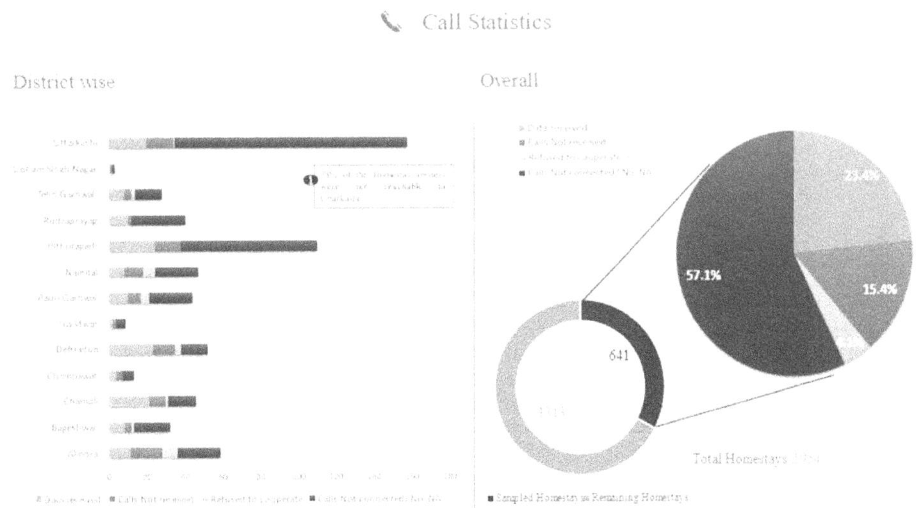

The table below details the data regarding employment loss based on the sample homestay in each of the 13 districts. We estimated the total number of people employed, both permanent and seasonal, by dividing the average expense per month (peak season and off-season) for all homestays by the average salary of employees. This was further adjusted according to the correction factor, and the total number of people employed turned out to be 2164 (Table 1). This number was used to calculate the total expense.

Figure 6. Socioeconomic details of average homestay of Uttrakhand

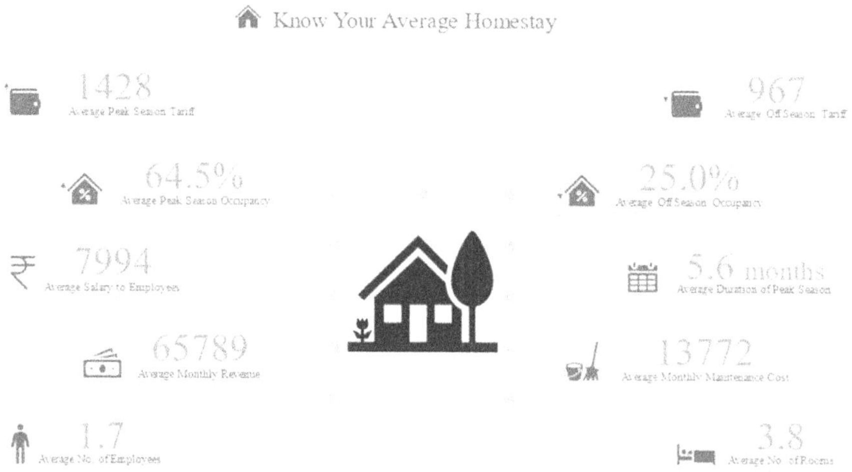

Based on the above variables, the total revenue loss assessed was INR7,63,69,282.95/- per month, the total expense calculated was INR 1,59,89,753.66 per month, and the number of unemployed people was estimated as 2164. Thus, the total monthly loss computed for homestay was 60 million INR, estimated to be 720 million INR annually. See Figure 7 for the district-wise division of revenue and expense.

Table 1. Calculation of the total people employed

District	No. of homestays	Average Expense per Month per Homestay (Peak)	Average Expense per Month per Homestay (Off-Season)	Peak season duration in months	Average Expense per Month (Peak)	Average Expense per Month (Off-Season)	Average salary per Month	No. of employed people (Total)	No. of employed people (Permanent)	No. of employed people (Adjusted)
Almora	128	15,181.80	454.54	5	19,43,270.40	58,181.12	7000	278	8	228
Bageshwar	56	6,400.00	-	3	3,58,400.00	-	6500	55	-	31
Chamoli	320	25,342.86	19,572.73	3	81,09,714.29	62,63,272.73	9000	901	696	609
Champawat	16	1,400.00	-	0	22,400.00	-	1500	15	-	9
Dehradun	397	19,065.20	13,173.90	10	75,68,884.40	52,30,038.30	9500	797	551	625
Haridwar	9	3,500.00	2,500.00	4	31,500.00	22,500.00	2500	13	9	9
Pauri Garhwal	75	21,400.00	1,242.86	3	16,05,000.00	93,214.29	6500	247	14	125
Nainital	240	10,178.57	8,275.00	6	24,42,857.14	19,86,000.00	9500	257	209	159
Pithoragarh	230	6,380.21	755.21	3	14,67,447.92	1,73,697.92	6500	226	27	82
Rudraprayag	112	6,250.00	1,250.00	3	7,00,000.00	1,40,000.00	6500	108	22	33
Tehri Garhwal	133	7,444.44	5,888.89	3	9,90,111.11	7,83,222.22	4000	248	196	197
Udham Singh Nagar	6	25,000.00	-	6	1,50,000.00	-	7500	20	-	16
Uttarkashi	232	4,078.57	-	6	9,46,228.57	-	6500	146	-	42
Total people employed										**2,164**

Figure 7. District-wise division of revenue and expense

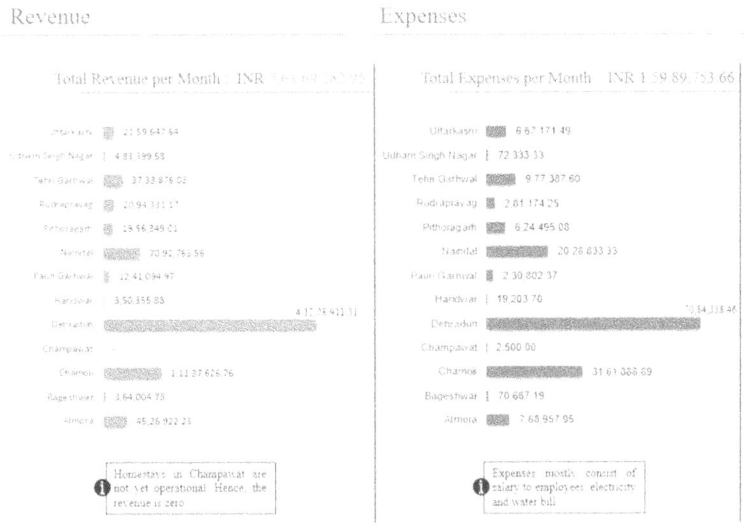

5. RESILIENT STRATEGIES ADOPTED BY HOMESTAY OWNERS

Uttarakhand's tourism sector was severely affected by the COVID-19 pandemic, and homestay businesses were especially distressed. New challenges emerged with the homestay business resurrection. The semi-structured interviews with homestay owners helped us to understand the strategies adopted post covid for revival. Six themes were identified based on the eleven interviews under resilience strategies applied by homestay businesses.

- **Untact hospitality:** Homestay owners realized that due to the pandemic, customer expectations have changed. Post-pandemic guests want less interaction and more isolation, covered in a concept known as "Untact hospitality." Tidy, functionable property and quality tangibles (Farmaki et al., 2020; Moon et al., 2019) demand for host friendliness, responsiveness, and hospitableness (Gunter, 2018; Xie and Mao, 2017) were always a key prerequisite for guest satisfaction (Priporas et al., 2017). However, post-COVID-19, the nature of hosting changed, as keeping a distance is the key to enhancing guest experience. Untact Hospitality as a phenomenon already exists in marketing and customer service literature as Gen Z is unwilling to establish unwanted contacts in consuming services in a digital era (Buhalis et al., 2019; Chadee, Ren, and Tang, 2021; de Kervenoael, Hasan, Schwob, and Goh, 2020). But now homestay owners recognize that the guests want more space and less intrusion to self-protect and feel safe (Ghadcri et al., 2022).
- **Upgrading homestay services with digital technologies:** Homestay owners used COVID-19 as downtime to upgrade their properties to suit specific audiences. Services were being modified with different segments of travelers kept in mind. For digital nomads, who prefer to work from anywhere with WiFi connectivity, WiFi was upgraded to ensure high-speed connections as work-from-home has evolved into work-from-any-home. Homestay owners added pet-friendly services for pet parents. Homestay owners made infrastructural improvements to serve the revenge travel-

ers, such as increasing closet sizes, improving kitchen equipment, and implementing tech tools that limit human contact, such as keyless entry solutions, mobile payments, F&B smart menu. These changes helped them attract new audiences, thus resulting in an inflow of revenue sooner than expected.

- **Digitally transforming booking policies:** To revive the business, some homestay owners simplified the booking process by tying up with booking websites. Unlike hotels, few owners had reported that host practices were often doubted in terms of fairness. Gaining trust over time for any medium enterprise is an uphill task; the pandemic further pushed hosts to revise the terms and make conditions more flexible. The process of homestay booking and cancellation policies were made flexible with a superior website and apps. Travelers were free to cancel their reservation at the very last moment without a penalty fee, given varying lockdowns and city restrictions. Digitally aligning platforms with hotel policies made travelers more confident about booking. It also attracted families and large groups to plan long-duration trips to compensate for canceled celebrations.
- **Sustainability with legal registration:** Before COVID, many homestays were opened mostly to misuse government grants. The negative impact of this disorganized growth was seen with COVID-19, as there was a lack of contingency plans for the sector's recovery. However, post-COVID-19, homestay owners reorganized their businesses with the help of technological transformations. This further helped them take advantage of the government schemes and generate sufficient income for themselves and the locals. As part of resilience strategies, many homestays got legal registration, which helped them access funds for service recovery by local authorities. This helped them sustain their businesses longer in times of crisis. Post covid, legal registration of all the homestays has given access to the funds and promotion. After gaining legal registration, homestays have an organized structure which has further aided them in working collaboratively with the tourism board. This has ushered growth due to awareness of schemes and aids by the government, such as formulating credit policies for homestays region-wise and a performance-based incentive scheme for the workforce. Using local resources and workforce was not just budget-friendly but also prevented unnecessary migration to other areas for jobs.
- **Supporting intrapreneurship:** Homestay owners reported making special efforts to keep staff morale positive with an open and effective communication system. The motivation of the frontline employees made them loyal to the organization and improved their effectiveness at work, as they felt valued in the workplace. Thus, minimum terminations were done, and temporary paid leaves were offered. Involving more locals as full-time and part-time employees brought cohesiveness to the business. Training programs were developed to enhance the preparedness of those working in the sector, increase awareness, and improve understanding of risks. These trainings promoted capacity building and ultimately changed attitudes and behaviors concerning security. Employees' protection was ensured by giving them proper gear, and they were trained to make the guests feel safe.
- **Follow the social entrepreneurship model:** To withstand COVID-19 difficulties, some homestay owners designed a social entrepreneurship model. They decided to work with the community by organizing guided tours and arranging local tour guides to let guests experience local scenery and culture, promoting artwork, exhibiting music and dance shows, and food festivals, and organizing haat/melas to develop a destination by involving the community. Technology and digitalization have helped make these community-based activities possible with ease. These pursuits were

propagated through various digital platforms portraying homestays as experiential products with particular emphasis on meditation or returning to nature. These efforts got more business and proved valuable.

Figure 8. Resilience strategies adopted by homestays

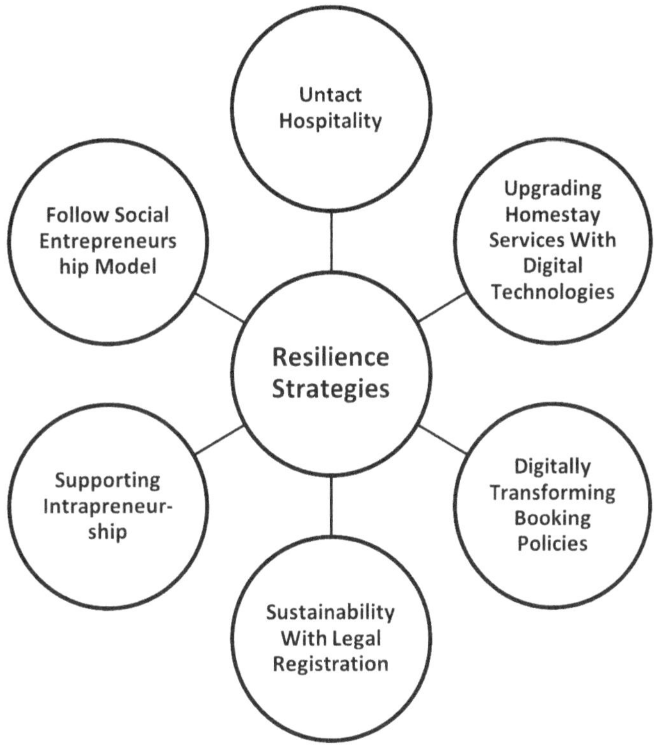

6. IMPLICATIONS

The COVID-19 epidemic wreaked havoc on the world's travel, tourist, and hospitality sectors, resulting in severe socioeconomic disruption worldwide (Bae and Chang, 2021; UNWTO, 2020). Its impact was seen in the Indian tourism sector as well and, especially, on homestays of Uttarakhand, which was the key focus of this study. Homestays in Uttarakhand are still at an infant stage, as they are small hospitality units notably managed by individual families. These units suffered huge losses due to the pandemic, resulting in severe socioeconomic disruption. Despite its apparent susceptibility to the effects of the pandemic (Hassanli et al., 2016; Lynch and MacWhannell, 2000), this sector has been neglected academically. Since much mainstream research has focused on large-scale hospitality sectors

In this research, we estimated the revenue loss incurred by homestay owners of 13 districts of the Uttarakhand region. Since the sector is unorganized and reliable data is unavailable, we used novel techniques and unique variables for estimation. We found that the total monthly loss computed for homestay was 60 million INR, which accounted for 720 million INR annually. We also provided district-wise data for the region. We also particularized the process and variables used in the study, which could be used in other emerging economies that lack organized tourism-related data. The Uttarakhand government also used this data to formulate mitigation strategies to overcome the loss. It helped them design schemes to support the registered homestay owners in making a forward plan and guidelines to overcome the loss during the pandemic. Studies in the past have estimated loss due to COVID-19; however, no study has estimated revenue loss for homestays using the variables in this study.

Along with estimating revenue loss, we also tried to uncover resilience strategies adopted by homestay owners to recover their businesses. We found six themes from the eleven interviews conducted with these homestay hosts. These themes were based on different tactics and strategies to recover their businesses during and post-COVID. We named these themes untact hospitality, upgrading homestay services with digital technologies, digitally transforming booking policies, offering sustainability to businesses with legal registration, supporting intrapreneurship, and following the social entrepreneurship model. Studies related to resilience have captured guest and travelers' behaviors, while homestay owners and host perspectives have not been accounted for (Farmaki et al. 2020). This study contributes to crisis management by evaluating the impact of the pandemic on the tourism sector in the state of Uttarakhand and thereafter elaborating on adaptation and recovery strategies essential for businesses to survive crises and disasters.

This study makes a novel contribution as it will help researchers and academicians estimate revenue loss due to any crisis. This study contributes to the evolving literature on assessing the impact of the COVID-19 pandemic on the development of the tourism sector. Also, the findings of this work can be used in advancing tourism resilience literature. Practitioners and policymakers can use the outcomes to promote homestays in India to address the increasing tourist accommodation deficit and stimulate the rural economy. The sector needs a collaborative and competitive culture to usher it to the next level. Policymakers need to re-evaluate the existing guidelines and improve them further concerning homestay operations while focusing on promoting the host-guest relationship with the new normal solitary concept. This study will also help homestay hosts improve their businesses since rural confinement leads to a lack of travel trade skills among homestay owners. These results could help homestay owners, guesthouse owners, small lodges, restaurant owners and other MSMEs in different regions formulate a forward plan and guidelines to overcome the loss during the pandemic.

7. LIMITATIONS AND FUTURE CHALLENGES

There are limitations and assumptions which should be taken into account before generalizing the results. There were intrinsic challenges in finding authentic data since some initial respondents exaggerated their loss. Thus, we used many nonconventional variables to estimate loss. Despite the inherent limitations, these assumptions have been made for estimating loss due to COVID-19. Such as, homestays registered till March 2018 have been considered to be active, the correction factor is calculated to get the number of active homestays, peak season has been suitably assumed based on discussion with homestay owners, and secondary research, utility bills are for the homestay and not for the household. Only the homestay owners' perspective is taken into account. Other stakeholders should be included in future studies.

Despite the resilience strategies adopted by homestay owners to revive their business in Uttrakhand, there are still challenges. These challenges can arise within the home-stay operators and host community or beyond the control of the host community or home-stay operators. Some internal challenges are financial assistance, sustainability training, and capacity building. Some external challenges are cross-cultural conflicts, land degradation, green certification, and improper solid waste management, which can hinder sustainable tourism development. Due to the increase in construction across hilly regions and unbalanced topography, there are multiple climate change issues. Although technology and internet connectivity have improved, external parties can always exploit data. Cyber crimes are increasing manifold, and with artificial intelligence, things have gone beyond the control of an average person. The essence of simplicity is lost due to over-commercialization and unaccounted-for development. The cultural value of each destination needs to be well-kept and restored by communities. Although progress has been made in the establishment and operation of homestays in many parts of the world, especially after the pandemic, there is still much room for progress and development.

REFERENCES

Bae, S. Y., & Chang, P. J. (2021). The effect of coronavirus disease-19 (COVID-19) risk perception on behavioural intention towards 'untact'tourism in South Korea during the first wave of the pandemic (March 2020). *Current Issues in Tourism*, *24*(7), 1017–1035. doi:10.1080/13683500.2020.1798895

Braun, V., & Clarke, V. (2006). Using thematic analysis in psychology. *Qualitative Research in Psychology*, *3*(2), 77–101. doi:10.1191/1478088706qp063oa

Bresciani, S., Ferraris, A., Santoro, G., Premazzi, K., Quaglia, R., Yahiaoui, D., & Viglia, G. (2021). The seven lives of Airbnb. The role of accommodation types. *Annals of Tourism Research*, *88*, 103170. doi:10.1016/j.annals.2021.103170

Brinkmann, S., & Kvale, S. (2015). *Interviews: Learning the craft of qualitative research interviewing*. Sage publications.

Bryman, A. (2016). *Social research methods*. Oxford university press.

Buhalis, D., Harwood, T., Bogicevic, V., Viglia, G., Beldona, S., & Hofacker, C. (2019). Technological disruptions in services: Lessons from tourism and hospitality. *Journal of Service Management*, *30*(4), 484–506. doi:10.1108/JOSM-12-2018-0398

Centeno, R. S., & Marquez, J. P. (2020). How much did the tourism industry lost? Estimating earning loss of tourism in the Philippines. *arXiv preprint arXiv:2004.09952*.

Chhabra, S. (2022). Why Does Privacy Paradox Exist?: A Qualitative Inquiry to Understand the Reasons for Privacy Paradox Among Smartphone Users. Journal of Electronic Commerce in Organizations (JECO), 20(1), 1-20.

Chadee, D., Ren, S., & Tang, G. (2021). Is digital technology the magic bullet for performing work at home? Lessons learned for post COVID-19 recovery in hospitality management. *International Journal of Hospitality Management*, *92*, 102718. doi:10.1016/j.ijhm.2020.102718 PMID:33071425

Creswell, J. W., & Tashakkori, A. (2007). Differing perspectives on mixed methods research. *Journal of Mixed Methods Research*, *1*(4), 303–308. doi:10.1177/1558689807306132

Dahles, H., & Susilowati, T. P. (2015). Business resilience in times of growth and crisis. *Annals of Tourism Research*, *51*, 34–50. doi:10.1016/j.annals.2015.01.002

de Kervenoael, R., Hasan, R., Schwob, A., & Goh, E. (2020). Leveraging human-robot interaction in hospitality services: Incorporating the role of perceived value, empathy, and information sharing into visitors' intentions to use social robots. *Tourism Management*, *78*, 104042. doi:10.1016/j.tourman.2019.104042

Di Domenico, M., & Lynch, P. A. (2007). Host/guest encounters in the commercial home. *Leisure Studies*, *26*(3), 321–338. doi:10.1080/02614360600898110

Eichelberger, S., & Peters, M. (2021). Family firm management in turbulent times: opportunities for responsible tourism. In *Resiliency Models and Addressing Future Risks for Family Firms in the Tourism Industry* (pp. 103–124). IGI Global. doi:10.4018/978-1-7998-7352-5.ch005

Farmaki, A., Miguel, C., Hadjielia, M., Aleksi, A., Ceh, A., & Efthymiadou, F. (2020). Impacts of Covid-19 on peer-to-peer accommodation platforms: Host perceptions and responses. *International Journal of Hospitality Management*, *91*(September), 102663. doi:10.1016/j.ijhm.2020.102663 PMID:32901166

Ghaderi, Z., Butler, R., & Béal, L. (2022). Exploring home-based accommodation operators' responses to Covid-19: Implications of untact hospitality adoption. *Tourism Management Perspectives*, *43*, 100979. doi:10.1016/j.tmp.2022.100979 PMID:35706982

Ghaderi, Z., King, B., & Hall, C. M. (2022). Crisis preparedness of hospitality managers: Evidence from Malaysia. *Journal of Hospitality and Tourism Insights*, *5*(2), 292–310. doi:10.1108/JHTI-10-2020-0199

Giousmpasoglou, C., Marinakou, E., & Zopiatis, A. (2021). Hospitality managers in turbulent times: The COVID-19 crisis. *International Journal of Contemporary Hospitality Management*, *33*(4), 1297–1318. doi:10.1108/IJCHM-07-2020-0741

Gossling, ̈. S., Scott, D., & Hall, C. M. (2020). Pandemics, tourism and global change: A rapid assessment of COVID-19. *Journal of Sustainable Tourism*, *29*(1), 1–20. doi:10.1080/09669582.2020.1758708

Günay, F., Bayraktaroğlu, E., & Özkul, K. (2020). Assessing the short-term impacts of COVID-19 pandemic on foreign visitor's demand for Turkey: A scenario analysis. *Journal of Ekonomi*, *2*(2), 80–85.

Gunter, U. (2018). What makes an Airbnb host a superhost? Empirical evidence from San Francisco and the Bay Area. *Tourism Management*, *66*, 26–37. doi:10.1016/j.tourman.2017.11.003

Hassanli, N., Gross, M. J., & Brown, G. (2016). The emergence of home-based accommodations in Iran: A study of self-organization. *Tourism Management*, *54*, 284–295. doi:10.1016/j.tourman.2015.11.018

Hinson, G., & Slade, R. (2011). Forty hard-won business continuity lessons from the 2011 earthquakes in New Zealand and Japan. *EDPACS*, *44*(3), 1–18. doi:10.1080/07366981.2011.608316

Hoang, T. G., Truong, N. T., & Nguyen, T. M. (2021). The survival of hotels during the COVID-19 pandemic: A critical case study in Vietnam. *Service Business*, *15*(2), 209–229. doi:10.1007/s11628-021-00441-0

Hussain, A. (2021). A future of tourism industry: Conscious travel, destination recovery and regenerative tourism. *Journal of Sustainability and Resilience*, *1*(1), 5.

Kaján, E., & Saarinen, J. (2013). Tourism, climate change and adaptation: A review. *Current Issues in Tourism*, *16*(2), 167–195. doi:10.1080/13683500.2013.774323

Kayat, K. (2010). The homestay program in Malaysia and its implications for the future. *Tourism research in Malaysia: What, which way and so what*, 293-316.

King, C., Iba, W., & Clifton, J. (2021). Reimagining resilience: COVID-19 and marine tourism in Indonesia. *Current Issues in Tourism*, *24*(19), 2784–2800. doi:10.1080/13683500.2021.1873920

Leh, F. C., & Hamzah, M. R. (2012). Homestay tourism and pro-poor tourism strategy in Banghuris Selangor, Malaysia. *Elixir International Journal*, *45*, 7602–7610.

Lynch, P. A. (2005). Sociological impressionism in a hospitality context. *Annals of Tourism Research*, *32*(3), 527–548. doi:10.1016/j.annals.2004.09.005

Lynch, P. A., & MacWhannell, D. (2000). Home and conceptualizing hospitality. *In search of hospitality: Theoretical perspectives and debates*, 1965-1982.

Malhotra, N. K., & Dash, S. J. M. R. (2010). An applied orientation. *Marketing Research, 2*.

Marco-Lajara, B., Úbeda-García, M., Ruiz-Fernández, L., Poveda-Pareja, E., & Sánchez-García, E. (2022). Rural hotel resilience during COVID-19: The crucial role of CSR. *Current Issues in Tourism*, *25*(7), 1121–1135. doi:10.1080/13683500.2021.2005551

Mayson, S., & Barrett, R. (2017). A new argument using embeddedness and sensemaking to explain small firms' responses to employment regulation. *Human Resource Management Journal*, *27*(1), 189–202. doi:10.1111/1748-8583.12141

McCartney, G. (2021). The impact of the coronavirus outbreak on Macao. From tourism lockdown to tourism recovery. *Current Issues in Tourism*, *24*(19), 2683–2692. doi:10.1080/13683500.2020.1762549

Mehta, K., & Jha, S. S. (2020). COVID-19: a nightmare for the Indian economy. *UGC Care Journal, 31*(20).

Ministry of Tourism. (2020). *Annual Report 2019-20*. Government of India. Available at: https://tourism.gov.in/annual-reports/annual-report-2019-20

Moon, H., & Han, H. (2019). Tourist experience quality and loyalty to an island destination: The moderating impact of destination image. *Journal of Travel & Tourism Marketing*, *36*(1), 43–59. doi:10.1080/10548408.2018.1494083

Nikos, N. (2022, January 2). Tourism and Hospitality industry resilience during the Covid-19 pandemic: Evidence from England. *Current Issues in Tourism*, *25*(1), 46–59. Advance online publication. doi:10.1080/13683500.2021.1883556

Njoroge, J. M. (2015). Climate change and tourism adaptation: Literature review. *Tourism and Hospitality Management*, *21*(1), 95–108. doi:10.20867/thm.21.1.7

Pappas, N. (2021). COVID19: Holiday intentions during a pandemic. *Tourism Management*, *84*, 104287. Advance online publication. doi:10.1016/j.tourman.2021.104287 PMID:33518882

Pramana Paramartha, Ermawan, Deli, & Srimulyani. (2021). Impact of COVID-19 pandemic on tourism in Indonesia. *Current Issues*.

Prayag, G. (2018). Symbiotic relationship or not? Understanding resilience and crisis management in tourism. *Tourism Management Perspectives*, *25*, 133–135. doi:10.1016/j.tmp.2017.11.012

Prayag, G. (2020). Time for reset? Covid-19 and tourism resilience. *Tourism Review International*, *24*(2/3), 179–184. doi:10.3727/154427220X15926147793595

Priporas, C. V., Stylos, N., Vedanthachari, L. N., & Santiwatana, P. (2017). Service quality, satisfaction, and customer loyalty in Airbnb accommodation in Thailand. *International Journal of Tourism Research*, *19*(6), 693–704. doi:10.1002/jtr.2141

Priyadarshini, I. (2020). *A Survey on some of the Global Effects of the COVID-19 Pandemic*. Academic Press.

Sarkar, S., & Clegg, S. R. (2021). Resilience in a time of contagion: Lessons from small businesses during the COVID-19 pandemic. *Journal of Change Management*, *21*(2), 242–267. doi:10.1080/14697017.2021.1917495

Sharma, G. D., Thomas, A., & Paul, J. (2021). Reviving tourism industry post-COVID-19: A resilience-based framework. *Tourism Management Perspectives*, *37*, 100786. doi:10.1016/j.tmp.2020.100786 PMID:33391988

Sigala, M. (2017). Collaborative commerce in tourism: Implications for research and industry. *Current Issues in Tourism*, *20*(4), 346–355. doi:10.1080/13683500.2014.982522

Stone, L. S., Stone, M. T., Mogomotsi, P. K., & Mogomotsi, G. E. (2021). The impacts of Covid-19 on nature-based tourism in Botswana: Implications for community development. *Tourism Review International*, *25*(2-3), 263–278. doi:10.3727/154427221X16098837279958

Su, L., Tang, B., & Nawijn, J. (2021). How tourism activity shapes travel experience sharing: Tourist well-being and social context. *Annals of Tourism Research*, *91*, 103316. doi:10.1016/j.annals.2021.103316

Ugur, N. G., & Akbıyık, A. (2020). Impacts of COVID-19 on global tourism industry: A cross-regional comparison. *Tourism Management Perspectives*, *36*, 189–194. doi:10.1016/j.tmp.2020.100744 PMID:32923356

UNWTO. (2020). *Impact assessment of the COVID-19 outbreak on international tourism*. www.unwto.org/impact-assessment-of-the-covid-19-outbreakon-international-tourism

Wen, J., Aston, J., Liu, X., & Ying, T. (2020). Effects of misleading media coverage on public health crisis: A case of the 2019 novel coronavirus outbreak in China. *Anatolia*, *31*(2), 331–336. doi:10.1080/13032917.2020.1730621

Worldometer. (2020). *COVID-19 coronavirus / cases*. Retrieved January 3, 2021, from https://www.worldometers.info/coronavirus/coronavirus-cases/#case-distribution

Xie, K., & Mao, Z. (2017). The impacts of quality and quantity attributes of Airbnb hosts on listing performance. *International Journal of Contemporary Hospitality Management*, *29*(9), 2240–2260. doi:10.1108/IJCHM-07-2016-0345

Yong, K., & Hassan, R. A. (2019). The relationships between social media marketing and entrepreneurial success: A conceptual study on homestay business in Sabah, Malaysia. *Review of Integrative Business and Economics Research*, *8*, 25–33.

Respondent	Designation	Number of Years in Business	Number of Rooms	Tourist Type	Native of Uttarakhand
1	Owner	8	6	Domestic and Foreign	Yes
2	Manager	5	4	Domestic and Foreign	No
3	Manager	4	3	Domestic	Yes
4	Owner	11	6	Domestic and Foreign	Yes
5	Owner	2	2	Domestic	Yes
6	Owner	6	4	Domestic and Foreign	Yes
7	Owner	9	5	Domestic and Foreign	Yes
8	Manager	7	4	Domestic and Foreign	Yes
9	Manager	6	3	Domestic and Foreign	No
10	Owner	8	4	Domestic and Foreign	Yes
11	Owner	7	4	Domestic and Foreign	Yes

APPENDIX

Interview Protocol

1) How did the COVID-19 pandemic affect the business?
2) What changes were made in the business strategy and tactics to adapt to the pandemic?
3) Which shift in strategy/tactics benefitted you the most?
4) Has the business recovered completely?
5) Any innovative idea/parameter that worked effectively for you?

Respondent Profile

Compilation of References

Aall, C. (2014). Sustainable tourism in practice: Promoting or perverting the quest for a sustainable development? *Sustainability (Basel)*, *6*(5), 2562–2583. doi:10.3390/su6052562

Abdel Maksoud, W. (2023). *Opportunities and challenges of applying "Block Chain" Technology in the Tourism Industry* (Master Thesis, Faculty of Tourism and Hotels, University of Sadat City).

Abdellah, M. (2023). The impact of blockchain technology on the efficiency and effectiveness of tourism services. *Journal of the Faculty of Tourism and Hotels-University of Sadat City*, *7*(1/2), 27–48.

Abduljabbar, R., Liyanage, S., & Dia, H. (2021). The role of micro-mobility in shaping sustainable cities: A systematic literature review. *Transportation Research Part D, Transport and Environment*, *92*, 102734. doi:10.1016/j.trd.2021.102734

Acharya, S., Mekker, M., & De Vos, J. (2023). Linking travel behavior and tourism literature : Investigating the impacts of travel satisfaction on destination satisfaction and revisit intention. *Transportation Research Interdisciplinary Perspectives*, *17*, 100745. doi:10.1016/j.trip.2022.100745

Ackermann, A. E., & Sellitto, M. A. (2022). Méthodes de prévision de la demande: Une revue de la littérature. *Innovar (Universidad Nacional de Colombia)*, *32*(85), 83–99. doi:10.15446/innovar.v32n85.100979

Adachi, Y., Takahashi, S., & Nakano, Y. (2020). The effect of virtual reality on emotional experience and cognitive assessment: A comparison of head-mounted displays and desktop 360° videos. *Computers in Human Behavior*, *103*, 106202.

Adongo, C. A., Taale, F., & Adam, I. (2018). Tourists' values and empathic attitude toward sustainable development in tourism. *Ecological Economics*, *150*, 251–263. doi:10.1016/j.ecolecon.2018.04.013

Agrawal, R., Wankhede, V. A., Kumar, A., Luthra, S., & Huisingh, D. (2022). Big data analytics and sustainable tourism : A comprehensive review and network based analysis for potential future research. *International Journal of Information Management Data Insights*, *2*(2), 100122. doi:10.1016/j.jjimei.2022.100122

Aguirre Montero, A., & López-Sánchez, J. A. (2021). Intersection of data science and smart destinations: A systematic review. *Frontiers in Psychology*, *12*, 712610. doi:10.3389/fpsyg.2021.712610 PMID:34393952

Agyeiwaah, E., McKercher, B., & Suntikul, W. (2017). Identifying core indicators of sustainable tourism: A path forward? *Tourism Management Perspectives*, *24*, 26–33. doi:10.1016/j.tmp.2017.07.005

Ahlert, G. (2006). Hosting the FIFA World Cup™ Germany 2006: Macroeconomic and regional economic impacts. *Journal of Sport Management*, *20*(1), 1–22.

Ahn, T., Ekinci, Y., & Li, G. (2013). Self-congruence, functional congruence, and destination choice. *Journal of Business Research*, *66*(6), 719–723. doi:10.1016/j.jbusres.2011.09.009

Ajzen, I. (1991). The theory of planned behaviours. *Organizational Behavior and Human Decision Processes*, *50*(2), 179–211. doi:10.1016/0749-5978(91)90020-T

Alahakoon, D., Nawaratne, R., Xu, Y., De Silva, D., Sivarajah, U., & Gupta, B. (2020). Self-Building Artificial Intelligence and Machine Learning to Empower Big Data Analytics in Smart Cities. *Information Systems Frontiers*. Advance online publication. doi:10.1007/s10796-020-10056-x

Alami, Y. H., Khazzan, B. E., & Souab, M. (2017). Heritage and cultural tourism in Fes (Morocco). *International Journal of Scientific Management and Tourism*, *3*(2). https://ojs.scientificmanagementjournal.com/ojs/index.php/smj/article/view/293

AlBanna, B., Sakr, M., Moussa, S., & Moawad, I. (2016). Interest Aware Location-Based Recommender System Using Geo-Tagged Social Media. *ISPRS International Journal of Geo-Information*, *5*(12), 245. Advance online publication. doi:10.3390/ijgi5120245

Alghizzawi, M., Habes, M., Al Assuli, A., & Ezmigna, A. (2023). Digital Marketing and Sustainable Businesses: As Mobile Apps in Tourism. In *Artificial Intelligence and Transforming Digital Marketing*. Springer Nature Switzerland.

AlKhader, W., Jayaraman, R., Salah, K., Sleptchenko, A., Antony, J., & Omar, M. (2023). Leveraging blockchain and NFTs for quality 4.0 implementation in digital manufacturing. *Journal of Manufacturing Technology Management*, *34*(7), 1208–1234. doi:10.1108/JMTM-05-2023-0172

Alla, L., Hmioui, A., & Taghzouti, A. (2018). Impact de la création de valeur pour le client sur la performance des entreprises touristiques marocaines: Etude exploratoire des pratiques des hotels de Fès. *Les Cahiers du CEDIMES*, 75.

Alla, L., Hmioui, A., & Taghzouti, A. (n.d.). Impact de la creation de valeur pour le client sur la performance des entreprises touristiques marocaines : Etude exploratoire des pratiques des hotels de fes. *Les Cahiers Du Cedimes*, 75.

AllaL.KamalM.BouhtatiN. (2022). Big data et efficacité marketing des entreprises touristiques : Une revue de littérature. *Alternatives Managériales Economiques, 4*(0). doi:10.48374/IMIST.PRSM/ame-v1i0.36928

Allam, Z., & Newman, P. (2018). Redefining the smart city: Culture, metabolism and governance. *Smart Cities*, *1*(1), 4–25. doi:10.3390/smartcities1010002

Al-Muezzin, M. (2008). *Principles of Marketing* (Vol. 1). Dar Althaqafa.

Al-Nafjan, A., Alrashoudi, N., & Alrasheed, H. (2022). Recommendation System Algorithms on Location-Based Social Networks: Comparative Study. *Information (Basel)*, *13*(4), 188. Advance online publication. doi:10.3390/info13040188

Alnajim, R., & Fakieh, B. (2023). A Tourist-Based Framework for Developing Digital Marketing for Small and Medium-Sized Enterprises in the Tourism Sector in Saudi Arabia. *Data*, *8*(12), 179. Advance online publication. doi:10.3390/data8120179

Al-Romeedy, B. (2024). AI and HRM in Tourism and Hospitality in Egypt: Inevitability, Impact and Future. In HRM, Artificial Intelligence and the Future of Work: Critical Insights from the Global South. Palgrave Macmillan.

Álvarez-García, J., González-Vázquez, E., Del Río-Rama, M. D. L. C., & Durán-Sánchez, A. (2019). Quality in customer service and its relationship with satisfaction: An innovation and competitiveness tool in sport and health centers. *International Journal of Environmental Research and Public Health*, *16*(20), 3942. doi:10.3390/ijerph16203942 PMID:31623254

Amalia, J., & Hanika, I. (2021). Tourism in Digital Era : The Influence of Digital Marketing on the Intention to Visit Museum. *Jurnal Komunikasi-Malaysian Journal of Communication*, *37*(3), 274–287. doi:10.17576/JKMJC-2021-3703-16

Amirkhanyan, A., & Meinel, C. (2016). Analysis of the Value of Public Geotagged Data from Twitter from the Perspective of Providing Situational Awareness. In. Lecture Notes in Computer Science: Vol. 9844. *Including Subseries Lecture Notes in Artificial Intelligence and Lecture Notes in Bioinformatics).* LNCS. doi:10.1007/978-3-319-45234-0_48

Andrews, D. C. (2002). Audience-specific online community design. *Communications of the ACM, 45*(4), 64–68. doi:10.1145/505248.505275

Ante, L. (2021). Smart contracts on the blockchain–A bibliometric analysis and review. *Telematics and Informatics, 57*, 101519. doi:10.1016/j.tele.2020.101519

Arachchi, H. D. M., & Samarasinghe, G. D. (2023). Impact of embedded AI mobile smart speech recognition on consumer attitudes towards AI and purchase intention across Generations X and Y. *European Journal of Management Studies.*

Arasli, H., Abdullahi, M., & Gunay, T. (2021). Social media as a destination marketing tool for a sustainable heritage festival in Nigeria: A moderated mediation study. *Sustainability (Basel), 13*(11), 6191. doi:10.3390/su13116191

Avraham, E., & Ketter, E. (2016). *Tourism marketing for developing countries: Battling stereotypes and crises in Asia, Africa and the Middle East.* Springer. doi:10.1057/9781137342157

Azman, N., Albattat, A., & Valeri, M. (2023). The Importance of Big Data Analysis: Developing Neuromarketing in Tourism Industry. In Tourism Innovation in the Digital Era: Big Data, AI and Technological Transformation (pp. 1-24). Emerald Publishing Limited.

Azuma, R. T. (1997). A Survey of Augmented Reality. *Presence (Cambridge, Mass.), 6*(4), 355–385. doi:10.1162/pres.1997.6.4.355

Bachouche, H., & Sabri, O. (2022). The impact of consumers' status-participant or nonparticipant-on the evaluation of empowerment strategies. *Recherche et Applications en Marketing, 37*(3), 149–175. doi:10.1177/20515707221118811

Bae, S. Y., & Chang, P. J. (2021). The effect of coronavirus disease-19 (COVID-19) risk perception on behavioural intention towards 'untact'tourism in South Korea during the first wave of the pandemic (March 2020). *Current Issues in Tourism, 24*(7), 1017–1035. doi:10.1080/13683500.2020.1798895

Baggio, R., & Valeri, M. (2022). Network science and sustainable performance of family businesses in tourism. *Journal of Family Business Management, 12*(2), 200–213. doi:10.1108/JFBM-06-2020-0048

Baik, J. (2020). The Geotagging Counterpublic: The Case of Facebook Remote Check-Ins to Standing Rock. *International Journal of Communication*, 4.

Baker, M. A., Davis, E. A., & Weaver, P. A. (2014). Eco-friendly attitudes, barriers to participation, and differences in behaviours at green hotels. *Cornell Hospitality Quarterly, 55*(1), 89–99. doi:10.1177/1938965513504483

Balasubramanian, K., Kunasekaran, P., Konar, R., & Sakkthivel, A. M. (2022). Integration of augmented reality (AR) and virtual reality (VR) as marketing communications channels in the hospitality and tourism service sector. In Marketing Communications and Brand Development in Emerging Markets Volume II: Insights for a Changing World (pp. 55-79). Cham: Springer International Publishing.

Balasubramanian, K., & Konar, R. (2022). Moving Forward with Augmented Reality Menu: Changes in Food Consumption Behaviour Patterns. *Asia-Pacific Journal of Innovation in Hospitality and Tourism, 11*(3), 91–96.

Balatska, N., Radkevych, L., Robul, Y., Vdovichena, O., & Strenkovska, A. (2022). Digital technologies and digital marketing: New opportunities for tourism and hospitality business. *Financial and Credit Activity-Problems of Theory and Practice, 6*(47), 424–432. doi:10.55643/fcaptp.6.47.2022.3924

Baloch, Q. B., Shah, S. N., Iqbal, N., Sheeraz, M., Asadullah, M., Mahar, S., & Khan, A. U. (2023). Impact of tourism development upon environmental sustainability: A suggested framework for sustainable ecotourism. *Environmental Science and Pollution Research International*, *30*(3), 5917–5930. doi:10.1007/s11356-022-22496-w PMID:35984561

Baloch, Q. B., Shah, S. N., Maher, S., Irshad, M., Khan, A. U., Kiran, S., & Shah, S. S. (2022). Determinants of evolving responsible tourism behavior : Evidences from supply chain. *Cogent Social Sciences*, *8*(1), 2099565. doi:10.1080/23311886.2022.2099565

Baltaci, A. (2021). Strategic Digital Marketing and Sustainability. In *Financial Ecosystem and Strategy in the Digital Era: Global Approaches and New Opportunities* (pp. 237–264). Springer International Publishing. doi:10.1007/978-3-030-72624-9_11

Banafa, A. (2022). *Blockchain technology and applications*. CRC Press. doi:10.1201/9781003337393

Băndoi, A., Jianu, E., Enescu, M., Axinte, G., Tudor, S., & Firoiu, D. (2020). The relationship between development of tourism, quality of life and sustainable performance in EU countries. *Sustainability (Basel)*, *12*(4), 1628. doi:10.3390/su12041628

Bank, A. D. (2021). *Big data for better tourism policy, management, and sustainable recovery from COVID-19*. Asian Development Bank.

Bano, M., & Siddiqui, S. (2022). Examining the role of perceived ease of use, perceived usefulness, and social influence on students' acceptance of learning management systems in higher education. *International Journal of Educational Technology in Higher Education*, *19*(1), 1–17. PMID:35013716

Bao, J., Zheng, Y., Wilkie, D., & Mokbel, M. (2015). Recommendations in Location-Based Social Networks: A Survey. *GeoInformatica*, *19*(3), 525–565. Advance online publication. doi:10.1007/s10707-014-0220-8

Baralla, G., Pinna, A., Tonelli, R., Marchesi, M., & Ibba, S. (2021). Ensuring transparency and traceability of food local products: A blockchain application to a Smart Tourism Region. *Concurrency and Computation*, *33*(1), e5857. doi:10.1002/cpe.5857

Barbeau, S. J., Winters, P. L., Georggi, N. L., Labrador, M. A., & Perez, R. (2010). Travel Assistance Device: Utilising Global Positioning System-Enabled Mobile Phones to Aid Transit Riders with Special Needs. IET Intelligent Transport Systems, 4. doi:10.1049/iet-its.2009.0028

Baruca, P. Z., & Civre, Z. (2012). How do guests choose a hotel. *Academica Turistica*, *5*(1), 75–84.

Başer, G. D. M. K. (2020). Digital marketing practices: Trends and challenges for the tourism industry. *Journal of Tourism and Hospitality Management*, *11*(1), 1–13. https://dergipark.org.tr/en/download/article-file/1120498

Bashir, I. (2017). *Mastering blockchain*. Packt Publishing Ltd.

Bassano, C., Barile, S., Piciocchi, P., Spohrer, J., Iandolo, F., & Fisk, R. (2019). Storytelling about places: Tourism marketing in the digital age. *Cities (London, England)*, *87*, 10–20. doi:10.1016/j.cities.2018.12.025

Bassano, C., Piciocchi, P., & Pietronudo, M. C. (2018). Managing value co-creation in consumer service systems within smart retail settings. *Journal of Retailing and Consumer Services*, *45*, 190–197. doi:10.1016/j.jretconser.2018.09.008

Basso, S., Campagnolo, D., & Castaldo, S. (2023). Sustainable design of digital technologies: A framework for integrating user needs and environmental responsibility. *Journal of Cleaner Production*, *390*, 133222. https://www.mdpi.com/2071-1050/15/21

Bauer, C., & Strauss, C. (2016). Location-Based Advertising on Mobile Devices: A Literature Review and Analysis. *Management Review Quarterly*, *66*(3), 159–194. Advance online publication. doi:10.1007/s11301-015-0118-z

Bayarsaikhan, T., Kim, S. T., & Gim, T. H. T. (2020). International tourists' destination choice differences according to Plog's personality types: Analyzing the case of Mongolia based on the recreation opportunity spectrum. *International Journal of Urban Sciences*, *24*(4), 485–515. doi:10.1080/12265934.2020.1771195

Bec, A., Moyle, B., Schaffer, V., & Timms, K. (2021). Virtual reality and mixed reality for second chance tourism. *Tourism Management*, *83*, 104256. doi:10.1016/j.tourman.2020.104256

Beirman, D. (2020). *Restoring tourism destinations in crisis: A strategic marketing approach*. Routledge. doi:10.4324/9781003117148

Bekele, M. K., Pierdicca, R., Frontoni, E., Malinverni, E. S., & Gain, J. (2018). A Survey of Augmented, Virtual, and Mixed Reality for Cultural Heritage. *Journal on Computing and Cultural Heritage*, *11*(2), 1–36. doi:10.1145/3145534

Belas, J., Jan, D., Hlawiczka, R., Smrcka, L., & Khan, K. A. (2024). SMEs sustainability: The role of human resource management, corporate social responsibility and financial management. *Oeconomia Copernicana*, *15*(1), 307–342. doi:10.24136/oc.2937

Bell, S., & Morse, S. (2008). *Sustainability indicators Measuring the immeasurable?* (2nd ed.). Earthscan Publication for Sustainable Future.

Benbba, B., Ismaili, F. E. A., & El Archi, Y. (2022). Smart tourism destination et développement durable: Quels apports à l'expérience touristique? *Alternatives Managériales Economiques, 4*, 1-18. https://revues.imist.ma/index.php/AME/article/view/36926

BenbbaB.IsmailiF. E. A.ArchiY. E. (2022). Smart tourism destination et développement durable : Quels apports à l'expérience touristique? *Alternatives Managériales Economiques, 4*(0), Article 0. doi:10.48374/IMIST.PRSM/ame-v1i0.36926

Bernard, M., & Parker, L. (2021). *The effect of conscious consumerism on purchasing behaviours: The example of greenwashing in the cosmetics industry*. Academic Press.

Bernini, C., Cracolici, M. F., & Nijkamp, P. (2017). Place-based attributes and spatial expenditure behavior in tourism. *Journal of Regional Science*, *57*(2), 218–244. doi:10.1111/jors.12308

Bhaskar, P., & Sharma, K. (2022). A Critical Insight into the Role of Artificial Intelligence (AI) in Tourism and Hospitality Industries. *A Refereed Monthly. International Journal of Management*, *15*, 76–85.

Bhuiyan, M., Darda, M. A., & Hasan, M. R. (2021). Tourist perception and satisfaction on safari tourism at Bangabandhu Sheikh Mujib Safari Park in Bangladesh. *International Journal of Geoheritage and Parks*, *9*(4), 430–440. doi:10.1016/j.ijgeop.2021.11.005

Bhushan, B., Sinha, P., Sagayam, K., & Andrew, J. (2021). Untangling blockchain technology: A survey on state of the art, security threats, privacy services, applications and future research directions. *Computers & Electrical Engineering*, *90*, 106897. doi:10.1016/j.compeleceng.2020.106897

Bhushan, N., & Rai, K. (2004). Strategic decision-making. In *Strategic Decision Making: Applying the Analytic Hierarchy Process* (pp. 3–10). Springer London. doi:10.1007/978-1-85233-864-0_1

Bibri, S. E. (2018). The IoT for smart sustainable cities of the future: An analytical framework for sensor-based big data applications for environmental sustainability. *Sustainable Cities and Society*, *38*, 230–253. doi:10.1016/j.scs.2017.12.034

Bibri, S. E. (2019a). Big Data Science and Analytics for Smart Sustainable Urbanism. In *Advances in Science, Technology & Innovation*. Springer International Publishing. doi:10.1007/978-3-030-17312-8

Bibri, S. E. (2019b). On the sustainability of smart and smarter cities in the era of big data: An interdisciplinary and transdisciplinary literature review. *Journal of Big Data*, *6*(1), 25. Advance online publication. doi:10.1186/s40537-019-0182-7

Bibri, S. E., & Krogstie, J. (2017). The core enabling technologies of big data analytics and context-aware computing for smart sustainable cities: A review and synthesis. *Journal of Big Data*, *4*(1), 38. Advance online publication. doi:10.1186/s40537-017-0091-6

Bieger, T., & Laesser, C. (2004). Information sources for travel decisions: Toward a source process model. *Journal of Travel Research*, *42*(4), 357–371. doi:10.1177/0047287504263030

Binns, T., & Nel, E. (2002). Tourism as a local development strategy in South Africa. *The Geographical Journal*, *168*(3), 235–247. doi:10.1111/1475-4959.00051

Bissonnette, M. M., & Contento, I. R. (2001). Adolescents' perspectives and food choice behaviours in terms of the environmental impacts of food production practices: Application of a psychosocial model. *Journal of Nutrition Education*, *33*(2), 72–82. doi:10.1016/S1499-4046(06)60170-X PMID:12031187

Biswas, B., & Gupta, R. (2019). Analysis of barriers to implement blockchain in industry and service sectors. *Computers & Industrial Engineering*, *136*, 225–241. doi:10.1016/j.cie.2019.07.005

Blanco-Moreno, S., González-Fernández, A. M., & Muñoz-Gallego, P. A. (2023). Big data in tourism marketing: past research and future opportunities. *Spanish Journal of Marketing-ESIC.*

Bodkhe, U., Bhattacharya, P., Tanwar, S., Tyagi, S., Kumar, N., & Obaidat, M. (2019). BloHosT: Blockchain enabled smart tourism and hospitality management. In 2019 international conference on computer, information and telecommunication systems (CITS) (pp. 1-5). IEEE.

Bodkhe, U., Bhattacharya, P., Tanwar, S., Tyagi, S., Kumar, N., & Obaidat, M. S. (2019). BloHosT : Blockchain enabled smart tourism and hospitality management. *2019 international conference on computer, information and telecommunication systems (CITS)*, 1-5. https://ieeexplore.ieee.org/abstract/document/8862001/

Bodkhe, U., Tanwar, S., Parekh, K., Khanpara, P., Tyagi, S., Kumar, N., & Alazab, M. (2020). Blockchain for industry 4.0: A comprehensive review. *IEEE Access : Practical Innovations, Open Solutions*, *8*, 79764–79800. doi:10.1109/ACCESS.2020.2988579

Bogicevic, V., Matic, D., & Kovacevic, V. (2019). The role of virtual reality in destination image formation. *Tourism Management Perspectives*, *32*, 100579.

Bogoch, I. I., Watts, A., Thomas-Bachli, A., Huber, C., Kraemer, M. U. G., & Khan, K. (2020). Pneumonia of Unknown Aetiology in Wuhan, China: Potential for International Spread via Commercial Air Travel. *Journal of Travel Medicine*, *27*(2), taaa008. doi:10.1093/jtm/taaa008 PMID:31943059

Bonink, C. (1992). *Cultural tourism development and government policy* (Unpublished ma dissertation). Rijksuniversiteit, Utrecht, The Netherlands.

Booking (n.d.). https://www.booking.com/

Borysova, O., Gryniuk, O., Mykhailenko, T., Shparaha, T., & Hryniuk, T. (2022). Global Trends in Tourism and Imperatives of the Circular Economy in the Context of Sustainable Environmental Development of States. *Revue d'Economie Financiere*, *20*(1), 196–202. Advance online publication. doi:10.55365/1923.x2022.20.23

Boto-Garcia, D., Mayor, M., & De la Vega, P. (2021). Spatial price mimicking on Airbnb: Multi-host vs single-host. *Tourism Management*, *87*, 104365. doi:10.1016/j.tourman.2021.104365

Bouaouinate, A. (2008). *Les acteurs locaux du tourisme de désert au Maroc: Cas de l'erg Chebbi et de Zagora-M'hamid*. https://www.academia.edu/89480928/Les_acteurs_locaux_du_tourisme_de_d%C3%A9sert_au_Maroc_Cas_de_lerg_Chebbi_et_de_Zagora_Mhamid

Bouaouinate, A. (2016). *La mise en tourisme des espaces oasiens du Maroc d'un tourisme de masse à un tourisme alternatif Rapport de Synthèse des Travaux scientifiques et pédagogiques* [Thesis, Université Hassan II Casablanca; Faculté des Lettres et des Sciences Humaines Mohammedia]. https://theses.hal.science/tel-01353766

Boukil, M. (2023). La mise en tourisme de la médina de Fès: le cas de la réhabilitation et transformation des Riads en maisons d'hôtes. *Journal of urban and territorial planning* (N°18, p149-164), Democratic Arabic Center http://democraticac.de/?page_id=61723

Boukil, M., Aarab.H, Serghini. B., Chaouki, M. (2021) Le tourisme international à Fès est-il vers un nouvel essor ? Revue territoire et développement numéro 9-10, Faculté des Lettres et des sciences Humaines Dhar el Mahraz Fès.

Boulhila, S., Alouat, M., Rezzaz, M. A., & Schmitz, S. (2022). Towards a development model of local cultural tourism through the involvement of local actors (Province of Constantine, Algeria). *Geo Journal of Tourism and Geosites*, *40*(1), 9–19. doi:10.30892/gtg.40101-797

Boulos, M. N., Anastasiou, A., Bekiaris, E., & Panou, M. (2011). Geo-Enabled Technologies for Independent Living: Examples from Four European Projects. *Technology and Disability*, *23*(1), 7–17. Advance online publication. doi:10.3233/TAD-2011-0300

Bouraga, S. (2021). A taxonomy of blockchain consensus protocols: A survey and classification framework. *Expert Systems with Applications*, *168*, 114384. doi:10.1016/j.eswa.2020.114384

Boyle, P., & Haggerty, K. (2009). Spectacular security: Mega-events and the security complex. *International Political Sociology*, *3*(3), 257–274. doi:10.1111/j.1749-5687.2009.00075.x

Brabham, D. C. (2008). Crowdsourcing as a Model for Problem Solving: An Introduction and Cases. *Convergence (London)*, *14*(1), 75–90. Advance online publication. doi:10.1177/1354856507084420

Braun, V., & Clarke, V. (2006). Using thematic analysis in psychology. *Qualitative Research in Psychology*, *3*(2), 77–101. doi:10.1191/1478088706qp063oa

Bresciani, S., Ferraris, A., Santoro, G., Premazzi, K., Quaglia, R., Yahiaoui, D., & Viglia, G. (2021). The seven lives of Airbnb. The role of accommodation types. *Annals of Tourism Research*, *88*, 103170. doi:10.1016/j.annals.2021.103170

Bricker, K. S., Black, R., & Cottrell, S. (Eds.). (2013). *Sustainable tourism & the millennium development goals: Effecting positive change*. Jones & Bartlett Publishers.

Brinkmann, S., & Kvale, S. (2015). *Interviews: Learning the craft of qualitative research interviewing*. Sage publications.

Brodie, R. J., Ilic, A., Juric, B., & Hollebeek, L. (2013). Consumer engagement in a virtual brand community: An exploratory analysis. *Journal of Business Research*, *66*(1), 105–114. doi:10.1016/j.jbusres.2011.07.029

Browning, V., So, K. K. F., & Sparks, B. (2013). The influence of online reviews on consumers' attributions of service quality and control for service standards in hotels. *Journal of Travel & Tourism Marketing*, *30*(1-2), 23–40. doi:10.1080/10548408.2013.750971

Bryman, A. (2016). *Social research methods*. Oxford university press.

Bucur, C. (2015). Using Opinion Mining Techniques in Tourism. *Procedia Economics and Finance*, *23*, 1666–1673. doi:10.1016/S2212-5671(15)00471-2

Budeanu, A., Miller, G., Moscardo, G., & Ooi, C. S. (2016). Sustainable tourism, progress, challenges and opportunities: An introduction. *Journal of Cleaner Production*, *111*, 285–294. doi:10.1016/j.jclepro.2015.10.027

Buhalis, D., & Amaranggana, A. (2014). Smart tourism destinations. In Z. Xiang & I. Tussyadiah (Eds.), *Information and communication technologies in tourism 2014* (pp. 553–564). Springer.

Buhalis, D., & Amaranggana, A. (2015). Smart Tourism Destinations Enhancing Tourism Experience Through Personalisation of Services. In I. Tussyadiah & A. Inversini (Eds.), *Information and Communication Technologies in Tourism 2015* (pp. 377–389). Springer International Publishing. doi:10.1007/978-3-319-14343-9_28

Buhalis, D., Harwood, T., Bogicevic, V., Viglia, G., Beldona, S., & Hofacker, C. (2019). Technological disruptions in services: Lessons from tourism and hospitality. *Journal of Service Management*, *30*(4), 484–506. doi:10.1108/JOSM-12-2018-0398

Buhalis, D., & Law, R. (2008). Progress in information technology and tourism management: 20 years on and 10 years after the Internet—The state of eTourism research. *Tourism Management*, *29*(4), 609–623. doi:10.1016/j.tourman.2008.01.005

Buhalis, D., & O'Connor, P. (2005). Role of information technology in tourism. *Tourism Management*, *26*(6), 999–1019.

Buhalis, D., O'Connor, P., & Leung, R. (2023). Smart hospitality: From smart cities and smart tourism towards agile business ecosystems in networked destinations. *International Journal of Contemporary Hospitality Management*, *35*(1), 369–393. doi:10.1108/IJCHM-04-2022-0497

Bulchand-Gidumal, J., & Melián-González, S. (2023). Fighting fake reviews with blockchain-enabled consumer-generated reviews. *Current Issues in Tourism*. Advance online publication. doi:10.1080/13683500.2023.2173054

Bulchand-Gidumal, J., Secin, E. W., O'Connor, P., & Buhalis, D. (2023). Artificial Intelligence's impact on hospitality and tourism marketing: Exploring key themes and addressing challenges. *Current Issues in Tourism*, 1–18. doi:10.1080/13683500.2023.2229480

Buonomano, A., Del Papa, G., Giuzio, G., Palombo, A., & Russo, G. (2023). Future pathways for decarbonization and energy efficiency of ports: Modelling and optimization as sustainable energy hubs. *Journal of Cleaner Production*, *420*, 138389. doi:10.1016/j.jclepro.2023.138389

Burguete, J. L., Hidalgo-Alcazár, C., & Lanero, A. (2020). Turismo sostenible y emprendimiento digital: Una aproximación a los nuevos perfiles de turistas. *Economía Industrial*, *417*, 59–68.

Butler, R. W. (1993). Tourism – An Evolutionary Perspective. In Tourism and Sustainable Development: Monitoring, Planing, Managing. University of Waterloo Press.

Butler, R. W. (1999). Sustainable tourism: A state-of-the-art review. *Tourism Geographies*, *1*(1), 7–25. doi:10.1080/14616689908721291

Caia, Y., Lib, G., Wena, L., & Liua, C. (2023). *Intellectual Landscape and Emerging Trends of Big Data Research in Tourism and Hospitality: A Scientometric Analysis*. Academic Press.

Cai, L. (2002). An examination of destination image formation model. *Tourism Management*, *23*(5), 537–545.

Camilleri, M. (2014). Advancing the sustainable tourism agenda through strategic CSR perspectives. *Tourism Planning & Development*, *11*(1), 42–56. doi:10.1080/21568316.2013.839470

Cao, K., Li, D., & Xie, L. (2019). Bearing-ratio-of-distance rigidity theory with application to directly similar formation control. *Automatica*, *109*, 108540. doi:10.1016/j.automatica.2019.108540

Cao, S., Johnson, H., & Tulloch, A. (2023b). Exploring blockchain-based traceability for food supply chain sustainability: Towards a better way of sustainability communication with consumers. *Procedia Computer Science*, *217*, 1437–1445. doi:10.1016/j.procs.2022.12.342

Cao, S., Xu, H., & Bryceson, K. (2023a). Blockchain Traceability for Sustainability Communication in Food Supply Chains: An Architectural Framework, Design Pathway and Considerations. *Sustainability (Basel)*, *15*(18), 13486. doi:10.3390/su151813486

Caratù, M., Cherubino, P., Menicocci, S., & Martinez-Levy, A. (2023). Does sustainable communication have an impact on international social media audiences? A neuromarketing explorative study between Finland and Italy. *Italian Journal of Marketing*. doi:10.1007/s43039-023-00086-z

Cardoso, L., Dias, F., de Araújo, A. F., & Marques, M. I. A. (2019). A destination imagery processing model: Structural differences between dream and favourite destinations. *Annals of Tourism Research*, *74*, 81–94. doi:10.1016/j.annals.2018.11.001

Carlisle, S., Ivanov, S., & Dijkmans, C. (2023). The digital skills divide : Evidence from the European tourism industry. *Journal of Tourism Futures, 9*(2), 240-266. Advance online publication. doi:10.1108/JTF-07-2020-0114

Carneiro, T., Picoto, W. N., & Pinto, I. (2023). Big Data Analytics and Firm Performance in the Hotel Sector. *Tourism and Hospitality*, *4*(2), 244–256. doi:10.3390/tourhosp4020015

Carrigan, M., & Attalla, A. (2001). The myth of the ethical consumer–do ethics matter in purchase behaviour? *Journal of Consumer Marketing*, *18*(7), 560–578. doi:10.1108/07363760110410263

Cater, E., & Goodall, B. (1992). Must Tourism Destroy Its Resource Base? In A. M. Mannion & S. R. Bowlby (Eds.), *Environmental Issues in the 1990s* (pp. 309–324). Wiley.

Centeno, R. S., & Marquez, J. P. (2020). How much did the tourism industry lost? Estimating earning loss of tourism in the Philippines. *arXiv preprint arXiv:2004.09952*.

Centobelli, P., Cerchione, R., Del Vecchio, P., Oropallo, E., & Secundo, G. (2022). Blockchain technology for bridging trust, traceability and transparency in circular supply chain. *Information & Management*, *59*(7), 103508. doi:10.1016/j.im.2021.103508

Chadee, D., Ren, S., & Tang, G. (2021). Is digital technology the magic bullet for performing work at home? Lessons learned for post COVID-19 recovery in hospitality management. *International Journal of Hospitality Management*, *92*, 102718. doi:10.1016/j.ijhm.2020.102718 PMID:33071425

Chandra, S., & Verma, S. (2023). Big data and sustainable consumption: A review and research agenda. *Vision (Basel)*, *27*(1), 11–23. doi:10.1177/09722629211022520 PMID:36810315

Chang, J. Y. S., Konar, R., Cheah, J. H., & Lim, X. J. (2024). Does privacy still matter in smart technology experience? A conditional mediation analysis. *Journal of Marketing Analytics*, *12*(1), 71–86. doi:10.1057/s41270-023-00240-8

Chang, V. (2021). An ethical framework for big data and smart cities. *Technological Forecasting and Social Change*, *165*(2), 120559. doi:10.1016/j.techfore.2020.120559

Chan, J., Marzuki, K., & Mohtar, T. (2021). Local community participation and responsible tourism practices in ecotourism destination: A case of lower Kinabatangan, Sabah. *Sustainability (Basel)*, *13*(23), 13302. doi:10.3390/su132313302

Chaturvedi, R., Verma, S., Ali, F., & Kumar, S. (2023). Reshaping Tourist Experience with AI-Enabled Technologies: A Comprehensive Review and Future Research Agenda. *International Journal of Human-Computer Interaction*, 1–17. doi:10.1080/10447318.2023.2238353

Chebli, A., Othmani, M. C., & Said, F. B. (2020). Market segmentation in urban tourism: Exploring the influence of personal factors on tourists' perception. *Journal of Tourism and Services*, *11*(20), 74–108. doi:10.29036/jots.v11i20.144

Chen, C. F., Wang, Y. M., & Li, S. (2022). Can virtual reality experiences promote sustainable tourism behavior? An empirical examination. *Journal of Sustainable Tourism*, *30*(8), 2202–2223.

Chen, D. (2018). Utility of the blockchain for climate mitigation. *The Journal of The British Blockchain Association*, *1*(1), 75–80. doi:10.31585/jbba-1-1-(6)2018

Cheng, M., Mackenzie, S. H., & Degarege, G. A. (2022). Airbnb impacts on host communities in a tourism destination: An exploratory study of stakeholder perspectives in Queenstown, New Zealand. In *Platform-mediated tourism* (pp. 194–212). Routledge. doi:10.4324/9781003230618-11

Cheong, F. H. (1995). Virtual reality: A new window on the world for tourism. *Tourism Management*, *16*(7), 551–558.

Chettiparamb, A., & Kokkranikal, J. (2012). Responsible tourism and sustainability: The case of Kumarakom in Kerala, India. *Journal of Policy Research in Tourism, Leisure & Events*, *4*(3), 302–326. doi:10.1080/19407963.2012.711088

Chhabra, S. (2022). Why Does Privacy Paradox Exist?: A Qualitative Inquiry to Understand the Reasons for Privacy Paradox Among Smartphone Users. Journal of Electronic Commerce in Organizations (JECO), 20(1), 1-20.

Chhabra, D. (2010). *Sustainable marketing of cultural and heritage tourism*. Routledge. doi:10.4324/9780203855416

Choi, J., Lee, H., & Kim, J. (2022). Understanding the acceptance of virtual reality tourism: A technology acceptance model perspective. *Journal of Hospitality & Tourism Research (Washington, D.C.)*, *46*(1), 1–23.

Chua, A., Servillo, L., Marcheggiani, E., & Vande Moere, A. (2016). Mapping Cilento: Using Geotagged Social Media Data to Characterize Tourist Flows in Southern Italy. *Tourism Management*, *57*, 295–310. Advance online publication. doi:10.1016/j.tourman.2016.06.013

Chuang, C.-M. (2023). The conceptualization of smart tourism service platforms on tourist value co-creation behaviours : An integrative perspective of smart tourism services. *Humanities & Social Sciences Communications*, *10*(1), 1–16. doi:10.1057/s41599-023-01867-9

Ciegis, R., Ramanauskiene, J., & Martinkus, B. (2009). The concept of sustainable development and its use for sustainability scenarios. *The Engineering Economist*, *62*(2), 28–37.

Cinnamon, J., & Gaffney, A. (2022). Do-it-yourself Street Views and the urban imaginary of Google Street View. *Journal of Urban Technology*, *29*(3), 95–116. doi:10.1080/10630732.2021.1910467

Close, P., Askew, D., & Xin, X. (2006). *The Beijing Olympiad: The political economy of a sporting megaevent*. Routledge. doi:10.4324/9780203003015

Coaffee, J., & Wood, S. M. (2006). Security is coming home: Rethinking scale and constructing resilience in the global urban response to terrorist risk. *International Relations*, *20*(4), 503–517. doi:10.1177/0047117806069416

Cobanoglu, C., Terrah, A., Hsu, M. J., Corte, V. D., & Gaudio, G. D. (2022). A systematic review of big data: Research approaches and future prospects. *Journal of Smart Tourism*, *2*(1), 21–31.

Cooper, C., Fletcher, J., Gilbert, D., & Wanhill, R. S. (2008). *Tourism principles and practice* (4th ed.). Longman Publishing.

Cooper, C., & Macneil, S. (2005). *Tourism: Principles and practice* (3rd ed.). Pearson Education Limited.

Coroş, M., Gică, O., Yallop, A., & Moisescu, O. (2017). Innovative and sustainable tourism strategies: A viable alternative for Romania's economic development. *Worldwide Hospitality and Tourism Themes*, *9*(5), 504–515. doi:10.1108/WHATT-07-2017-0033

Corrales, M., Fenwick, M., & Haapio, H. (2019). *Legal tech, smart contracts and blockchain*. Springer. doi:10.1007/978-981-13-6086-2

Cousin, S., Chareyron, G., & Jacquot, S. (2017). Big data and tourism. In L. L. Lowry (Ed.), *The SAGE international encyclopedia of travel and tourism*. Sage publications.

Cranmer, E. E., tom Dieck, M. C., & Fountoulaki, P. (2020). Exploring the Value of Augmented Reality for Tourism. *Tourism Management Perspectives*, *35*, 100672. Advance online publication. doi:10.1016/j.tmp.2020.100672

Cranmer, E. E., tom Dieck, M. C., & Jung, T. (2023). The role of augmented reality for sustainable development : Evidence from cultural heritage tourism. *Tourism Management Perspectives*, *49*, 101196. doi:10.1016/j.tmp.2023.101196

Creevey, D., Kidney, E., & Mehta, G. (2019). From dreaming to believing: A review of consumer engagement behaviours with brands' social media content across the holiday travel process. *Journal of Travel & Tourism Marketing*, *36*(6), 679–691. doi:10.1080/10548408.2019.1624242

Creswell, J. W., & Tashakkori, A. (2007). Differing perspectives on mixed methods research. *Journal of Mixed Methods Research*, *1*(4), 303–308. doi:10.1177/1558689807306132

Crouch, G. I., Huybers, T., & Oppewal, H. (2016). *On the Choice of Tourism Destination versus Tourism Experience: Insights from an Analysis of Past Choice and Future Interest*. Academic Press.

Cunha, C. R., Morais, E. P., & Martins, C. A. (2018). Can we use big data analytics to leverage tourism in rural tourism destinations? In *Proceedings of the 32nd International Business Information Management Association Conference*. International Business Information Management Association (IBIMA).

Cuomo, M. T., Tortora, D., Foroudi, P., Giordano, A., Festa, G., & Metallo, G. (2021). Digital transformation and tourist experience co-design : Big social data for planning cultural tourism. *Technological Forecasting and Social Change*, *162*, 120345. doi:10.1016/j.techfore.2020.120345

D'Alessandro, F. (2016). Green Building for a Green Tourism. A new model of eco-friendly agritourism. *Agriculture and Agricultural Science Procedia*, *8*, 201–210. doi:10.1016/j.aaspro.2016.02.094

D'Arco, M., Presti, L. L., Marino, V., & Maggiore, G. (2021). Is sustainable tourism a goal that came true? The Italian experience of the Cilento and Vallo di Diano National Park. *Land Use Policy*, *101*, 105198. doi:10.1016/j.landusepol.2020.105198

D'Eusanio, M., & Zamagni, A., & Petti, L. (2019). Social sustainability and supply chain management : Methods and tools. *Journal of Cleaner Production*, *235*, 178-189. Advance online publication. doi:10.1016/j.jclepro.2019.06.323

da Costa Liberato, P. M., Alén-González, E., & de Azevedo Liberato, D. F. V. (2018). Digital technology in a smart tourist destination: The case of Porto. *Journal of Urban Technology*, *25*(1), 75–97. doi:10.1080/10630732.2017.1413228

Dahles, H., & Susilowati, T. P. (2015). Business resilience in times of growth and crisis. *Annals of Tourism Research*, *51*, 34–50. doi:10.1016/j.annals.2015.01.002

Dai, Y., & Wang, T. (2021). Prediction of customer engagement behaviour response to marketing posts based on machine learning. *Connection Science*, *33*(4), 891–910. doi:10.1080/09540091.2021.1912710

Davis, F. D., Bagozzi, R. P., & Warshaw, P. R. (1989). User acceptance of computer technology: A comparison of two theoretical models. *Management Science*, *35*(8), 982–1003. doi:10.1287/mnsc.35.8.982

De Carlo, M., Ferilli, G., d'Angella, F., & Buscema, M. (2020). Artificial Intelligence to design collaborative strategy: An application to urban destinations. *Journal of Business Research*. Advance online publication. doi:10.1016/j.jbusres.2020.09.013

de Kervenoael, R., Hasan, R., Schwob, A., & Goh, E. (2020). Leveraging human-robot interaction in hospitality services: Incorporating the role of perceived value, empathy, and information sharing into visitors' intentions to use social robots. *Tourism Management*, *78*, 104042. doi:10.1016/j.tourman.2019.104042

De Oliveira, P. S. G., Wada, E. K., Lopes, A. S., & da Silva, L. F. (2023). *Machine learning and sentiment analysis to assess the evolution of the COVID-19 pandemic and the impacts on tourism.* Anais Brasileiros de Estudos Turísticos.

Decrop, A. (2010). Destination choice sets: An inductive longitudinal approach. *Annals of Tourism Research*, *37*(1), 93–115. doi:10.1016/j.annals.2009.08.002

Del Vecchio, P., Mele, G., Ndou, V., & Secundo, G. (2018). Open Innovation and Social Big Data for Sustainability: Evidence from the Tourism Industry. *Sustainability (Basel)*, *10*(9), 3215. doi:10.3390/su10093215

Della Corte, V., & Del Gaudio, G. (2022). A Systematic Review of Big Data: Research Approaches and Future Prospects. *Journal of Smart Tourism*.

DeLone, W. H., & McLean, E. R. (2003). The DeLone and McLean model of information systems success: A ten-year update. *Journal of Management Information Systems*, *19*(4), 9–37. doi:10.1080/07421222.2003.11045748

Devinney, T. M., Auger, P., & Eckhardt, G. M. (2010). *The myth of the ethical consumer hardback with DVD*. Cambridge University Press.

Devkota, N., Gajdka, K., Siwakoti, R., Klímová, M., & Dhakal, K. (2023). Promoting Sustainable Tourist Behavior through Promotional Marketing. *Journal of Tourism and Services*, *14*(26), 219–241. doi:10.29036/jots.v14i26.512

Dewailly, J. M. (1999). Virtual reality for disability and rehabilitation applications. *Journal of Neuroengineering and Rehabilitation*, *4*(1), 1–8.

Dharel, Y. (2020). Online reviews and their impact in tourism business. Centria University of Applied Sciences.

Di Domenico, M., & Lynch, P. A. (2007). Host/guest encounters in the commercial home. *Leisure Studies*, *26*(3), 321–338. doi:10.1080/02614360600898110

Di Vaio, A., & Varriale, L. (2020). Blockchain technology in supply chain management for sustainable performance: Evidence from the airport industry. *International Journal of Information Management*, *52*, 102014. doi:10.1016/j.ijinfomgt.2019.09.010

Dickinger, A., Scharl, A., Stern, H., Weichselbraun, A., & Wöber, K. (2008). Acquisition and Relevance of Geotagged Information in Tourism. In Information and Communication Technologies in Tourism 2008. doi:10.1007/978-3-211-77280-5_48

Dimitrios. (2021). The Use of Big Data in Tourism: Current Trends and Directions for Future Research. Diverses institutions incluant l'University of Thessaly, Arab Open University - Kuwait, Panteion University of Social and Political Sciences, et le Metropolitan College.

Dionysis, S. (2023). *Blockchain and distributed ledger technologies for supply chain traceability: industry considerations and consumer preferences* (Doctoral dissertation, University of Nottingham).

Divisekera, S. (2010). Economics of tourist's consumption behaviour: Some evidence from Australia. *Tourism Management, 31*(5), 629–636. doi:10.1016/j.tourman.2009.07.001

Dodds, R., & Graci, S. (2012). *Sustainable tourism in island destinations*. Routledge. doi:10.4324/9781849776660

Doe, J., & Hinson, R. (2023). AI-driven sustainability brand activism for family businesses: a future-proofing perspective article. *Journal of Family Business Management*. doi:10.1108/JFBM-10-2023-0217

Donthu, N., Kumar, S., Mukherjee, D., Pandey, N., & Lim, W. M. (2021). How to conduct a bibliometric analysis : An overview and guidelines. *Journal of Business Research, 133*, 285–296. doi:10.1016/j.jbusres.2021.04.070

Douglas, A., Hoogendoorn, G., & Richards, G. (2024). Activities as the critical link between motivation and destination choice in cultural tourism. *Journal of Hospitality and Tourism Insights, 7*(1), 249–271. doi:10.1108/JHTI-09-2022-0442

Doyle, J., & Kelliher, F. (2023). Bringing the past to life : Co-creating tourism experiences in historic house tourist attractions. *Tourism Management, 94*, 104656. doi:10.1016/j.tourman.2022.104656

Dreamstime. (n.d.). https://www.dreamstime.com/social-media-background-network-people-connecting-modern-technology-devices-image108829611

Drezner, J. A., Drezner, S. M., Magner, K. N., & Ayala, J. T. (2021). COVID-19 Surveillance in Youth Soccer during Small Group Training: A Safe Return to Sports Activity. *Sports Health, 13*(1), 15–17. doi:10.1177/1941738120964458 PMID:32986534

Dubihlela, J., & Ngxukumeshe, T. (2016). Eco-friendly retail product attributes, customer attributes and the repurchase intentions of South African consumers. *The International Business & Economics Research Journal, 15*(4), 163–174. doi:10.19030/iber.v15i4.9754

Durif, F., François-Lecompte, A., & Prim-Allaz, I. (2017). Pratiques de tourisme durable: proposition d'une typologie croisée avec les comportements de consommation. *Téoros. Revue de recherche en tourisme, 36*(1).

Durmaz, Y., & Başer, M. Y. (2023). *A Systematic Literature Review on Artificial İntelligence Applications in Tourism Marketing*. Academic Press.

Dutta, P., Choi, T., Somani, S., & Butala, R. (2020). Blockchain technology in supply chain operations: Applications, challenges and research opportunities. *Transportation research part e: Logistics and transportation review, 142*, 102067.

Edgell, D. (2019). *Managing sustainable tourism: A legacy for the future*. Routledge.

Edwards, D. (2012). Using GPS to Track Tourists Spatial Behaviour in Urban Destinations. SSRN *Electronic Journal*. doi:10.2139/ssrn.1905286

Eichelberger, S., & Peters, M. (2021). Family firm management in turbulent times: opportunities for responsible tourism. In *Resiliency Models and Addressing Future Risks for Family Firms in the Tourism Industry* (pp. 103–124). IGI Global. doi:10.4018/978-1-7998-7352-5.ch005

El Archi, Y., & Benbba, B. (2022). Ecotourisme et développement durable: Étude empirique sur la perception des con sommateurs. *Revue Internationale des Sciences de Gestion, 5*(2).

El Archi, Y., & Benbba, B. (2023). The Applications of Technology Acceptance Models in Tourism and Hospitality Research: A Systematic Literature Review. *Journal of Environmental Management & Tourism, 14*(2), 379–391. doi:10.14505/jemt.v14.2(66).08

El Archi, Y., Benbba, B., Kabil, M., & Dávid, L. D. (2023a). Digital Technologies for Sustainable Tourism Destinations: State of the Art and Research Agenda. *Administrative Sciences, 13*(8), 184. doi:10.3390/admsci13080184

El Archi, Y., Benbba, B., Nizamatdinova, Z., Issakov, Y., Vargáné, G. I., & Dávid, L. D. (2023b). Systematic Literature Review Analysing Smart Tourism Destinations in Context of Sustainable Development: Current Applications and Future Directions. *Sustainability (Basel)*, *15*(6), 5086. doi:10.3390/su15065086

El Archi, Y., Benbba, B., Zhu, K., El Andaloussi, Z., Pataki, L., & Dávid, L. D. (2023). Mapping the Nexus between Sustainability and Digitalization in Tourist Destinations: A Bibliometric Analysis. *Sustainability (Basel)*, *15*(12), 9717. doi:10.3390/su15129717

El Jaouhari, S., & Lhoussaine, A. L. L. A. (2021). Approche participative en tourisme et gouvernance territoriale, quel apport pour le développement territorial? *Alternatives Managériales Economiques*, *3*(2), 257–277.

El Ouarrak, Y., Hmioui, A., Rajouani, B., Alla, L., & Benjelloun, A. (2024). A Systematic Review of the Relationship Between Supply Chain Risk Management and Sustainability. In A. Iglesias, J. Shin, B. Patel, and A. Joshi (Éds.), *Proceedings of World Conference on Information Systems for Business Management* (p. 41-52). Springer Nature. 10.1007/978-981-99-8349-0_5

El País. (2023). *Bono Cultural Joven: qué es, cómo solicitarlo y en qué tipo de eventos se puede utilizar.* Disponible online: https://elpais.com/escaparate/2023-08-02/todo-sobre-el-bono-cultural-joven.html

El, Youssef, Özgöker, & Ben Slimene. (2021). Big Data for Entrepreneurship and Sustainable Development. CRC Press.

El-Khadrawy, R. K., Attia, A. A., Khalifa, M., & Rashed, R. (2020). Sustainable Tourism and Culture: A Symbiotic Relationship. *International Journal of Environmental Science & Sustainable Development*, *5*(1), 54–67. doi:10.21625/essd.v5i1.717

El-Sherbiny, M., & Al-Romeedy, B. (2023). Does loyalty moderate the impact of digital marketing on the purchase intention of customers in airlines? *Journal of the Faculty of Tourism and Hotels-University of Sadat City*, *7*(2), 1–16.

Emami, A., & Ranjbarian, B. (2019). The perceived risk of iran as a tourism destination (A mixed method approach). *Iranian Journal of Management Studies*, *12*(1), 45–67.

Ergül, B. (2023). *Digital marketing in tourism destinations*. ResearchGate. https://www.researchgate.net/publication/366014740_DIGITAL_MARKETING_IN_TOURISM_DESTINATIONS

Erol, I., Neuhofer, I. O., Dogru, T., Oztel, A., Searcy, C., & Yorulmaz, A. C. (2022). Improving sustainability in the tourism industry through blockchain technology: Challenges and opportunities. *Tourism Management*, *93*, 104628. doi:10.1016/j.tourman.2022.104628

Fails, J. A., Herbert, K., Hill, E., De Stefano, A., Hesse, B., Cushman, P., Gant, T., & ... 2015). Geotagger: A Collaborative Environmental Inquiry Platform. *2015 International Conference on Collaboration Technologies and Systems, CTS 2015*. 10.1109/CTS.2015.7210453

Falcão, D., & Roseira, C. (2022). Mapping the socially responsible consumption gap research : Review and future research agenda. *International Journal of Consumer Studies*, *46*(5), 1718–1760. doi:10.1111/ijcs.12803

Fall Diallo, M., Diop-Sall, F., Leroux, E., & Valette-Florence, P. (2015). Comportement responsable des touristes: Le rôle de l'engagement social. *Recherche et Applications en Marketing*, *30*(3), 88–108. doi:10.1177/0767370115571048

Farahat, E. (2023). *Applications of artificial intelligence as a marketing tool and their impact on the competitiveness of the Egyptian tourist destination* (Doctoral Dissertation, Minia University).

Farahat, E., Mohamed, M., & Al-Romeedy, B. (2022). Artificial intelligence applications and its impact on the competitiveness of the Egyptian tourist destination. *Research Journal of the Faculty of Tourism and Hotels Mansoura University*, *11*(2), 57–93.

Farhangi, S., & Alipour, H. (2021). Social media as a catalyst for the enhancement of destination image : Evidence from a mediterranean destination with political conflict. Sustainability (Switzerland), 13(13). *Sustainability (Basel)*, *13*(13), 7276. Advance online publication. doi:10.3390/su13137276

Farmaki, A., Miguel, C., Hadjielia, M., Aleksi, A., Ceh, A., & Efthymiadou, F. (2020). Impacts of Covid-19 on peer-to-peer accommodation platforms: Host perceptions and responses. *International Journal of Hospitality Management*, *91*(September), 102663. doi:10.1016/j.ijhm.2020.102663 PMID:32901166

Fazel, H. (2024). Digital marketing and sustainability in the era of climate change: PLS-structural equation modeling approach. *Journal of Statistics Applications & Probability*, *13*(1), 491–505. doi:10.18576/jsap/130134

Feldmann, C., & Hamm, U. (2015). Consumers' perceptions and preferences for local food: A review. *Food Quality and Preference*, *40*, 152–164. doi:10.1016/j.foodqual.2014.09.014

Femenia-Serra, F., & Neuhofer, B. (2018). *Smart tourism experiences : Conceptualisation, key dimensions and research agenda*. Academic Press.

Fennell, D. A. (2020). *Ecotourism*. Routledge. doi:10.4324/9780429346293

Ferrer-Rosell, B., Massimo, D., & Berezina, K. (2023). *Information and Communication Technologies in Tourism 2023: Proceedings of the ENTER 2023 eTourism Conference*. Springer Nature.

Filimonau, V., & Naumova, E. (2020). The blockchain technology and the scope of its application in hospitality operations. *International Journal of Hospitality Management*, *87*, 102383. doi:10.1016/j.ijhm.2019.102383

Fiorentino, S., & Bartolucci, S. (2021). Blockchain-based smart contracts as new governance tools for the sharing economy. *Cities (London, England)*, *117*, 103325. doi:10.1016/j.cities.2021.103325

Font, X., & McCabe, S. (2017). Sustainability and marketing in tourism: Its contexts, paradoxes, approaches, challenges and potential. *Journal of Sustainable Tourism*, *25*(7), 869–883. doi:10.1080/09669582.2017.1301721

Foris, D., Crihalmean, N., & Foris, T. (2020). Exploring the environmental practices in hospitality through booking websites and online tourist reviews. *Sustainability (Switzerland)*, *12*(24), 1-18. Advance online publication. doi:10.3390/su122410282

François-Lecompte, A., Prim-Allaz, I., & Durif, F. (2013). Le tourisme durable: Les sept clés du succès 1. *Gestion*, *38*(3), 84–96. doi:10.3917/riges.383.0084

Franzoni, S. (2015). Measuring the sustainability performance of the tourism sector. *Tourism Management Perspectives*, *16*, 22–27. doi:10.1016/j.tmp.2015.05.007

Frey, N., & George, R. (2010). Responsible tourism management: The missing link between business owners' attitudes and behaviour in the Cape Town tourism industry. *Tourism Management*, *31*(5), 621–628. doi:10.1016/j.tourman.2009.06.017

Frias, C., Silva, C., Amante, M. J., & Seabra, C. (2019). Storytelling tourist experiences in the websites of world heritage historical centers. *Experiential Consumption and Marketing in Tourism within a Cross-Cultural Context*, 174.

García, F. C., Valverde, M. I., Mascuñano, P. J., & Gimeno, V. M. (2018). Quality Implications of the use of Big Data in Tourism Statistics: Three exploratory examples. *Boletin de estadistica e Investigacion Operativa, 34*, 138-148.

García-Madurga, M.-Á., & Grilló-Méndez, A.-J. (2023). Artificial Intelligence in the Tourism Industry: An Overview of Reviews. *Administrative Sciences*, *13*(8), 172. doi:10.3390/admsci13080172

Garrido, R. H., Grávalos-Gastaminza, A., Escobar-Rodríguez, T., & Pérez-Calañas, C. (2022). Determining factors influencing use of caravanning tourism apps. *Journal of Tourism Analysis, 29*(1), 160-184. Advance online publication. doi:10.53596/jta.v29i1.399

Gaur, L., Afaq, A., Singh, G., & Dwivedi, Y. K. (2021). Role of artificial Intelligence and robotics to foster the touchless travel during a pandemic: a review and research agenda. *International Journal of Contemporary Hospitality Management*. doi:10.1108/IJCHM-11-2020-1246

Gayvoronskaya, T., & Meinel, C. (2021). *Blockchain*. Springer International Publishing. doi:10.1007/978-3-030-61559-8

Gemar, G., Sánchez-Teba, E. M., & Soler, I. P. (2022). Factors determining cultural city tourists' length of stay. *Cities (London, England), 130*, 103938. doi:10.1016/j.cities.2022.103938

Genç, R. (2018). *The Impact of Augmented Reality*. Technology on Tourist Satisfaction., doi:10.1007/978-3-319-64027-3_8

Ghaderi, Z., Butler, R., & Béal, L. (2022). Exploring home-based accommodation operators' responses to Covid-19: Implications of untact hospitality adoption. *Tourism Management Perspectives, 43*, 100979. doi:10.1016/j.tmp.2022.100979 PMID:35706982

Ghaderi, Z., King, B., & Hall, C. M. (2022). Crisis preparedness of hospitality managers: Evidence from Malaysia. *Journal of Hospitality and Tourism Insights, 5*(2), 292–310. doi:10.1108/JHTI-10-2020-0199

Gheorghe, G., Tudorache, P., & Roşca, I. (2023). The contribution of green marketing in the development of a sustainable destination through advanced clustering methods. *Sustainability (Basel), 15*(18), 13691. doi:10.3390/su151813691

Giesler, M., & Veresiu, E. (2014). Creating the responsible consumer: Moralistic governance regimes and consumer subjectivity. *The Journal of Consumer Research, 41*(3), 840–857. doi:10.1086/677842

Gigauri, I., Palazzo, M., & Ferri, M. (2023). *Handbook of Research on Achieving Sustainable Development Goals with Sustainable Marketing*. IGI Global. doi:10.4018/978-1-6684-8681-8

Giousmpasoglou, C., Marinakou, E., & Zopiatis, A. (2021). Hospitality managers in turbulent times: The COVID-19 crisis. *International Journal of Contemporary Hospitality Management, 33*(4), 1297–1318. doi:10.1108/IJCHM-07-2020-0741

Giuliano, R., & Eloy, F. (2021). Le comportement responsable des consommateurs face à l'agriculture belge. *Question (s) de management*, (2), 143-156.

Glasson, J., & Therivel, R. (2013). *Introduction to environmental impact assessment*. Routledge.

GNFI. (2020, March 27). *Dampak virus Corona terhadap pariwisata Indonesia* [The impact of the Coronavirus on Indonesian tourism]. Guinness World Records Indonesia.

Gohil, N. (2015). Role and impact of social media in tourism: A case study on the initiatives of Madhya Pradesh State Tourism. *International Journal of Research in Economics and Social Sciences, 5*(4), 8–15.

Goh, K. Y., Heng, C. S., & Lin, Z. (2013). Social media brand community and consumer behavior: Quantifying the relative impact of user-and marketer-generated content. *Information Systems Research, 24*(1), 88–107. doi:10.1287/isre.1120.0469

Gomes, S., & Lopes, J. M. (2023). Insights for pro-sustainable tourist behavior: The role of sustainable destination information and pro-sustainable tourist habits. *Sustainability (Basel), 15*(11), 8856. doi:10.3390/su15118856

Gonda, M. (2013). Supporters' movement. Against modern football and sport mega events: European and Polish contexts. *Przegląd Socjologiczny, 62*(3), 85–106.

González, A. (2020). La influencia de los medios de comunicación en el turismo. *Proceedings Book, 1*(4), 6.

Goodwin, H. (2011). Taking responsibility for tourism. Goodfellow Publishers Limited.

Goossens, C. (2000). Experiential marketing and its impact on post-purchase product related emotions. *Journal of Marketing Management*, *16*(1-2), 145–167.

Gossling, ̈. S., Scott, D., & Hall, C. M. (2020). Pandemics, tourism and global change: A rapid assessment of COVID-19. *Journal of Sustainable Tourism*, *29*(1), 1–20. doi:10.1080/09669582.2020.1758708

Gössling, S. (2020). Technology, ICT and tourism: From big data to the big picture. *Journal of Sustainable Tourism*, 1–10. doi:10.1080/09669582.2020.1865387

Gössling, S., Scott, D., & Hall, C. M. (2019). *Tourism and development in an era of climate change*. Routledge.

Government Decree 241/2020 (V.27.) on Outdoor Events (241/2020. (V. 27.)

Greenwald, A., Hampel, G., Phadke, C., & Poosala, V. (2011). An Economically Viable Solution to Geofencing for Mass-Market Applications. *Bell Labs Technical Journal*, *16*(2), 21–38. Advance online publication. doi:10.1002/bltj.20500

Gretzel, U., Fesenmaier, D. R., & O'Leary, J. T. (2006). The transformation of consumer behaviour. *Tourism business frontiers: Consumers, products and industry*, *9*, 18.

Gretzel, U., Hwang, Y., & Gibson, H. (2022). Virtual reality for sustainable tourism: Literature review and future research directions. *Journal of Sustainable Tourism*, *30*(10), 2465–2488.

Gretzel, U., Kim, Y., & Li, X. (2020). Smart tourism: A sustainability perspective. *Journal of Sustainable Tourism*, *28*(1), 7–22.

Gretzel, U., Sigala, M., Xiang, Z., & Koo, C. (2015). Smart tourism : Foundations and developments. *Electronic Markets*, *25*(3), 179–188. doi:10.1007/s12525-015-0196-8

Guagnano, G. A., Stern, P. C., & Dietz, T. (1995). Influences on attitude-behaviours relationships: A natural experiment with curbside recycling. *Environment and Behavior*, *27*(5), 699–718. doi:10.1177/0013916595275005

Gumede, T. (2019). Heritage tourism as a strategy for the local economic development in KwaZulu-Natal, South Africa. *African Journal of Hospitality, Tourism and Leisure*, *8*(3), 1–15.

Gummerus, J., Liljander, V., Weman, E., & Pihlström, M. (2012). Customer engagement in a Facebook brand community. *Management Research Review*, *35*(9), 857–877. doi:10.1108/01409171211256578

Günay, F., Bayraktaroğlu, E., & Özkul, K. (2020). Assessing the short-term impacts of COVID-19 pandemic on foreign visitor's demand for Turkey: A scenario analysis. *Journal of Ekonomi*, *2*(2), 80–85.

Gunduz, E., & Agayi, C. (2021). Assessment of the State and Impact of Tourism Activities in Kenya. *Kent Akademisi*, *14*(1), 174–185. doi:10.35674/kent.822245

Gunter, U. (2018). What makes an Airbnb host a superhost? Empirical evidence from San Francisco and the Bay Area. *Tourism Management*, *66*, 26–37. doi:10.1016/j.tourman.2017.11.003

Gunter, U., & Önder, I. (2021). An Exploratory Analysis of Geotagged Photos From Instagram for Residents of and Visitors to Vienna. *Journal of Hospitality & Tourism Research (Washington, D.C.)*, *45*(2), 373–398. Advance online publication. doi:10.1177/1096348020963689

Guo, L., Zhang, B., & Zhao, X. (2021). A Consumer Behaviours Prediction Model Based on Multivariate Real-Time Sequence Analysis. Hindawi Mathematical Problems in Engineering.

Guo, Y., Liu, H., & Chai, Y. (2014). The embedding convergence of smart cities and tourism internet of things in China: An advance perspective. *Advances in Hospitality and Tourism Research, 2*(1), 54–69.

Gupta, A., Dash, S., & Mishra, A. (2019). All that glitters is not green: Creating trustworthy ecofriendly services at green hotels. *Tourism Management, 70*, 155–169. doi:10.1016/j.tourman.2018.08.015

Gupta, S., Modgil, S., Bhattacharyya, S., & Bose, I. (2021). Artificial intelligence for decision support systems in the field of operations research: Review and future scope of research. *Annals of Operations Research*. Advance online publication. doi:10.1007/s10479-020-03856-6

Gupta, S., Modgil, S., Lee, C.-K., & Sivarajah, U. (2022). The future is yesterday: Use of AI-driven facial recognition to enhance value in the travel and tourism industry. *Information Systems Frontiers*. Advance online publication. doi:10.1007/s10796-022-10271-8 PMID:35529102

Gutierriz, I., Ferreira, J. J., & Fernandes, P. O. (2023). Digital transformation and the new combinations in tourism : A systematic literature review. *Tourism and Hospitality Research*. Advance online publication. doi:10.1177/14673584231198414

Guttentag, D. A. (2010). Virtual reality: Applications and implications for tourism. *Tourism Recreation Research, 35*(2), 137–144.

Hagemann, A. (2008). *"From stadium to "Fanzone": The architecture of control.* Paper presented at the conference Security and Surveillance at Mega Sport Events, Durham University, Durham, UK.

Hair, J. F., Black, W. C., Babin, B. J., & Anderson, R. E. (2019). *Multivariate data analysis* (8th ed.). Pearson Education Limited.

Hair, J. F., Ringle, C. M., & Sarstedt, M. (2019). *A primer on partial least squares structural equation modeling (PLS-SEM)*. Sage Publications Limited. doi:10.3926/oss.37

Halkiopoulos, C., Antonopoulou, H., Papadopoulos, D., Giannoukou, I., & Gkintoni, E. (2020). Online reservation systems in e-Business: Analyzing decision making in e-Tourism. *Journal of Tourism Heritage & Services Marketing, 6*(1), 9–16.

Hall, C. M. (2006). Urban entrepreneurship, corporate interests and sports mega-events: The thin policies of competitiveness within the hard outcomes of neoliberalism. *The Sociological Review, 54*(2), 59–70. doi:10.1111/j.1467-954X.2006.00653.x

Hallem, Y., Hikkerova, L., & Sahut, J.-M. (2017). Commercialisation de la durabilité dans le tourisme au travers de l'expérience client : Le cas d'un parc naturel: *Management & Avenir, 96*(6), 187-205. doi:10.3917/mav.096.0187

Hallo, J. C., Beeco, J. A., Goetcheus, C., McGee, J., McGehee, N. G., & Norman, W. C. (2012). GPS as a Method for Assessing Spatial and Temporal Use Distributions of Nature-Based Tourists. *Journal of Travel Research, 51*(5), 591–606. Advance online publication. doi:10.1177/0047287511431325

Hameduddin, T., & Engbers, T. (2022). Leadership and public service motivation : A systematic synthesis. *International Public Management Journal, 25*(1), 86-119. Advance online publication. doi:10.1080/10967494.2021.1884150

Hamouda, M. (2022). Mobile Applications in Tourism : Examining the Determinants of Intention to Use. *International Journal of Technology and Human Interaction, 18*(1), 1–13. Advance online publication. doi:10.4018/IJTHI.293198

Han, D. I. D., Jung, T., & Claudia Tom Dieck, M. (2019). Translating Tourist Requirements into Mobile AR Application Engineering Through QFD. *International Journal of Human-Computer Interaction, 35*(19), 1842–1858. Advance online publication. doi:10.1080/10447318.2019.1574099

Han, D. I., tom Dieck, M. C., & Jung, T. (2018). User Experience Model for Augmented Reality Applications in Urban Heritage Tourism. *Journal of Heritage Tourism*, *13*(1), 46–61. Advance online publication. doi:10.1080/1743873X.2016.1251931

Hardy, A., Hyslop, S., Booth, K., Robards, B., Aryal, J., Gretzel, U., & Eccleston, R. (2017). Tracking Tourists' Travel with Smartphone-Based GPS Technology: A Methodological Discussion. *Information Technology & Tourism*, *17*(3), 255–274. Advance online publication. doi:10.1007/s40558-017-0086-3

Harris, R., Williams, P., & Griffin, T. (2012). *Sustainable tourism*. Routledge. doi:10.4324/9780080498362

Hartmann, K., & Lederer, M. (2021). The current state of Big Data research in tourism: Results of a systematic literature analysis. *Zeitschrift für Tourismuswissenschaft*, *13*(2), 209–226. doi:10.1515/tw-2021-0015

Hassan, H. N., Zakaria, Z., & Musa, N. A. (2023). The effects of virtual reality on tourist interest in Malaysian cultural heritage sites. *Tourism Management Perspectives*, *55*, 100932.

Hassanli, N., Gross, M. J., & Brown, G. (2016). The emergence of home-based accommodations in Iran: A study of self-organization. *Tourism Management*, *54*, 284–295. doi:10.1016/j.tourman.2015.11.018

Hays, S., Page, S. J., & Buhalis, D. (2013). Social Media as a Destination Marketing Tool: Its Use by National Tourism Organisations. *Current Issues in Tourism*, *16*(3), 211–239. doi:10.1080/13683500.2012.662215

Hernigou, P., & Scarlat, M. (2023). Two minutes of orthopaedics with ChatGPT: It is just the beginning; it's going to be hot, hot, hot! *International Orthopaedics*, *47*(8), 1887–1893. doi:10.1007/s00264-023-05887-7 PMID:37428231

He, S., Tang, Q., Wu, C., & Shen, X. (2019). Decentralizing IoT management systems using blockchain for censorship resistance. *IEEE Transactions on Industrial Informatics*, *16*(1), 715–727. doi:10.1109/TII.2019.2939797

Higgins-Desbiolles, F. (2010). The elusiveness of sustainability in tourism: The culture-ideology of consumerism and its implications. *Tourism and Hospitality Research*, *10*(2), 116–129. doi:10.1057/thr.2009.31

Hinson, G., & Slade, R. (2011). Forty hard-won business continuity lessons from the 2011 earthquakes in New Zealand and Japan. *EDPACS*, *44*(3), 1–18. doi:10.1080/07366981.2011.608316

Histouria. (n.d.). *Unlocking History, One Chat at a Time*. Retrieved March 15, 2024, from https://historia-landing.vercel.app/

Hmioui, A., Alla, L., & Bentalha, B. (2022). *Perception of sustainable tourism by international clients: Case of the visit to a hotel establishment*. Academic Press.

Hoang, H., & Trang, P. H. (2023). Navigating the Rise of Smart Tourism : Implications of Technology and Data for Sustainable Industry Growth. *Brawijaya Journal of Social Science*, *3*(1), 1–18. doi:10.21776/ub.bjss.2023.003.01.1

Hoang, T. G., Truong, N. T., & Nguyen, T. M. (2021). The survival of hotels during the COVID-19 pandemic: A critical case study in Vietnam. *Service Business*, *15*(2), 209–229. doi:10.1007/s11628-021-00441-0

Hoffman, R. R., Klein, G., & Mueller, S. T. (2018, September). Explaining explanation for "explainable AI". *Proceedings of the Human Factors and Ergonomics Society Annual Meeting*, *62*(1), 197–201. doi:10.1177/1541931218621047

Holloway, J. C. (2004). *The business of tourism* (5th ed.). Pearson Education Limited.

Holloway, J. C. (2009). *Marketing in tourism* (4th ed.). Pearson Education Limited.

Holotescu, C. (2018). Understanding blockchain technology and how to get involved. *The 14thInternational Scientific ConferenceeLearning and Software for EducationBucharest*, 20.

Hosseini, M., Shahri, A., Phalp, K., Taylor, J., & Ali, R. (2015). Crowdsourcing: A Taxonomy and Systematic Mapping Study. *Computer Science Review*, *17*, 43–69. Advance online publication. doi:10.1016/j.cosrev.2015.05.001

Hu, F. H., & Wei, G. (2013). *The impact of the knowledge sharing in social media on consumer behaviour*. Academic Press.

Huang, M. H., Law, R., & Yuan, Y. (2022). The impact of virtual reality (VR) destination marketing on millennials' travel intention: The role of emotional experience and perceived authenticity. *International Journal of Hospitality Management*, *107*, 103382.

Huang, M. H., Ryu, K., & Law, R. (2013). The influence of virtual reality on destination image formation: A moderated mediation model. *Tourism Management*, *38*(1), 227–241.

Huang, Y., Li, Y., Li, S., Prendergast, M., & Gretzel, U. (2023). Understanding the intention to use virtual reality tourism: A social exchange theory perspective. *Journal of Travel Research*, *62*(2), 332–347.

Hudson, S., & Thal, K. (2013). The impact of social media on the consumer decision process: Implications for tourism marketing. *Journal of Travel & Tourism Marketing*, *30*(1-2), 156–160. doi:10.1080/10548408.2013.751276

Hussain, A. (2021). A future of tourism industry: Conscious travel, destination recovery and regenerative tourism. *Journal of Sustainability and Resilience*, *1*(1), 5.

Hyland, J. (2000). *New measures against football hooligans undermine civil liberties in Britain*. Available from https://www.wsws.org/en/articles/2000/08/act-a29. html

Hysa, B., Karasek, A., & Zdonek, I. (2021). Social media usage by different generations as a tool for sustainable tourism marketing in society 5.0 idea. *Sustainability (Basel)*, *13*(3), 1018. doi:10.3390/su13031018

Hyun, S. Y., & O'Keeffe, G. M. (2012). Tourists' experiential perceptions of destination branding. *Journal of Travel Research*, *51*(1), 78–91.

Imre, D. (2018). Professionalism, modesty, diligence: A festive volume in honour of 65-year- old József Boda. Budapest: Dialóg Campus Publishing House.

Inge, C. (2022). *Marketing metrics: leverage analytics and data to optimize marketing strategies*. Kogan Page Publishers.

Ioannides, D., & Hajibabaei, M. (2020). Augmented reality and virtual reality in tourism: A review. *Annals of Tourism Research*, *81*, 102855.

Iqbal, R., Doctor, F., More, B., Mahmud, S., & Yousuf, U. (2018). Big data analytics: Computational intelligence techniques and application areas. *Technological Forecasting and Social Change*, *153*, 119253. doi:10.1016/j.techfore.2018.03.024

Isabelle, D. A., & Westerlund, M. (2022). A Review and Categorization of Artificial Intelligence-Based Opportunities in Wildlife, Ocean and Land Conservation. *Sustainability (Basel)*, *14*(4), 1979. doi:10.3390/su14041979

Issa, E., Tsou, M. H., Nara, A., & Spitzberg, B. (2017). Understanding the Spatio-Temporal Characteristics of Twitter Data with Geotagged and Non-Geotagged Content: Two Case Studies with the Topic of Flu and Ted (Movie). *Annals of GIS*, *23*(3), 219–235. Advance online publication. doi:10.1080/19475683.2017.1343257

Ivanov, S., & Webster, C. (2019). Conceptual Framework of the Use of Robots, Artificial Intelligence and Service Automation in Travel, Tourism, and Hospitality Companies. *Robots, Artificial Intelligence, and Service Automation in Travel, Tourism and Hospitality*, 7–37. doi:10.1108/978-1-78756-687-320191001

IvanovS. H.UmbrelloS. (2021, December 31). *The ethics of artificial Intelligence and robotization in tourism and hospitality – a conceptual framework and research agenda*. https://papers.ssrn.com/sol3/papers.cfm?abstract_id=4030669

J, P. A., & Md, R. K. (2024). Revolutionizing the Tourism Industry through Artificial Intelligence: A Comprehensive Review of AI Integration, Impact on Customer Experience, Operational Efficiency, and Future Trends. *International Journal for Multidimensional Research Perspectives, 2*(2), 1-14. https://www.chandigarhphilosophers.com/index.php/ijmrp/article/view/115

Jabbar, S., Lloyd, H., Hammoudeh, M., Adebisi, B., & Raza, U. (2021). Blockchain-enabled supply chain: Analysis, challenges, and future directions. *Multimedia Systems*, *27*(4), 787–806. doi:10.1007/s00530-020-00687-0

Jacobsen, J. K. S., & Munar, A. M. (2012). Tourist information search and destination choice in a digital age. *Tourism Management Perspectives*, *1*, 39–47. doi:10.1016/j.tmp.2011.12.005

Jamal, T., & Camargo, B. (2014). Sustainable tourism, justice and an ethic of care: Toward the just destination. *Journal of Sustainable Tourism*, *22*(1), 11–30. doi:10.1080/09669582.2013.786084

Jasrotia, A., & Gangotia, A. (2018). Smart cities to smart tourism destinations: A review paper. *Journal of Tourism Intelligence and Smartness*, *1*(1), 47–56.

Jin, X., Xiang, Y., Weber, K., & Liu, Y. (2019). Motivation and involvement in adventure tourism activities: A Chinese tourists' perspective. *Asia Pacific Journal of Tourism Research*, *24*(11), 1066–1078. doi:10.1080/10941665.2019.1666152

Jones, P., & Wynn, M. (2019). The circular economy, natural capital and resilience in tourism and hospitality. *International Journal of Contemporary Hospitality Management*, *31*(6), 2544–2563. doi:10.1108/IJCHM-05-2018-0370

Jude, S. E., & Ukekwe, C. N. (2020). The impacts of technology on tourism development in developing countries: A review of the literature. *Journal of Hospitality & Tourism Research (Washington, D.C.)*, *44*(2), 381–403.

Judmayer, A., Stifter, N., Krombholz, K., & Weippl, E. (2022). *Blocks and chains: introduction to bitcoin, cryptocurrencies, and their consensus mechanisms*. Springer Nature.

Kadri, M., Khalloufi, H., & Azough, A. (2020). V-Museum: A Virtual Museum Based on Augmented and Virtual Realities for Cultural Heritage Mediation. *2020 International Conference on Intelligent Systems and Computer Vision (ISCV)*, 1–5. 10.1109/ISCV49265.2020.9204253

Kaján, E., & Saarinen, J. (2013). Tourism, climate change and adaptation: A review. *Current Issues in Tourism*, *16*(2), 167–195. doi:10.1080/13683500.2013.774323

Kalıpçı, M. B., & Şimşek, E. K. (2023). Exploring The Intersection of Big Data and Hospitality: A Bibliometric Analysis. *Journal of Business Research-Turk*, *15*(3), 2096–2110.

Kamyab, H., Khademi, T., Chelliapan, S., SaberiKamarposhti, M., Rezania, S., Yusuf, M., Farajnezhad, M., Abbas, M., Hun Jeon, B., & Ahn, Y. (2023). The latest innovative avenues for the utilization of artificial Intelligence and big data analytics in water resource management. *Results in Engineering*, *20*, 101566–101566. doi:10.1016/j.rineng.2023.101566

Kamyuka, D., Carlin, L., McPherson, G., & Misener, L. (2020). Access to physical activity and sport and the effects of isolation and cordon sanitaire during COVID-19 for people with disabilities in Scotland and Canada. *Frontiers in Sports and Active Living*, *2*, 594501. doi:10.3389/fspor.2020.594501 PMID:33426519

Kanyam, D. A., Kostandini, G., & Ferreira, S. (2017). The mobile phone revolution: Have mobile phones and the internet reduced corruption in Sub-Saharan Africa? *World Development*, *99*, 271–284. doi:10.1016/j.worlddev.2017.05.022

Kapil, S., & Kapil, K. (2022). Blockchain in hospitality and tourism industry way forward. *International Journal of Business and Economics*, *6*(2), 289–298.

Karamitsos, I., Papadaki, M., Al-Hussaeni, K., & Kanavos, A. (2023). Transforming Airport Security: Enhancing Efficiency through Blockchain Smart Contracts. *Electronics (Basel), 12*(21), 4492. doi:10.3390/electronics12214492

Karasakal, S., & Albayrak, T. (2022). How to create flow experience during travel : The role of destination attributes. *Journal of Vacation Marketing, 28*(3), 303–318. doi:10.1177/13567667211053386

Karl, M., Muskat, B., & Ritchie, B. W. (2020). Which travel risks are more salient for destination choice? An examination of the tourist's decision-making process. *Journal of Destination Marketing & Management, 18*, 100487. doi:10.1016/j.jdmm.2020.100487

Karode, T., & Werapun, W. (2022). Robustness against fraudulent activities of a blockchain-based online review system. *Peer-to-Peer Networking and Applications, 15*(1), 92–106. doi:10.1007/s12083-021-01225-z

Kaspar, C. (1973). Fremderverkehrsoekologie—eine neue Dimension der Fremdenverkehslehre [Tourism Ecology—A New Dimension of Tourism Science]. *Festscrift zur Vollendung des, 65.*

Kaufman, I., Horton, C., & Soltanifar, M. (2023). *Digital Marketing: Integrating Strategy, Sustainability, and Purpose.* Taylor & Francis. doi:10.4324/9781351019187

Kayat, K. (2010). The homestay program in Malaysia and its implications for the future. *Tourism research in Malaysia: What, which way and so what*, 293-316.

Kaźmierczak, R., Szczepańska, A., Kowalczyk, C., Grunwald, G., & Janowski, A. (2021). Using Ar Technology in Tourism Based on the Example of Maritime Educational Trips—A Conceptual Model. *Sustainability (Basel), 13*(13), 7172. Advance online publication. doi:10.3390/su13137172

Ketter, E., & Avraham, E. (2021). # StayHome today so we can# TravelTomorrow: Tourism destinations' digital marketing strategies during the Covid-19 pandemic. *Journal of Travel & Tourism Marketing, 38*(8), 819–832. doi:10.1080/10548408.2021.1921670

Kew, T., Shaitarova, A., Meraner, I., Goldzycher, J., Clematide, S., & Volk, M. (2019). Geotagging a Diachronic Corpus of Alpine Texts: Comparing Distinct Approaches to Toponym Recognition. *Proceedings of the Workshop on Language Technology for Digital Historical Archives*, 11–18. 10.26615/978-954-452-059-5_003

Khalloufi, H., Zaifri, M., Kadri, M., Benlahbib, A., Kaghat, F. Z., & Azough, A. (2024). El-FnaVR: An Immersive Virtual Reality Representation of Jemaa El-Fna in Marrakech for Intangible Cultural Heritage Experiences. *IEEE Access : Practical Innovations, Open Solutions, 12*, 9331–9349. doi:10.1109/ACCESS.2023.3347195

Khan, M. N., Gul, A., Khan, F., Khan, M. W., & Naz, A. (2024). Unlocking the potential : Artificial intelligence (AI) applications in sustainable tourism. *Remittances Review, 9*(1). https://remittancesreview.com/menu-script/index.php/remittances/article/view/1258

Khan, A., Wang, M., Ehsan, S., Nurunnabi, M., & Hashmi, M. (2019). Linking sustainability-oriented marketing to social media and web atmospheric cues. *Sustainability (Basel), 11*(9), 2663. doi:10.3390/su11092663

Khandai, S., Mathew, J., Yadav, R., Kataria, S., & Kohli, H. (2023). Ensuring brand loyalty for firms practising sustainable marketing: A roadmap. *Society and Business Review, 18*(2), 219–243. doi:10.1108/SBR-10-2021-0189

Khan, S., Loukil, F., Ghedira-Guegan, C., Benkhelifa, E., & Bani-Hani, A. (2021). Blockchain smart contracts: Applications, challenges, and future trends. *Peer-to-Peer Networking and Applications, 14*(5), 2901–2925. doi:10.1007/s12083-021-01127-0 PMID:33897937

Khelifi, K. (2023, October). *Conceptualisation des déterminants du choix d'une destination touristique par la valeur perçue. 9èmes Rendez-vous Champlain.*

Kilipiri, E., Papaioannou, E., & Kotzaivazoglou, I. (2023). Social Media and Influencer Marketing for Promoting Sustainable Tourism Destinations: The Instagram Case. *Sustainability (Basel)*, *15*(8), 6374. doi:10.3390/su15086374

Kim, H., Ahn, S.-K., & Forney, J. A. (2014). Shifting paradigms for fashion : From total to global to smart consumer experience. *Fashion and Textiles*, *1*(1), 1. Advance online publication. doi:10.1186/s40691-014-0015-4

Kim, J., Jeong, S., & Goo, H. (2022). The role of virtual reality (VR) technology in tourism during the COVID-19 pandemic. *Journal of Travel Research*, *61*(7), 1427–1441.

Kim, J., Kim, Y., & Park, S. (2021). The effects of virtual reality travel experience on destination attitude and visit intention: Focusing on the roles of presence and escape motivation. *Journal of Travel Research*, *50*(2), 552–567.

Kim, J., Kim, Y., & Wang, Y. (2020). Understanding the role of authenticity in virtual reality travel experiences: A stimulus–organism–response framework. *Tourism Management*, *83*, 104212.

Kim, J., Lee, H., & Jung, T. (2020). The effects of virtual reality on destination image and purchase intention: The mediating role of emotional experience. *Journal of Travel Research*, *59*(8), 1674–1689.

Kim, M. J., Lee, C. K., & Jung, T. (2020). Exploring consumer behavior in virtual reality tourism using an extended stimulus-organism-response model. *Journal of Travel Research*, *59*(1), 69–89. doi:10.1177/0047287518818915

Kindzule-Millere, I., & Zeverte-Rivza, S. (2022). Digital transformation in tourism: Opportunities and challenges. *Economic Science for Rural Development Conference Proceedings*, *56*, 476–486. https://doi.org/10.22616/ESRD.2022.56.047

King, C., Iba, W., & Clifton, J. (2021). Reimagining resilience: COVID-19 and marine tourism in Indonesia. *Current Issues in Tourism*, *24*(19), 2784–2800. doi:10.1080/13683500.2021.1873920

Kirtil, İ. G., & Aşkun, V. (2021, June 1). Artificial Intelligence in Tourism: A Review and Bibliometrics Research. *Advances in Hospitality and Tourism Research*, *9*(1), 205–233. Advance online publication. doi:10.30519/ahtr.801690

Knani, M., Echchakoui, S., & Ladhari, R. (2022). Artificial intelligence in tourism and hospitality: Bibliometric analysis and research agenda. *International Journal of Hospitality Management*, *107*, 103317. doi:10.1016/j.ijhm.2022.103317

Kompatsiaris, P. (2024). *Curation in the Age of Platform Capitalism: The Value of Selection, Narration, and Expertise in New Media Cultures*. Taylor & Francis. doi:10.4324/9781003182801

Konar, R., Ali, F., & Hussain, K. (2018). Empowerment in Hospitality Service Leadership: A Moderated Mediation Approach. *Asia-Pacific Journal of Innovation in Hospitality and Tourism*, *7*(2), 21–38.

Koščak, M., & O'Rourke, T. (2023). *Ethical and responsible tourism: managing sustainability in local tourism destinations*. Taylor & Francis.

Kotler, P., Kartajaya, H., Setiawan, I., & Vandercammen, M. (2022). Chapitre 9. Le marketing prédictif. *Marketing*, 169-183.

Kotler, P., Kartajaya, H., & Setiawan, I. (2023). *Marketing 6.0: The Future Is Immersive*. John Wiley & Sons.

Kounavis, C. D., Kasimati, A. E., & Zamani, E. D. (2012). Enhancing the Tourism Experience through Mobile Augmented Reality: Challenges and Prospects. *International Journal of Engineering Business Management*, *4*(1), 10. Advance online publication. doi:10.5772/51644

Krause, D. (1993). Environmental consciousness: An empirical study. *Environment and Behavior*, *25*(1), 126–142. doi:10.1177/0013916593251007

Krisna, D. F., Handayani, P. W., & Azzahro, F. (2019). The Antecedents of Hashtag and Geotag Use in Smart Tourism: Case Study in Indonesia. *Asia Pacific Journal of Tourism Research*, *24*(12), 1141–1154. Advance online publication. doi:10.1080/10941665.2019.1665559

Kumar, K. S., Kaur, J., & Ramani, S. (2017). Consumer Behvior Analysis using Big Data Analytic. *International Journal of Control Theory and Applications, 10*(30).

Kumar, N. R., Pavithra, R., Yuvaraj, V., & Kumar, T. M. (2023). The role of smart tourism in India for optimizing visitor experiences through technology. *JIMS8I International Journal of Information Communication and Computing Technology, 11*(2), 692-701.

Kumar, P. (2021). Digital marketing in hospitality and tourism. In C. Cobanoglu, S. Dogan, K. Berezina, & G. Collins (Eds.), *Hospitality & tourism information technology* (pp. 1–23). USF M3 Publishing. https://www.doi.org/10.5038/9781732127593

Kumar, A. P., Shankar, S. G., Praveen, K. R. M., Gadekallu, T. R., Al-Ahmari, A., & Abidi, M. H. (2020). Location Based Business Recommendation Using Spatial Demand. *Sustainability (Basel)*, *12*(10), 4124. Advance online publication. doi:10.3390/su12104124

Kumar, S. (2024). The Role of Digital Marketing on Customer Engagement in the Hospitality Industry. In *Leveraging ChatGPT and Artificial Intelligence for Effective Customer Engagement*. IGI Global. doi:10.4018/979-8-3693-0815-8.ch010

Kuo, C. M., Chen, L. C., & Tseng, C. Y. (2017). Investigating an innovative service with hospitality robots. *International Journal of Contemporary Hospitality Management*, *29*(5), 1305–1321. doi:10.1108/IJCHM-08-2015-0414

Kurowksa, K., Marks-Bielska, R., Bielski, S., Aleknavičius, A., & Kowalczyk, C. (2021). Geographic Information Systems and the Sustainable Development of Rural Areas. *Land (Basel)*. Advance online publication. doi:10.3390/land10010006

Labre, M. (2022, April 8). *Leveraging Digital Marketing to Boost Your Tourism Business*. LinkedIn. https://www.linkedin.com/pulse/future-travel-leveraging-digital-marketing-boost-your-marco-labre

Lahcene, S. A., & Naili, N. (2019). *Les attentes des consommateurs touristiques en matière de tourisme durable* (Doctoral dissertation, Université Mouloud Mammeri).

Lai, I. K. W., & Hitchcock, M. (2017). Local reactions to mass tourism and community tourism development in Macau. *Journal of Sustainable Tourism*, *25*(4), 451–470. doi:10.1080/09669582.2016.1221413

Lai, I. K. W., Hitchcock, M., Lu, D., & Liu, Y. (2018). The influence of word of mouth on tourism destination choice: Tourist–resident relationship and safety perception among Mainland Chinese tourists visiting Macau. *Sustainability (Basel)*, *10*(7), 2114. doi:10.3390/su10072114

Lanquar, R. (2023). Tourisme, créativité et intelligence artificielle. *Marché et organisations, 47*, 185-203. doi:10.3917/maorg.047.0185

Laroche, M., Habibi, M. R., & Richard, M. O. (2013). To be or not to be in social media: How brand loyalty is affected by social media? *International Journal of Information Management*, *33*(1), 76–82. doi:10.1016/j.ijinfomgt.2012.07.003

Lashkari, B., & Musilek, P. (2021). A comprehensive review of blockchain consensus mechanisms. *IEEE Access : Practical Innovations, Open Solutions*, *9*, 43620–43652. doi:10.1109/ACCESS.2021.3065880

Lau, S.-Y. J., & Agius, H. (2021). A framework and immersive serious game for mild cognitive impairment. *Multimedia Tools and Applications*, *80*(20), 31183–31237. doi:10.1007/s11042-021-11042-4

Lazić, M., Bradić-Martinović, A., & Banović, J. (2023). Digital skills in tourism and hospitality as a precondition for the sector resilient growth: The case of Serbia. *Менаџмент у хотелијерству и туризму, 11*(1), 25-40.

Leal, F., Malheiro, B., Veloso, B., & Burguillo, J. C. (2020). Responsible processing of crowdsourced tourism data. *Journal of Sustainable Tourism*, 1–21. doi:10.1080/09669582.2020.1778011

Lee, H. S., & Lim, J. H. (2013). Statistical package for the social sciences. *JypHyunJae Publication.*

Lee, C. K., & Gretzel, U. (2023). Virtual reality and overtourism: A conceptual framework for sustainable destination management. *Journal of Sustainable Tourism*, *31*(2), 381–400.

Lee, C.-K., Olya, H., Ahmad, M. S., Kim, K. H., & Oh, M.-J. (2021). Sustainable intelligence, destination social responsibility, and pro-environmental behaviour of visitors : Evidence from an eco-tourism site. *Journal of Hospitality and Tourism Management*, *47*, 365–376. doi:10.1016/j.jhtm.2021.04.010

Lee, S., Sun, K. A., Wu, L., & Xiao, Q. (2018). A moderating role of green practices on the relationship between service quality and customer satisfaction: Chinese hotel context. *Journal of China Tourism Research*, *14*(1), 42–60. doi:10.1080/19388160.2017.1419897

Lee, T. H., & Jan, F.-H. (2023). How do smart tourism experiences affect visitors' environmentally responsible behavior? Influence analysis of nature-based tourists in Taiwan. *Journal of Hospitality and Tourism Management*, *55*, 1–10. doi:10.1016/j.jhtm.2023.02.016

Legrand, W., Chen, J. S., & Laeis, G. C. (2022). *Sustainability in the hospitality industry: Principles of sustainable operations*. Routledge.

Leh, F. C., & Hamzah, M. R. (2012). Homestay tourism and pro-poor tourism strategy in Banghuris Selangor, Malaysia. *Elixir International Journal*, *45*, 7602–7610.

Lenskyj, H. (2002). *The best Olympics ever? The social impacts of Sydney 2000*. SUNY Press. doi:10.1353/book4571

Lenzen, M., Moran, D., Bhaduri, A., Christie, P., Foran, B., Hegarty, K., ... Wiedmann, T. (2023). The environmental footprint of global tourism. *Nature Climate Change*, *13*(4), 325–332.

Levac, D., Colquhoun, H., & O'brien, K. K. (2010). Scoping studies: Advancing the methodology. *Implementation Science : IS*, *5*(1), 1–9. doi:10.1186/1748-5908-5-69 PMID:20854677

Lew, A. A. (2011). Tourism's Role in the Global Economy. *Tourism Geographies*, *13*(1), 148–151. Advance online publication. doi:10.1080/14616688.2010.531046

Lhoussaine, A. L. L. A., Kamal, M., & Bouhtati, N. (2022). Big data et efficacité marketing des entreprises touristiques: Une revue de littérature. *Alternatives Managériales Economiques*, *1*, 39–58.

Lhoussaine, A., Kamal, M., & Bouhtati, N. (2022). Big data et efficacité marketing des entreprises touristiques : Une revue de littérature. *Alternatives Managériales Economiques*, *1*, 39–58.

Li, J., & Cao, B. (2022). Study on Tourism Consumer Behaviours and Countermeasures Based on Big Data. Hindawi Computational Intelligence and Neuroscience.

Liberati, A., Moher, D., Tetzlaff, J., & Altman, D. G. (2009). Preferred reporting items for systematic reviews and meta-analyses: The PRISMA statement. *PLoS Medicine*, *6*(7), e1000097. doi:10.1371/journal.pmed.1000100 PMID:19621072

Li, C., Zheng, W., Zhuang, X., & Chen, F. (2023). Intelligent transport design with a dual focus : Tourist experience and operating cost. *Annals of Tourism Research*, *101*, 103597. doi:10.1016/j.annals.2023.103597

Li, F., Yu, X., Ge, R., Wang, Y., Cui, Y., & Zhou, H. (2021). BCSE: Blockchain-based trusted service evaluation model over big data. *Big Data Mining and Analytics*, *5*(1), 1–14. doi:10.26599/BDMA.2020.9020028

Lifewire. (2024). https://www.lifewire.com/best-travel-planner-apps-4175910

Light, D. (2017). Progress in dark tourism and thanatourism research: An uneasy relationship with heritage tourism. *Tourism Management, 61*, 275–301. doi:10.1016/j.tourman.2017.01.011

Lignon-Darmaillac, S. (2016). *Œnotourisme et éthique de la consommation*. Academic Press.

Li, H., Meng, F., & Zhang, Z. (2016b). Non-participation of Domestic Tourism: Analyzing the Influence of Discouraging Factors. *International Journal of Tourism Research, 18*(6), 567–578. doi:10.1002/jtr.2074

Li, J., & Cao, B. (2022). Study on tourism consumer behaviours and countermeasures based on big data. *Computational Intelligence and Neuroscience*.

Li, J., & Kassem, M. (2021). Applications of distributed ledger technology (DLT) and Blockchain-enabled smart contracts in construction. *Automation in Construction, 132*, 103955. doi:10.1016/j.autcon.2021.103955

Li, K., Chen, X., Jiang, H., & Bertino, E. (2019). *Essentials of blockchain technology*. CRC Press. doi:10.1201/9780429674457

Lima, N. T. Y., Haro-Zea, K. L., & Saavedra, R. A. D. (2024). Digital marketing and customer orientation as predictors of sustainability in tourism SMEs. *Innovative Marketing, 20*(1), 160-171. Advance online publication. doi:10.21511/im.20(1).2024.14

Lin, M., Sharma, A., Pan, B., & Quadri-Felitti, D. (2023). Information asymmetry in the innovation adoption decision of tourism and hospitality SMEs in emerging markets: A mixed-method analysis. *Tourism Management, 99*, 104793. doi:10.1016/j.tourman.2023.104793

Lin, Y. C., & Chen, T. C. T. (2022). Type-II fuzzy approach with explainable artificial intelligence for nature-based leisure travel destination selection amid the COVID-19 pandemic. *Digital Health, 8*, 20552076221106322. doi:10.1177/20552076221106322 PMID:35707268

Li, Q., Liao, F., Timmermans, H. J., & Zhou, J. (2016a). A user equilibrium model for combined activity–travel choice under prospect theoretical mechanisms of decision-making under uncertainty. *Transportmetrica A: Transport Science, 12*(7), 629–649. doi:10.1080/23249935.2016.1189718

Lisi, A., De Salve, A., Mori, P., Ricci, L., & Fabrizi, S. (2021). Rewarding reviews with tokens: An ethereum-based approach. *Future Generation Computer Systems, 120*, 36–54. doi:10.1016/j.future.2021.02.003

Liu, C. H., Horng, J. S., Chou, S. F., Yu, T. Y., Huang, Y. C., & Lin, J. Y. (2023). Integrating big data and marketing concepts into tourism, hospitality operations and strategy development. *Quality & Quantity, 57*(2), 1905–1922. doi:10.1007/s11135-022-01426-5 PMID:35729961

Liu, C., Horng, J., Chou, S., Yu, T., Lee, M., & Lapuz, M. (2023). Discovery sustainable servicescape on behavioural intention practices and nationality: The moderating role of parasocial interaction. *Journal of Retailing and Consumer Services, 71*, 103213. doi:10.1016/j.jretconser.2022.103213

Liu, Y., Lehdonvirta, V., Alexandrova, T., & Nakajima, T. (2012). Drawing on Mobile Crowds via Social Media. *Multimedia Systems, 18*(1), 53–67. Advance online publication. doi:10.1007/s00530-011-0242-0

Liu, Y., Naveed, R. T., Kanwal, S., Tahir Khan, M., Dalain, A. F., & Lan, W. (2023). Psychology in action: Social media communication, CSR, and consumer behavior management in banking. *PLoS One, 18*(8), e0289281. doi:10.1371/journal.pone.0289281 PMID:37590276

Lomas, K. R., & Trujillo, C. A. (2018). Environmental educational model for community tourism of fakcha llakta community-Ecuador. *International Journal of Professional Business Review: Int. J. Prof. Bus. Rev., 3*(1), 95–110.

Loureiro, S. M. C., Guerreiro, J., & Ali, F. (2020). 20 years of research on virtual reality and augmented reality in tourism context: A text-mining approach. *Tourism Management*, *77*, 104028. doi:10.1016/j.tourman.2019.104028

Low, S., Ullah, F., Shirowzhan, S., Sepasgozar, S., & Lin Lee, C. (2020). Smart digital marketing capabilities for sustainable property development: A case of Malaysia. *Sustainability (Basel)*, *12*(13), 5402. doi:10.3390/su12135402

Lv, H., Shi, S., & Gursoy, D. (2021). A look back and a leap forward: A review and synthesis of big data and artificial intelligence literature in hospitality and tourism. *Journal of Hospitality Marketing & Management*, *31*(2), 1–31. doi:10.1080/19368623.2021.1937434

Lynch, P. A., & MacWhannell, D. (2000). Home and conceptualizing hospitality. *In search of hospitality: Theoretical perspectives and debates*, 1965-1982.

Lynch, P. A. (2005). Sociological impressionism in a hospitality context. *Annals of Tourism Research*, *32*(3), 527–548. doi:10.1016/j.annals.2004.09.005

Lyu, J., Khan, A., Bibi, S., Chan, J. H., & Qi, X. (2022). Big data in action: An overview of big data studies in tourism and hospitality literature. *Journal of Hospitality and Tourism Management*, *51*, 346–360. doi:10.1016/j.jhtm.2022.03.014

Maach, I., Azough, A., & Meknassi, M. (2018). Development of a use case for virtual reality to visit a historical monument. *2018 International Conference on Intelligent Systems and Computer Vision (ISCV)*, 1–4. 10.1109/ISACV.2018.8354052

Maggioni, I., Sands, S. J., Ferraro, C. R., Pallant, J. I., Pallant, J. L., Shedd, L., & Tojib, D. (2020). Consumer cross-channel behaviour: Is it always planned? *International Journal of Retail & Distribution Management*, *48*(12), 1357–1375. doi:10.1108/IJRDM-03-2020-0103

Majeed, S., Zhou, Z., Lu, C., & Ramkissoon, H. (2020). Online tourism information and tourist behaviours: A structural equation modeling analysis based on a self-administered survey. *Frontiers in Psychology*, *11*, 599. doi:10.3389/fpsyg.2020.00599 PMID:32373008

Majid, A., Chen, L., Chen, G., Mirza, H. T., Hussain, I., & Woodward, J. (2013). A Context-Aware Personalized Travel Recommendation System Based on Geotagged Social Media Data Mining. *International Journal of Geographical Information Science*, *27*(4), 662–684. Advance online publication. doi:10.1080/13658816.2012.696649

Mak, J. (2008). *Developing a dream destination: Tourism and tourism policy planning in Hawaii*. University of Hawaii Press. doi:10.21313/hawaii/9780824832438.001.0001

Makransky, G., Petersen, D., & Göbel, M. (2022). Virtual Reality (VR) in education: Benefits and challenges for STEM learning. *Computers & Education*, *180*, 102597.

Makridis, I. (2018). *Big data analytics and knowledge discovery through location-based social networks*. LBSN.

Ma, L., & Cao, J. (2019). How perceptions mediate the effects of the built environment on travel behavior? *Transportation*, *46*(1), 175–197. doi:10.1007/s11116-017-9800-4

Malhotra, N. K., & Dash, S. J. M. R. (2010). An applied orientation. *Marketing Research, 2.*

Manning, E. W. (1990). Geographic Information Systems and Sustainable Development. *Government Information Quarterly*, *7*(3), 329–342. Advance online publication. doi:10.1016/0740-624X(90)90028-M

Mansfeld, Y. (1992). Group-differentiated perceptions of social impacts related to tourism development. *The Professional Geographer*, *44*(4), 377–392. doi:10.1111/j.0033-0124.1992.00377.x

Mansfield, E. (1998). Academic research and industrial innovation: An update of empirical findings. *Research Policy*, *26*(7-8), 773–776. doi:10.1016/S0048-7333(97)00043-7

Mao, X., Meng, J., & Wang, Q. (2014). Modeling the effects of tourism and land regulation on land-use change in tourist regions: A case study of the Lijiang River Basin in Guilin, China. *Land Use Policy*, *41*, 368–377. doi:10.1016/j.landusepol.2014.06.018

Maquera, G., da Costa, B., Mendoza, O., Salinas, R., & Haddad, A. (2022). Intelligent Digital Platform for Community-Based Rural Tourism-A Novel Concept Development in Peru. *Sustainability (Basel)*, *14*(13), 7907. Advance online publication. doi:10.3390/su14137907

Marchi, V., Marasco, A., & Apicerni, V. (2023). Sustainability communication of tourism cities : A text mining approach. *Cities (London, England)*, *143*, 104590. doi:10.1016/j.cities.2023.104590

Marco-Lajara, B., Úbeda-García, M., Ruiz-Fernández, L., Poveda-Pareja, E., & Sánchez-García, E. (2022). Rural hotel resilience during COVID-19: The crucial role of CSR. *Current Issues in Tourism*, *25*(7), 1121–1135. doi:10.1080/13683500.2021.2005551

Mariani, M. M., & Baggio, R. (2022). Big data and analytics in hospitality and tourism: a systematic literature review. *International Journal of Contemporary Hospitality Management, 34*(1), 231-278. doi:10.1108/IJCHM-03-2021-0301

Mariani, M., Baggio, R., Fuchs, M., & Höepken, W. (2018). Business intelligence and big data in hospitality and tourism: A systematic literature review. *International Journal of Contemporary Hospitality Management*, *30*(12), 3514–3554. doi:10.1108/IJCHM-07-2017-0461

Mariani, M., Bresciani, S., & Dagnino, G. B. (2021). The competitive productivity (CP) of tourism destinations: An integrative conceptual framework and a reflection on big data and analytics. *International Journal of Contemporary Hospitality Management*, *33*(9), 2970–3002. doi:10.1108/IJCHM-09-2020-1102

Martin, D., Alzua, A., & Lamsfus, C. (2011). A Contextual Geofencing Mobile Tourism Service. Information and Communication Technologies in Tourism 2011. doi:10.1007/978-3-7091-0503-0_16

Martinez-Ferran, M., de la Guía-Galipienso, F., Sanchis-Gomar, F., & Pareja-Galeano, H. (2020). Metabolic Impacts of Confinement during the COVID-19 Pandemic due to Modified Diet and Physical Activity Habits. *Nutrients*, *12*(6), 1549. doi:10.3390/nu12061549 PMID:32466598

Martínez, P., Herrero, Á., & García de los Salmones, M. D. M. (2022). An examination of the determining factors of users' intentions to share corporate CSR content on Facebook. *Current Issues in Tourism*, *25*(13), 2159–2176. doi:10.1080/13683500.2021.1946019

Masiero, L., & Qiu, R. T. (2018). Modeling reference experience in destination choice. *Annals of Tourism Research*, *72*, 58–74. doi:10.1016/j.annals.2018.06.004

Mateo, S. (2020). Procédure pour conduire avec succès une revue de littérature selon la méthode PRISMA. *Kinésithérapie, la Revue*, *20*(226), 29–37. doi:10.1016/j.kine.2020.05.019

Mathew, V., & Soliman, M. (2021). Does digital content marketing affect tourism consumer behavior? An extension of t echnology acceptance model. *Journal of Consumer Behaviour*, *20*(1), 61–75. doi:10.1002/cb.1854

Mátyás, Sz., Tokodi, P. V., & Vince, T. M. (2021). Investigation of Certain Components of Tourism Safety in the Visegrád Countries. Internal Security, 13(1), 89-97.

Mátyás, Sz., Nagy-Tóth, N. A., Dávid, L. D., Gogo, A. F. C., & Bujdosó, Z. (2022). The Role of Sports Policing and Tourism Safety at the Summer Olympics. *Sustainability*, *14*(10), 1–16.

Maurer, C. (2021). Digital marketing in tourism. In *Handbook of e-Tourism*. Springer International Publishing. doi:10.1007/978-3-030-05324-6_82-1

Mayson, S., & Barrett, R. (2017). A new argument using embeddedness and sensemaking to explain small firms' responses to employment regulation. *Human Resource Management Journal*, *27*(1), 189–202. doi:10.1111/1748-8583.12141

McCartney, G. (2021). The impact of the coronavirus outbreak on Macao. From tourism lockdown to tourism recovery. *Current Issues in Tourism*, *24*(19), 2683–2692. doi:10.1080/13683500.2020.1762549

McCool, S., & Bosak, K. (2016). *Reframing sustainable tourism*. Springer. doi:10.1007/978-94-017-7209-9

McGuine, T. A., Biese, K. M., Petrovska, L., Hetzel, S. J., Reardon, C., Kliethermes, S., Bell, D. R., Brooks, A., & Watson, A. M. (2021). Mental Health, Physical Activity, and Quality of Life of US Adolescent Athletes during COVID-19-Related School Closures and Sport Cancellations: A Study of 13,000 Athletes. *Journal of Athletic Training*, *56*(1), 11–19. doi:10.4085/1062-6050-0478.20 PMID:33290516

McKeown-Ice, R. (1994). Environmental Education: A Geographical Perspective. *The Journal of Geography*, *93*(1), 40–42. Advance online publication. doi:10.1080/00221349408979684

McNiven, A., & Harris, J. (2021). Can community football clubs in Scotland develop the local partnership support they need? *Managing Sport and Leisure*, *26*(5), 346–361. doi:10.1080/23750472.2020.1838940

Mehrabian, A., & Russell, J. A. (1974). An approach to environmental psychology. *Psychological Bulletin*, *81*(12), 703–710.

Mehraliyev, F., Choi, Y., & Köseoglu, M. A. (2019). Progress on smart tourism research. *Journal of Hospitality and Tourism Technology*, *10*(4), 522–538. doi:10.1108/JHTT-08-2018-0076

Mehta, K., & Jha, S. S. (2020). COVID-19: a nightmare for the Indian economy. *UGC Care Journal, 31*(20).

Miah, S. J., Vu, H. Q., Gammack, J., & McGrath, M. (2017). A big data analytics method for tourist behaviour analysis. *Information & Management*, *54*(6), 771–785. doi:10.1016/j.im.2016.11.011

Micić, J., Denda, S., & Popescu, M. (2019). The significance of the risk-related challenges in tourist destination choice. *Journal of the Geographical Institute "Jovan Cvijić". SASA*, *69*(1), 39–52.

Mihajlovic, I. (2012). The impact of information and communication technology (ICT) as a key factor of tourism development on the role of Croatian travel agencies. *International Journal of Business and Social Science*, *3*(24).

Mihalič, T. (2006). *Tourism and its environments: ecological, economic and political sustainability issues*. Ekonomska fakulteta.

Miller, G. P. (2017). *The Law of Governance, Risk Management and Compliance*. Wolters Kluwer Law & Business.

Miller, H. H. (2000). Mega-events, urban boosterism and growth strategies: An analysis of the objectives and legitimations of the Cape Town 2004 Olympic bid. *International Journal of Urban and Regional Research*, *24*, 439–458.

Ministry of Tourism and Wildlife. (2022). *New Tourism Strategy for Kenya 2021-2025*. Republic of Kenya.

Ministry of Tourism. (2020). *Annual Report 2019-20*. Government of India. Available at: https://tourism.gov.in/annual-reports/annual-report-2019-20

Minkwitz, A. (2018). TripAdvisor as a source of data in the planning process of tourism development on a local scale. *Turyzm (Lódz)*, *28*(2), 49–55. doi:10.2478/tour-2018-0014

Modi, R. (2018). *Solidity Programming Essentials: A beginner's guide to build smart contracts for Ethereum and blockchain*. Packt Publishing Ltd.

Mohamed, M., Farahat, E., & Al-Romeedy, B. (2022). The Mediating Role of Marketing Effectiveness in the Relationship between Artificial Intelligence and Destination Competitiveness. *Minia Journal of Tourism and Hospitality Research, 14*(1), 113–129.

Moher, D., Liberati, A., Tetzlaff, J., & Altman, D. G. (2009). Preferred Reporting Items for Systematic Reviews and Meta-Analyses : The PRISMA Statement. *PLoS Medicine, 6*(7), e1000097. doi:10.1371/journal.pmed.1000097 PMID:19621072

Monaco, S., & Sacchi, G. (2023). Travelling the Metaverse : Potential Benefits and Main Challenges for Tourism Sectors and Research Applications. *Sustainability (Basel), 15*(4), 3348. Advance online publication. doi:10.3390/su15043348

Mondino, E., & Beery, T. (2019). Ecotourism as a learning tool for sustainable development. The case of Monviso Transboundary Biosphere Reserve, Italy. *Journal of Ecotourism, 18*(2), 107–121. doi:10.1080/14724049.2018.1462371

Monnot, E., & Reniou, F. (2013). Ras le bol d'entendre parler d'écologie! comprendre la contestation des discours écologiques par les consommateurs. *Décisions Marketing*, 93-109.

Moon, H., & Han, H. (2019). Tourist experience quality and loyalty to an island destination: The moderating impact of destination image. *Journal of Travel & Tourism Marketing, 36*(1), 43–59. doi:10.1080/10548408.2018.1494083

Morrison, A. M., Taylor, S., Morrison, A. J., & Morrison, A. D. (1999). Marketing small hotels on the World Wide Web. *Information Technology & Tourism, 2*(2), 97–113.

Morrongiello, C. (2014). *De l'empowerment à l'engagement du client sur les plateformes en ligne: ou comment favoriser l'activité des clients sur Internet* (Doctoral dissertation, Université de Grenoble).

Moser, M. B., & Liu, A. (2005). Pre-testing of questionnaires: Developmental and psychometric considerations. *Journal of Nursing Measurement, 13*(1), 18–24.

Moser, M. B., & Liu, A. (2005). Pre-testing questionnaires: Tips and best practices. *AAOHN Journal, 53*(6), 243–248.

Mousavi, S. S., Doratli, N., Mousavi, S. N., & Moradiahari, F. (2016, December). Defining cultural tourism. In *International Conference on Civil, Architecture and Sustainable Development* (Vol. 1, No. 2, pp. 70-75). Academic Press.

Mousavian, S., Miah, S. J., & Zhong, Y. (2023). A design concept of big data analytics model for managers in hospitality industries. *Personal and Ubiquitous Computing, 27*(4), 1–11. doi:10.1007/s00779-023-01714-3 PMID:36818420

Mousazadeh, H., Ghorbani, A., Azadi, H., Almani, F. A., Zangiabadi, A., Zhu, K., & Dávid, L. D. (2023). Developing sustainable behaviors for underground heritage tourism management: The case of Persian Qanats, a UNESCO world heritage property. *Land (Basel), 12*(4), 808. doi:10.3390/land12040808

Moustafa, S. M., Ismail, Z., & Al-Deghim, Y. M. (2020). The impact of virtual reality (VR) technology on students' achievement in geometry: A systematic review and meta-analysis. *Computers & Education, 151*, 103868.

Mpofu, F. Y. (2022). Green Taxes in Africa: Opportunities and challenges for environmental protection, sustainability, and the attainment of sustainable development goals. *Sustainability (Basel), 14*(16), 10239. doi:10.3390/su141610239

MTAESS. (n.d.). *Tourisme – Ministère du Tourisme, de l'Artisanat et de l'Economie Sociale et Solidaire*. Retrieved March 31, 2024, from https://mtaess.gov.ma/fr/tourisme/

Mu, S., Piwowarczyk, M., Kutrzyński, M., Trawiński, B., & Telec, Z. (2020). Comparative Analysis of Selected Geotagging Methods. Communications in Computer and Information Science, 1178. doi:10.1007/978-981-15-3380-8_29

Müller, A., Bíró, M., Bodolai, M., Hidvégi, P., Váczi, P., Dávid, L., & Szántó, Á. (2017). A 2016-os fitnesztrendek helye és szerepe a rekreációban [The 2016 fitness trends place and role of recreation]. *ACTA Academiae Paedagogicae Agriensis Nova Series: Sectio Sport, 44*, 91–102.

Munar, A. M., & Jacobsen, J. K. S. (2014). Motivations for sharing tourism experiences through social media. *Tourism Management*, *43*, 46–54. doi:10.1016/j.tourman.2014.01.012

Munn, Z., Peters, M. D., Stern, C., Tufanaru, C., McArthur, A., & Aromataris, E. (2018). Systematic review or scoping review? Guidance for authors when choosing between a systematic or scoping review approach. *BMC Medical Research Methodology*, *18*(1), 1–7. doi:10.1186/s12874-018-0611-x PMID:30453902

Munn, Z., Stern, C., Aromataris, E., Lockwood, C., & Jordan, Z. (2018). What kind of systematic review should I conduct? A proposed typology and guidance for systematic reviewers in the medical and health sciences. *BMC Medical Research Methodology*, *18*(1), 1–9. doi:10.1186/s12874-017-0468-4 PMID:29316881

Mura, A., Franco, M., Russo, T., & Ciffoli, R. (2017). Virtual reality for enhancing cultural heritage: A user experience evaluation. *Journal of Cultural Heritage*, *28*, 126–132.

Mura, P. (2010). 'Scary… but I like it!' Young tourists' perceptions of fear on holiday. *Journal of Tourism and Cultural Change*, *8*(1-2), 30–49. doi:10.1080/14766825.2010.482209

Musil, L., & Pigel, J. (1994). Virtual tourism - a contradiction in terms? *MTT Schriftenreihe*, *15*, 133–139.

Mutanga, C., Kolawole, O., Gondo, R., & Mbaiwa, J. (2024). A review and SWOC analysis of natural heritage tourism in sub-Saharan Africa. *Journal of Heritage Tourism*, *19*(1), 49–67. doi:10.1080/1743873X.2023.2255689

Mutinda, R., & Mayaka, M. (2012). Application of destination choice model: Factors influencing domestic tourists destination choice among residents of Nairobi, Kenya. *Tourism Management*, *33*(6), 1593–1597. doi:10.1016/j.tourman.2011.12.008

Myat, A. A., Sharkasi, N., & Rajasekera, J. (2023). Myanmar's tourism : Sustainability of ICT to support hotel sector for online booking and digital marketing. *Benchmarking, 30*(7), 2486-2508. Advance online publication. doi:10.1108/BIJ-07-2017-0200

Nadkarni, S., & Teare, R. (2019). Reflections on the theme issue outcomes: Expo 2020: what will be the impact on Dubai? *Worldwide Hospitality and Tourism Themes*, *11*(3), 346–350. doi:10.1108/WHATT-03-2019-0013

Naif, M. R. K. (2023). *Big Data Analytics for Predicting Consumer Behaviour.* Academic Press.

Nascimento, J., & Loureiro, S. M. C. (2023). The Impact of Augmented and Virtual Reality for Sustainable Tourism. In Springer Proc. Bus. Econ. (p. 148-156). Springer Nature. doi:10.1007/978-3-031-25390-4_14

National Government of South Africa. (2024). *South African Tourism Annual Report 2021/2022*. South African Tourism Board.

Navío-Marco, L., Ruiz-Gómez, A. B., & Sevilla-Sevilla, J. (2018). The role of social media on tourist experiential quality: Mediating effects of information search, online engagement and travel planning. *Journal of Hospitality & Tourism Research (Washington, D.C.)*, *42*(4), 942–962.

Nayyar, A., Mahapatra, B., Le, D., & Suseendran, G. (2018). Virtual Reality (VR) & Augmented Reality (AR) technologies for tourism and hospitality industry. *International Journal of Engineering & Technology, 7*(2.21), 156-160.

Naznin, K. (2017). *Humanoid Educational Service Robots in Primary Classrooms: Design-based Strategies for Pre-implementation Planning*. https://researchbank.swinburne.edu.au/file/2876c640-2ed9-4aca-b649-779b9754973f/1/kaberi_naznin_thesis.pdf

Neidhardt, J., & Wörndl, W. (2020). *Information and Communication Technologies in Tourism 2020.* Springer International Publishing. doi:10.1007/978-3-030-36737-4

Nguyen, C., Hoang, D., Nguyen, D., Niyato, D., Nguyen, H., & Dutkiewicz, E. (2019). Proof-of-stake consensus mechanisms for future blockchain networks: Fundamentals, applications and opportunities. *IEEE Access : Practical Innovations, Open Solutions*, *7*, 85727–85745. doi:10.1109/ACCESS.2019.2925010

Nguyen, T. T., Camacho, D., & Jung, J. E. (2017). Identifying and Ranking Cultural Heritage Resources on Geotagged Social Media for Smart Cultural Tourism Services. *Personal and Ubiquitous Computing*, *21*(2), 267–279. Advance online publication. doi:10.1007/s00779-016-0992-y

Nicola-Gavrilă, L. (2023). Empirical Approaches Regarding Interdependency between Technology and Sustainable Tourism. *Journal of Environmental Management and Tourism*, *14*(4 (68)), 2140–2147. doi:10.14505/jemt.v14.4(68).25

Niemczyk, A. (2013). Cultural tourists: "An attempt to classify them". *Tourism Management Perspectives*, *5*, 24–30. doi:10.1016/j.tmp.2012.09.006

Nikitas, A., Michalakopoulou, K., Njoya, E. T., & Karampatzakis, D. (2020). Artificial Intelligence, Transport and the Smart City: Definitions and Dimensions of a New Mobility Era. *Sustainability (Basel)*, *12*(7), 2789. doi:10.3390/su12072789

Nikos, N. (2022, January 2). Tourism and Hospitality industry resilience during the Covid-19 pandemic: Evidence from England. *Current Issues in Tourism*, *25*(1), 46–59. Advance online publication. doi:10.1080/13683500.2021.1883556

Nittel, S. (2009). A Survey of Geosensor Networks: Advances in Dynamic Environmental Monitoring. *Sensors (Basel)*, *9*(7), 5664–5678. Advance online publication. doi:10.3390/s90705664 PMID:22346721

Nittel, S., Labrinidis, A., & Stefanidis, A. (2008). Introduction to Advances in Geosensor Networks. In. Lecture Notes in Computer Science: Vol. 4540. *Including Subseries Lecture Notes in Artificial Intelligence and Lecture Notes in Bioinformatics)*. LNCS. doi:10.1007/978-3-540-79996-2_1

Njoroge, J. M. (2015). Climate change and tourism adaptation: Literature review. *Tourism and Hospitality Management*, *21*(1), 95–108. doi:10.20867/thm.21.1.7

Noby, O., Rady, A., & Abd Eljalil, S. G. (2021). Big Data Analytics' Utilization in Egyptian Hotels' Decision-Making: What are The Challenges of Applying BDA? *Journal of Association of Arab Universities for Tourism and Hospitality*, *21*(5), 156–170. doi:10.21608/jaauth.2021.106301.1265

Norton, E., Li, Y., Mason, L. R., & Washington-Allen, R. A. (2019). Assessing the Impact of a Geospatial Data Collection App on Student Engagement in Environmental Education. *Education Sciences*, *9*(2), 118. Advance online publication. doi:10.3390/educsci9020118

Novas, N., Gázquez, J. A., MacLennan, J., García, R. M., Fernández-Ros, M., & Manzano-Agugliaro, F. (2017). A real-time underground environment monitoring system for sustainable tourism of caves. *Journal of Cleaner Production*, *142*, 2707–2721. doi:10.1016/j.jclepro.2016.11.005

Nowak, M. M., Słupecka, K., & Jackowiak, B. (2021). Geotagging of Natural History Collections for Reuse in Environmental Research. *Ecological Indicators*, *131*, 108131. Advance online publication. doi:10.1016/j.ecolind.2021.108131

Nwulu, C. S., & Ateke, B. W. (2014). The impact of marketing information credibility on consumer purchase behaviour in the pharmaceutical industry in Port Harcourt Nigeria. *West African Journal of Business and Management Science*, *3*(3), 41–49.

Nyman, E., Westin, K., & Carson, D. (2018). Tourism destination choice sets for families with wheelchair-bound children. *Tourism Recreation Research*, *43*(1), 26–38. doi:10.1080/02508281.2017.1362172

Ogilvie, D. M. (1987). The undesired self: A neglected variable in personality research. *Journal of Personality and Social Psychology*, *52*(2), 379–385. doi:10.1037/0022-3514.52.2.379

Oncioiu, I., & Priescu, I. (2022). The use of virtual reality in tourism destinations as a tool to develop tourist behavior perspective. *Sustainability (Basel)*, *14*(7), 4191. doi:10.3390/su14074191

Onder, I. (2023). Enhancing Sustainable Tourism through Blockchain-Based Tokenization: A Novel Approach for Revenue Generation. In *Blockchain for Tourism and Hospitality Industries*. Routledge. doi:10.4324/9781003351917-8

Ong, S. K., Yuan, M. L., & Nee, A. Y. C. (2008). Augmented Reality Applications in Manufacturing: A Survey. *International Journal of Production Research*, *46*(10), 2707–2742. Advance online publication. doi:10.1080/00207540601064773

Osburg, T. (2017). *Sustainability in a digital world needs trust*. Springer International Publishing. doi:10.1007/978-3-319-54603-2

Ottman, J. (2017). *The new rules of green marketing: Strategies, tools, and inspiration for sustainable branding*. Routledge. doi:10.4324/9781351278683

Özçağlar-Toulouse, N. (2009). Quel sens les consommateurs responsables donnent-ils à leur consommation? Une approche par les récits de vie. *Recherche et Applications en Marketing*, *24*(3), 3–23. doi:10.1177/076737010902400306

Page, M. J., McKenzie, J. E., Bossuyt, P. M., Boutron, I., Hoffmann, T. C., Mulrow, C. D., Shamseer, L., Tetzlaff, J. M., Akl, E. A., & Brennan, S. E. (2021). The PRISMA 2020 statement : An updated guideline for reporting systematic reviews. *BMJ (Clinical Research Ed.)*, *372*. https://www.bmj.com/content/372/bmj.n71.short PMID:33782057

Pai, C.-K., Liu, Y., Kang, S., & Dai, A. (2020). The Role of Perceived Smart Tourism Technology Experience for Tourist Satisfaction, Happiness and Revisit Intention. *Sustainability (Basel)*, *12*(16), 16. Advance online publication. doi:10.3390/su12166592

Palomo, J., Figueroa-Domecq, C., Flecha-Barrio, D., & Segovia-Pérez, M. (2017). The Use of New Data Analysis Techniques in Tourism: A Bibliometric Analysis in Data Mining, Big Data and Structural Equations Models. *Ereview of Tourism Research*, 8.

Pan, S. Y., Gao, M., Kim, H., Shah, K. J., Pei, S. L., & Chiang, P. C. (2018). Advances and challenges in sustainable tourism toward a green economy. *The Science of the Total Environment*, *635*, 452–469. doi:10.1016/j.scitotenv.2018.04.134 PMID:29677671

Pappas, N. (2021). COVID19: Holiday intentions during a pandemic. *Tourism Management*, *84*, 104287. Advance online publication. doi:10.1016/j.tourman.2021.104287 PMID:33518882

Park, C., Lee, S., Lee, C.-K., & Reisinger, Y. (2022). Volunteer tourists' environmentally friendly behavior and support for sustainable tourism development using Value-Belief-Norm theory : Moderating role of altruism. *Journal of Destination Marketing & Management*, *25*, 100712. doi:10.1016/j.jdmm.2022.100712

Park, S., Xu, Y., Jiang, L., Chen, Z., & Huang, S. (2020). Spatial structures of tourism destinations: A trajectory data mining approach leveraging mobile big data. *Annals of Tourism Research*, *84*, 102973. doi:10.1016/j.annals.2020.102973

Park, Y.-J. (2014). Tourism Information System Using Geofencing and NFC Technology. *Journal of Digital Convergence*, *12*(8), 211–217. Advance online publication. doi:10.14400/JDC.2014.12.8.211

Parmentola, A., Petrillo, A., & Tutore, I., & De Felice, F. (. (2022). Is blockchain able to enhance environmental sustainability? A systematic review and research agenda from the perspective of Sustainable Development Goals (SDGs). *Business Strategy and the Environment, 31*(1), 194-217. Advance online publication. doi:10.1002/bse.2882

Paul, A., Chauhan, S., & Dutta, D. (2023). Mobile-Based Image Interpretation and Geotagging Using Artificial Intelligence and Open-Source Geospatial Technology. *Applied Geomatics*, *15*(4), 795–805. Advance online publication. doi:10.1007/s12518-023-00522-x

Peceny, U. S., Urbančič, J., Mokorel, S., Kuralt, V., & Ilijaš, T. (2019). Tourism 4.0: Challenges in marketing a paradigm shift. In Consumer behavior and marketing. IntechOpen.

Peceny, U. S., Urbančič, J., Mokorel, S., Kuralt, V., & Ilijaš, T. (2019). Tourism 4.0: Challenges in marketing a paradigm shift. In Consumer behavior and marketing. IntechOpen.

Pencarelli, T. (2019). The digital revolution in the travel and tourism industry. *Information Technology & Tourism*, *22*(3), 455–476. doi:10.1007/s40558-019-00160-3

Peng, Y., & Huang, W. (2022). Using blockchain technology and sharing culture to promote sustainable forest management in tribal communities. *Journal of Environmental and Public Health*, *2022*, 1–12. Advance online publication. doi:10.1155/2022/1529407 PMID:35815248

Perkins, R., Khoo-Lattimore, C., & Arcodia, C. (2020). Understanding the contribution of stakeholder collaboration towards regional destination branding: A systematic narrative literature review. *Journal of Hospitality and Tourism Management*, *43*, 250–258. doi:10.1016/j.jhtm.2020.04.008

Pestana, M. H., Parreira, A., & Moutinho, L. (2020). Motivations, emotions and satisfaction: The keys to a tourism destination choice. *Journal of Destination Marketing & Management*, *16*, 100332. doi:10.1016/j.jdmm.2018.12.006

Peters, M. D., Godfrey, C. M., Khalil, H., McInerney, P., Parker, D., & Soares, C. B. (2015). Guidance for conducting systematic scoping reviews. *JBI Evidence Implementation*, *13*(3), 141–146. PMID:26134548

Pike, S. (2015). *Destination marketing: essentials*. Routledge. doi:10.4324/9781315691701

Platt, M., Sedlmeir, J., Platt, D., Xu, J., Tasca, P., Vadgama, N., & Ibañez, J. (2021). The energy footprint of blockchain consensus mechanisms beyond proof-of-work. In *2021 IEEE 21st International Conference on Software Quality, Reliability and Security Companion (QRS-C)* (pp. 1135-1144). IEEE. 10.1109/QRS-C55045.2021.00168

Politou, E., Alepis, E., Patsakis, C., Casino, F., & Alazab, M. (2020). Delegated content erasure in IPFS. *Future Generation Computer Systems*, *112*, 956–964. doi:10.1016/j.future.2020.06.037

Popova, P., Petrova, M., Popov, V., Marinova, K., & Sushchenko, O. (2023). Potential of the digital ecosystem for the sustainable development of the tourist destination. *IOP Conference Series. Earth and Environmental Science*, *1126*(1), 012021. Advance online publication. doi:10.1088/1755-1315/1126/1/012021

Porras-Bueno, N., Plaza-Mejía, M. D. L. Á., & Vargas-Sánchez, A. (2019). Quality of life and perception of the effects of tourism: A contingent approach. *Best Practices in Hospitality and Tourism Marketing and Management: A Quality of Life Perspective*, 109-132.

Prados-Castillo, J., Guaita Martínez, J., Zielińska, A., & Gorgues Comas, D. (2023). A Review of Blockchain Technology Adoption in the Tourism Industry from a Sustainability Perspective. *Journal of Theoretical and Applied Electronic Commerce Research*, *18*(2), 814–830. doi:10.3390/jtaer18020042

Prahadeeswaran, R. (2023). A Comprehensive Review: The Convergence of Artificial Intelligence and Tourism. *International Journal for Multidimensional Research Perspectives*, *1*(2), 12–24. https://www.chandigarhphilosophers.com/index.php/ijmrp/article/view/73

Pramana Paramartha, Ermawan, Deli, & Srimulyani. (2021). Impact of COVID-19 pandemic on tourism in Indonesia. *Current Issues*.

Pranita, D., Sarjana, S., Musthofa, B., Kusumastuti, H., & Rasul, M. (2023). Blockchain Technology to Enhance Integrated Blue Economy: A Case Study in Strengthening Sustainable Tourism on Smart Islands. *Sustainability (Basel)*, *15*(6), 5342. doi:10.3390/su15065342

Prayag, G. (2018). Symbiotic relationship or not? Understanding resilience and crisis management in tourism. *Tourism Management Perspectives*, *25*, 133–135. doi:10.1016/j.tmp.2017.11.012

Prayag, G. (2020). Time for reset? Covid-19 and tourism resilience. *Tourism Review International*, *24*(2/3), 179–184. doi:10.3727/154427220X15926147793595

Prideaux, B. (2002). The choreographic tourist: A poststructuralist analysis of tourism and performance. *Journal of Leisure Research*, *34*(2), 169–192.

Priporas, C. V., Stylos, N., Vedanthachari, L. N., & Santiwatana, P. (2017). Service quality, satisfaction, and customer loyalty in Airbnb accommodation in Thailand. *International Journal of Tourism Research*, *19*(6), 693–704. doi:10.1002/jtr.2141

Priyadarshini, I. (2020). *A Survey on some of the Global Effects of the COVID-19 Pandemic*. Academic Press.

Puntoni, S., Reczek, R. W., Giesler, M., & Botti, S. (2021). Consumers and artificial intelligence: An experiential perspective. *Journal of Marketing*, *85*(1), 131–151. doi:10.1177/0022242920953847

Purwaningwulan, M., & Ramdan, T. (2022). Digital promotion of local tourist destinations in the new normal era and its effect on the economy in Indonesia. *Journal of Eastern European and Central Asian Research*, *9*(1), 29–40. doi:10.15549/jeecar.v9i2.873

Quiroz-Fabra, J., Valencia-Arias, A., Londoño-Celis, W., & García-Pineda, V. (2022). Technological Tools for Knowledge Apprehension and Promotion in the Cultural and Heritage Tourism Sector : Systematic Literature Review. *Human Behavior and Emerging Technologies*, *2022*, 1–12. Advance online publication. doi:10.1155/2022/2851044

Quynh, T. D., & Dung, H. T. T. (2021). Prediction of Customer Behaviours using Machine Learning: A Case Study. *Proceedings of the 2nd International Conference on Human-centered Artificial Intelligence (Computing4Human 2021).*

Rahmadian, E., Feitosa, D., & Zwitter, A. (2021). A systematic literature review on the use of big data for sustainable tourism. *Current Issues in Tourism*, *25*(11), 1–20. doi:10.1080/13683500.2021.1974358

Rahman, M., Chamikara, M., Khalil, I., & Bouras, A. (2022). Blockchain-of-blockchains: An interoperable blockchain platform for ensuring IoT data integrity in smart city. *Journal of Industrial Information Integration*, *30*, 100408. doi:10.1016/j.jii.2022.100408

Rakotomalala-Ramandimbiarison, L. M. (2019). Les déterminants dans l'adhésion à la Charte du Tourisme Durable. *La Revue des Sciences de Gestion*, (5-6), 99–112. doi:10.3917/rsg.299.0099

Ramon-Rodriguez, A. B., Such-Devesa, M. J., Perles-Ribes, J. F., & Moreno-Izquierdo, L. (2021). Competitiveness, intelligence and sustainability in consolidated destinations : What we know and pending research challenges. *Tourism,* *69*(2), 281-299. Advance online publication. doi:10.37741/t.69.2.8

Rane, N., Choudhary, S., & Rane, J. (2023). Sustainable tourism development using leading-edge Artificial Intelligence (AI), Blockchain, Internet of Things (IoT), Augmented Reality (AR) and Virtual Reality (VR) technologies. *Blockchain, Internet of Things (IoT), Augmented Reality (AR) and Virtual Reality (VR) technologies.*

Rashideh, W. (2020). Blockchain technology framework: Current and future perspectives for the tourism industry. *Tourism Management*, *80*, 104125. doi:10.1016/j.tourman.2020.104125

Rasoolimanesh, S., Ramakrishna, S., Hall, C., Esfandiar, K., & Seyfi, S. (2023). A systematic scoping review of sustainable tourism indicators in relation to the sustainable development goals. *Journal of Sustainable Tourism*, *31*(7), 1497–1517. doi:10.1080/09669582.2020.1775621

Rathore, B. (2017). Aligning Profitability and Environmental Responsibility: A Study on Sustainable Marketing Strategies. *Eduzone: International Peer Reviewed/Refereed Multidisciplinary Journal, 6*(2), 7-15.

Rauscher, M., Humpe, A., & Brehm, L. (2020). Virtual reality in tourism : Is it « Real » enough? *Academica Turistica, 13*(2), 127-138. Advance online publication. doi:10.26493/2335-4194.13.127-138

Rehman Khan, S. A., Yu, Z., Sarwat, S., Godil, D. I., Amin, S., & Shujaat, S. (2022). The role of block chain technology in circular economy practices to improve organisational performance. *International Journal of Logistics, 25*(4-5), 605–622. doi:10.1080/13675567.2021.1872512

Rehman, F., Khalid, O., & Madani, S. A. (2016). A Comparative Study of Location-Based Recommendation Systems. *The Knowledge Engineering Review, 32*, e7. Advance online publication. doi:10.1017/S0269888916000308

Rejeb, A., Zailani, S., Rejeb, K., Treiblmaier, H., & Keogh, J. G. (2022). Modeling enablers for blockchain adoption in the circular economy. *Sustainable Futures : An Applied Journal of Technology, Environment and Society, 4*, 100095. doi:10.1016/j.sftr.2022.100095

Ribeiro, V., Pereira, M., Matos, O., Alves, J., Remoaldo, P., Gôja, R., & Freitas, I. (2020). Geo-Crowdsourcing Contributions for Cultural Mapping. *Journal of Information Science Theory and Practice, 8*(1). Advance online publication. doi:10.1633/JISTaP.2020.8.1.5

Ribes, J. F. P., & Baidal, J. I. (2018). Smart sustainability : A new perspective in the sustainable tourism debate. *Investigaciones Regionales-Journal of Regional Research, 42*, 151–170.

Richards, G. (2007). *Cultural tourism: Global and local perspectives*. Psychology Press.

Roberts, D., Bradley, E., Roth, K., Eckmann, T., & Still, C. (2010). Linking Physical Geography Education and Research through the Development of an Environmental Sensing Network and Project-Based Learning. *Journal of Geoscience Education, 58*(5), 262–274. Advance online publication. doi:10.5408/1.3559887

Rocha, J. (2020). Smart Tourism and Smart Destinations for a Sustainable Future. In W. Leal Filho, A. M. Azul, L. Brandli, A. Lange Salvia, & T. Wall (Eds.), *Decent Work and Economic Growth* (pp. 1–10). Springer International Publishing. doi:10.1007/978-3-319-71058-7_88-1

Rodrigues, V., Eusébio, C., & Breda, Z. (2023). Vitor Rodrigues, Celeste Eusébio, Zélia Breda. (2022), "Enhancing sustainable development through tourism digitalisation: A systematic literature review. *Information Technology & Tourism, 25*(1), 13–45. doi:10.1007/s40558-022-00241-w

Rosário, A. T., & Dias, J. C. (2024). Exploring the Landscape of Smart Tourism : A Systematic Bibliometric Review of the Literature of the Internet of Things. *Administrative Sciences, 14*(2), 22. doi:10.3390/admsci14020022

Rosário, A., & Dias, J. (2023). How has data-driven marketing evolved: Challenges and opportunities with emerging technologies. *International Journal of Information Management Data Insights, 3*(2), 100203. doi:10.1016/j.jjimei.2023.100203

Saarinen, J. (2020). *Tourism and sustainable development goals: Research on sustainable tourism geographies*. Routledge.

Saayman, M., & Saayman, A. (2012). Shopping tourism or tourists shopping? A case study of South Africa's African tourism market. *Tourism Economics, 18*(6), 1313–1329. doi:10.5367/te.2012.0169

Said, S. (2023). The Role of Artificial Intelligence (AI) and Data Analytics in Enhancing Guest Personalization in Hospitality. *Journal of Modern Hospitality, 2*(1), 1–13. doi:10.47941/jmh.1556

Sakas, D. P., Reklitis, D. P., Terzi, M. C., & Vassilakis, C. (2022). Multichannel Digital Marketing Optimizations through Big Data Analytics in the Tourism and Hospitality Industry. *Journal of Theoretical and Applied Electronic Commerce Research, 17*(4), 1383-1408. Advance online publication. doi:10.3390/jtaer17040070

Saleh, M. (2023). Advantaging Tourism Through Influencers: Applying Transaction Cost Theory to Recognize Top Hero, Hub, and Hygiene Content Tactics for Tourism Marketing. *Journal of Travel Research*, 00472875231214727. Advance online publication. doi:10.1177/00472875231214727

Samaddar, K., & Mondal, S. (2023). AR and VR-based travel : A responsible practice towards sustainable tourism. *International Journal of Tourism Cities*. https://www.emerald.com/insight/content/doi/10.1108/IJTC-05-2022-0135/full/html

Samara, D., Magnisalis, I., & Peristeras, V. (2020). Artificial intelligence and big data in tourism: A systematic literature review. *Journal of Hospitality and Tourism Technology, 11*(2), 343–367. doi:10.1108/JHTT-12-2018-0118

Samatas, M. (2007). Security and surveillance in the Athens 2004 Olympics: Some lessons from a troubled story. *International Criminal Justice Review, 17*(3), 220–238. doi:10.1177/1057567707306649

Sarfraz, M., Khawaja, K., Han, H., Ariza-Montes, A., & Arjona-Fuentes, J. (2023). Sustainable supply chain, digital transformation, and blockchain technology adoption in the tourism sector. *Humanities & Social Sciences Communications, 10*(1), 1–13. doi:10.1057/s41599-023-02051-9

Sarkar, S., & Clegg, S. R. (2021). Resilience in a time of contagion: Lessons from small businesses during the COVID-19 pandemic. *Journal of Change Management, 21*(2), 242–267. doi:10.1080/14697017.2021.1917495

Sarmento, E. M., & Loureiro, S. M. C. (2021). Exploring the role of norms and habit in explaining pro-environmental behavior intentions in situations of use robots and AI agents as providers in tourism sector. *Sustainability (Basel), 13*(24), 13928. doi:10.3390/su132413928

Saseanu, A., Ghita, S., Albastroiu, I., & Stoian, C. (2020). Aspects of Digitalization and Related Impact on Green Tourism in European Countries. *Information (Basel), 11*(11), 507. Advance online publication. doi:10.3390/info11110507

Savchuk & Pasichnyk. (2015). Modern Tendention in the Use of Gps Technology in Tourism Industry. *Journal on Economics of Technology*.

Scarano, M. C., Regany, F., & Özçağlar-Toulouse, N. (2023). L'engagement du consommateur envers la marque: Perspective socio-culturelle et voies de recherche. *Recherche et Applications en Marketing, 38*(4), 71–97. doi:10.1177/07673701231180409

Scholz, J., & Smith, A. N. (2016). Augmented Reality: Designing Immersive Experiences That Maximize Consumer Engagement. *Business Horizons, 59*(2), 149–161. Advance online publication. doi:10.1016/j.bushor.2015.10.003

Schreiber, S., Faude, O., Gärtner, B., Meyer, T., & Egger, F. (2022). Risk of SARS-CoV-2 Transmission from on-Field Player Contacts in Amateur, Youth and Professional Football (Soccer). *British Journal of Sports Medicine, 56*(3), 158–164. doi:10.1136/bjsports-2021-104441 PMID:34663570

Scott, D., Hall, C. M., & Gössling, S. (2021). *COVID-19 and tourism: Impacts, adaptation & recovery*. Routledge.

Scott, G., & Rajabifard, A. (2017). Sustainable Development and Geospatial Information: A Strategic Framework for Integrating a Global Policy Agenda into National Geospatial Capabilities. *Geo-Spatial Information Science, 20*(2), 59–76. doi:10.1080/10095020.2017.1325594

Scrioşteanu, A., Barbu, C. M., & Criveanu, M. M. (2022). Sustainable tourism in the current European context. In Contemp. Stud. Econ. Financ. Anal. (Vol. 108B, p. 17-36). Emerald Group Holdings Ltd.; Scopus. doi:10.1108/S1569-37592022000108B031

Sempels, C., & Lille, E. S. C. (2009). Vers une meilleure compréhension des écarts entre discours et comportements dans le cadre d'une consommation responsable: apports de la littérature. *International Marketing Trends Congress.*

Séré de Lanauze, G., Lallement, J., & De Ferran, F. (2021). Représentations sociales du consommateur responsable: les autres consommateurs ont-ils vraiment envie de le suivre? *Revue management et avenir*, (5), 141-160.

Séré de Lanauze, G., Lallement, J., & de Ferran, F. (2021). Représentations sociales du consommateur responsable : Les autres consommateurs ont-ils vraiment envie de le suivre? *Management & Avenir*, *125*(5), 141–160. doi:10.3917/mav.125.0141

Shafiee, S., Rajabzadeh Ghatari, A., Hasanzadeh, A., & Jahanyan, S. (2021). Smart tourism destinations: A systematic review. *Tourism Review*, *76*(3), 505–528. doi:10.1108/TR-06-2019-0235

Shamim, S., Yang, Y., Zia, N. U., & Shah, M. H. (2021). Big data management capabilities in the hospitality sector : Service innovation and customer generated online quality ratings. *Computers in Human Behavior*, *121*, 106777. doi:10.1016/j.chb.2021.106777

Sharma, G. D., Thomas, A., & Paul, J. (2021). Reviving tourism industry post-COVID-19: A resilience-based framework. *Tourism Management Perspectives*, *37*, 100786. doi:10.1016/j.tmp.2020.100786 PMID:33391988

Sheldon, P. J. (1997). 18 Tourism information technology. *International handbook on the economics of tourism, 399.*

Shen, S., Sotiriadis, M., & Zhou, Q. (2020). Could smart tourists be sustainable and responsible as well? The contribution of social networking sites to improving their sustainable and responsible behavior. *Sustainability (Basel)*, *12*(4), 1470. doi:10.3390/su12041470

Sheth, J. N., Sethia, N. K., & Srinivas, S. (2011). Mindful consumption: A customer-centric approach to sustainability. *Journal of the Academy of Marketing Science*, *39*(1), 21–39. doi:10.1007/s11747-010-0216-3

Shi, T. (2019). Spatial Data Mining and Big Data Analysis of Tourist Travel Behaviours. *Ingenierie Des Systemes D'information, 24*(2).

Shi, B., Zhao, J., & Po, J. C. (2017). Exploring Urban Tourism Crowding in Shanghai via Crowdsourcing Geospatial Data. *Current Issues in Tourism*, *20*(11), 1186–1209. Advance online publication. doi:10.1080/13683500.2016.1224820

Shin, H. H., Jeong, M., So, K. K. F., & DiPietro, R. (2022). Consumers' experience with hospitality and tourism technologies : Measurement development and validation. *International Journal of Hospitality Management*, *106*, 103297. doi:10.1016/j.ijhm.2022.103297

Shoval, N., Isaacson, M., & Chhetri, P. (2014). GPS, Smartphones, and the Future of Tourism Research. The Wiley Blackwell Companion to Tourism. doi:10.1002/9781118474648.ch20

Shoval, N. (2018). Sensing Tourists: Geoinformatics and the Future of Tourism Geography Research. *Tourism Geographies*, *20*(5), 910–912. Advance online publication. doi:10.1080/14616688.2018.1437768

Shoval, N., & Ahas, R. (2016). The Use of Tracking Technologies in Tourism Research: The First Decade. *Tourism Geographies*, *18*(5), 587–606. Advance online publication. doi:10.1080/14616688.2016.1214977

Shukla, A., Choudhury, T., Benara, N., Garg, P., Tiwari, A., & Um, J. S. (2023). Geo-Based Recommendation System Utilising Geo Tagging and K-Means Clustering. *Spatial Information Research*, *31*(3), 253–263. Advance online publication. doi:10.1007/s41324-022-00495-w

Sigala, M. (2017). Collaborative commerce in tourism: Implications for research and industry. *Current Issues in Tourism*, *20*(4), 346–355. doi:10.1080/13683500.2014.982522

Sigala, M. (2020). Tourism and COVID-19: Impacts and implications for advancing and resetting industry and research. *Journal of Business Research*, *117*, 312–321. doi:10.1016/j.jbusres.2020.06.015 PMID:32546875

Sigala, M., Christou, E., & Gretzel, U. (Eds.). (2012). *Social media in travel, tourism and hospitality: Theory, practice and cases*. Ashgate Publishing, Ltd.

Sin, K.-Y., Mohamad, A. A., & Lo, M.-C. (2020). A critical review of literature in the rising tide of social media towards promoting tourism. *Enlightening Tourism, 10*(2), 270-305. Advance online publication. doi:10.33776/et.v10i2.4887

Sivarajah, U., Irani, Z., Gupta, S., & Mahroof, K. (2020). Role of big data and social media analytics for business to business sustainability: A participatory web context. *Industrial Marketing Management*, *86*, 163–179. doi:10.1016/j.indmarman.2019.04.005

Sixt. (n.d.). https://www.sixt.com/magazine/tips/top-free-navigation-apps/

Slocum, S. L., Aidoo, A., & McMahon, K. (2020). *The business of sustainable tourism development and management*. Routledge. doi:10.4324/9781351031066

Small, D. A., & Cryder, C. (2016). Prosocial consumer behaviours. *Current Opinion in Psychology*, *10*, 107–111. doi:10.1016/j.copsyc.2016.01.001

Snyder, H. (2019). Literature review as a research methodology: An overview and guidelines. *Journal of Business Research*, *104*, 333–339. doi:10.1016/j.jbusres.2019.07.039

Song, J. (2017). *A Systematic Literature Review: Existing Hospitality & Tourism Research on Big Data*. Academic Press.

Sonko, S. M., & Deac, A. L. (2020). From simple tourism to smart tourism : The bet for the development of a responsible and sustainable tourism. cases in the Ziguinchor Region (Senegal). *Annals of the University of Oradea, Geography Series/Analele Universitatii Din Oradea, Seria Geografie, 30*(1). https://www.academia.edu/download/103324740/auog.301102-838.pdf

Soomro, K., Bhutta, M. N. M., Khan, Z., & Tahir, M. A. (2019). Smart city big data analytics: An advanced review. *Wiley Interdisciplinary Reviews. Data Mining and Knowledge Discovery*, *9*(5), e1319. Advance online publication. doi:10.1002/widm.1319

Sotomayor Granda, M. F. (2019). Propuesta de difusión turística mediante herramientas web y estrategias de marketing digital. Caso de estudio: Cantón Loja, Ecuador. *Siembra*, *6*(1), 68–84. doi:10.29166/siembra.v6i1.1717

Sousa, A., & Paniza Prados, J. L. (2020). Visitor management in world heritage destinations before and after Covid-19, Angkor. *Sustainability (Basel)*, *12*(23), 9929. doi:10.3390/su12239929

Spadaro, I., Pirlone, F., Bruno, F., Saba, G., Poggio, B., & Bruzzone, S. (2023). Stakeholder Participation in Planning of a Sustainable and Competitive Tourism Destination: The Genoa Integrated Action Plan. *Sustainability (Basel)*, *15*(6), 5005. doi:10.3390/su15065005

Sparks, P., & Shepherd, R. (1992). Self-identity and the theory of planned behaviours: Assesing the role of identification with" green consumerism. *Social Psychology Quarterly*, *55*(4), 388–399. doi:10.2307/2786955

Stankov, U., & Gretzel, U. (2020). Tourism 4.0 technologies and tourist experiences : A human-centered design perspective. *Information Technology & Tourism*, *22*(3), 477–488. Advance online publication. doi:10.1007/s40558-020-00186-y

Steenbosch, M. J., Van der Vegt, S. C., & Van der Mast, C. A. (2021). User interaction in virtual reality: A review of literature from a psychological perspective. *Computers in Human Behavior*, *117*, 106680. doi:10.1016/j.chb.2021.106680

Steene-Johannessen, J., Hansen, B. H., Dalene, K. E., Kolle, E., Northstone, K., Møller, N. C., Grøntved, A., Wedderkopp, N., Kriemler, S., Page, A. S., Puder, J. J., Reilly, J. J., Sardinha, L. B., van Sluijs, E. M. F., Andersen, L. B., van der Ploeg, H., Ahrens, W., Flexeder, C., Standl, M., ... Ekelund, U. (2020). Variations in Accelerometry Measured Physical Activity and Sedentary Time across Europe— Harmonized Analyses of 47,497 Children and Adolescents. *The International Journal of Behavioral Nutrition and Physical Activity*, *17*(1), 38. doi:10.1186/s12966-020-00930-x PMID:32183834

Stephan, E., Liberman, N., & Trope, Y. (2011). The effects of time perspective and level of construal on social distance. *Journal of Experimental Social Psychology*, *47*(2), 397–402. doi:10.1016/j.jesp.2010.11.001 PMID:21836728

Stern, P. C., Dietz, T., Abel, T., Guagnano, G. A., & Kalof, L. (1999). A value-belief-norm theory of support for social movements: The case of environmentalism. *Human Ecology Review*, 81–97.

Stock adobe (n.d.). https://stock.adobe.com/

Stojanovic, I., Andreu, L., & Curras-Perez, R. (2022). Social media communication and destination brand equity. *Journal of Hospitality and Tourism Technology*, *13*(4), 650–666. doi:10.1108/JHTT-11-2020-0302

Stone, L. S., Stone, M. T., Mogomotsi, P. K., & Mogomotsi, G. E. (2021). The impacts of Covid-19 on nature-based tourism in Botswana: Implications for community development. *Tourism Review International*, *25*(2-3), 263–278. doi: 10.3727/154427221X16098837279958

Stott, C. J., Hutchison, P., & Drury, J. (2001). Hooligans abroad? Inter-group dynamics, social identity and participation in collective disorder at the 1998 World Cup finals. *British Journal of Social Psychology*, *40*(3), 359–384. doi:10.1348/014466601164876 PMID:11593939

Stroumpoulis, A., Kopanaki, E., & Varelas, S. (2022). Role of artificial intelligence and big data analytics in smart tourism: A resource-based view approach. *WIT Transactions on Ecology and the Environment*, *256*, 99–108. doi:10.2495/ST220091

Stuit, A., Brockington, D., & Corbera, E. (2022). Smart, commodified and encoded: Blockchain technology for environmental sustainability and nature conservation. *Conservation & Society*, *20*(1), 12–23. doi:10.4103/cs.cs_41_21

Stylos, N., & Zwiegelaar, J. (2019). *Big data as a game changer: how does it shape business intelligence within a tourism and hospitality industry context?* Springer Singapore.

Su, L., Tang, B., & Nawijn, J. (2021). How tourism activity shapes travel experience sharing: Tourist well-being and social context. *Annals of Tourism Research*, *91*, 103316. doi:10.1016/j.annals.2021.103316

Sümer, S. I., & Parilti, N. (Eds.). (2023). *Social Media Analytics in Predicting Consumer Behavior*. CRC Press. doi:10.1201/9781003200154

Sun, Y., Fan, H., Bakillah, M., & Zipf, A. (2015). Road-Based Travel Recommendation Using Geo-Tagged Images. *Computers, Environment and Urban Systems*, *53*, 110–122. Advance online publication. doi:10.1016/j.compenvurbsys.2013.07.006

Šuput, D. (2021). FIFA and UEFA regulations against violence at football matches – the problem of overlapping competencies between the state and sports organizations. *Bezbednost*, *1/2012*, 166.

Suresh, K., Wilson, C., Quayle, A., Managi, S., & Khanal, U. (2022). Can a tourist levy protect national park resources and compensate for wildlife crop damage? An empirical investigation. *Environmental Development*, *42*, 100697. doi:10.1016/j.envdev.2021.100697

Sustacha, I., Baños-Pino, J. F., & Del Valle, E. (2023). The role of technology in enhancing the tourism experience in smart destinations : A meta-analysis. *Journal of Destination Marketing & Management*, *30*, 100817. doi:10.1016/j.jdmm.2023.100817

Sutarman, A., Rahardja, U., Oganda, F. P., Millah, S., & Azizah, N. N. (2023). The Role of Information Technology in Empowering the Creative Economy for Sustainable Tourism. [ATT]. *Aptisi Transactions on Technopreneurship*, *5*(2sp), 175–185. doi:10.34306/att.v5i2sp.352

Swarbrooke, J. (1999). *Sustainable Tourism Management*. Cabi. doi:10.1079/9780851993140.0000

Sykes, G. M., & Matza, D. (1957). Techniques of Neutralization: A Theory of Delinquency'. *American Sociological Review*, *22*(6), 664. doi:10.2307/2089195

Taha & El-Mawardy. (2022). *The effect of applying geofencing technology on the Egyptian customers' purchasing intention to tour services.* . doi:10.21608/mkaf.2022.254115

Tahiri, A., Kovaçi, I., & Trajkovska Petkoska, A. (2022). Sustainable tourism as a potential for promotion of regional heritage, local food, traditions, and diversity—Case of Kosovo. *Sustainability (Basel)*, *14*(19), 12326. doi:10.3390/su141912326

Tao, T. C. H., & Wall, G. (2009). Tourism as a sustainable livelihood strategy. *Tourism Management*, *30*(1), 90–98. doi:10.1016/j.tourman.2008.03.009

Tarazona-Montoya, R., Peris-Ortiz, M., & Devece, C. (2020). The value of cluster association for digital marketing in tourism regional development. *Sustainability (Switzerland)*, *12*(23), 1-18. Advance online publication. doi:10.3390/su12239887

Tariq, E., Alshurideh, M., Akour, I., Al-Hawary, S., & Kurdi, B. (2022). The role of digital marketing, CSR policy and green marketing in brand development. *International Journal of Data and Network Science*, *6*(3), 995–1004. doi:10.5267/j.ijdns.2022.1.012

Tavakoli, M., & Wijesinghe, A. M. (2019). Rethinking destination marketing in a disruptive digital environment. *Journal of Travel Research*, *58*(8), 1502–1522.

Tavitiyaman, P., Qu, H., Tsang, W. L., & Lam, C. R. (2021). The influence of smart tourism applications on perceived destination image and behavioral intention : The moderating role of information search behavior. *Journal of Hospitality and Tourism Management*, *46*, 476–487. doi:10.1016/j.jhtm.2021.02.003

Tawafak, J., Kim, J., & Liu, Z. (2023). The effects of virtual reality (VR) experience on tourists' behavioral intention, satisfaction, and destination image: A moderated mediation model. *Journal of Travel Research*, *62*(3), 822–842.

Tawafak, R. M., Al-Rahmi, W. M., Almogren, A. S., Al Adwan, M. N., Safori, A., Attar, R. W., & Habes, M. (2023). Analysis of e-learning system use using combined TAM and ECT factors. *Sustainability (Basel)*, *15*(14), 11100. doi:10.3390/su151411100

Tham, A., & Sigala, M. (2020). Road block (chain): Bit (coin) s for tourism sustainable development goals? *Journal of Hospitality and Tourism Technology*, *11*(2), 203–222. doi:10.1108/JHTT-05-2019-0069

Thangam, D., & Chavadi, C. (2023). Impact of Digital Marketing Practices on Energy Consumption, Climate Change, and Sustainability. *Climate and Energy*, *39*(7), 11–19. doi:10.1002/gas.22329

Thommandru, A., Espinoza-Maguiña, M., Ramirez-Asis, E., Ray, S., Naved, M., & Guzman-Avalos, M. (2023). Role of Tourism and Hospitality Business in Economic Development. *Materials Today: Proceedings*, *80*, 2901–2904. Advance online publication. doi:10.1016/j.matpr.2021.07.059

Tien, N. H., Viet, P. Q., Duc, N. M., & Tam, V. T. (2021). Sustainability of tourism development in Vietnam's coastal provinces. *World Review of Entrepreneurship, Management and Sustainable Development*, *17*(5), 579–598. doi:10.1504/WREMSD.2021.117443

Tjilen, A., Ririhena, S., & Manuhutu, F. (2021). Model Ecotourism-Based Tourism Development Strategy and Empowerment of Local Communities in Merauke Regency. *Journal of Economics, Trade and Marketing Management*, *3*(01), 48–59. doi:10.22158/jetmm.v3n1p48

Tong, Y., Zhou, Z., Zeng, Y., Chen, L., & Shahabi, C. (2020). Spatial Crowdsourcing: A Survey. *The VLDB Journal*, *29*(1), 217–250. Advance online publication. doi:10.1007/s00778-019-00568-7

Toros, A., & Özkan, O. A. (2024). Understanding users' acceptance of virtual reality applications in tourism: A technology acceptance model perspective. *International Journal of Tourism Research*, *26*(2), 391–406.

Toros, E., Asiksoy, G., & Sürücü, L. (2024). Refreshment students' perceived usefulness and attitudes towards using technology: A moderated mediation model. *Humanities & Social Sciences Communications*, *11*(1), 1–10. doi:10.1057/s41599-024-02839-3

Torre, C., Guazzo, G. M., Çekani, V., & Bacco, V. (2022). The relationship between big data and decision making. A systematic literature review. *Journal of Service Science and Management*, *15*(2), 89–107. doi:10.4236/jssm.2022.152007

Tóth, N. A. (2019). Sports Policing in the System of Sports Administration. University of Public Service, Faculty of Public Governance and International Studies.

Tóth, N.A. (2021). The Application of Innovative Sports Policing Tools. *Časopis Policajna Teoria a Prax, 2*, 102-114.

Tóth, N. A. (2020). Thoughts about the current issues of sports policing. *Belügyi Szemle*, *68*(Special Issue), 77–93. doi:10.38146/BSZ.SPEC.2020.3.6

Treiblmaier, H. (2022). Blockchain and tourism. In *Handbook of e-Tourism* (pp. 475–495). Springer International Publishing. doi:10.1007/978-3-030-48652-5_28

Trentmann, F. (2008). Le consommateur en tant que citoyen: Synergies et tensions entre bien-être et engagement civique. *Economics and Politics*, 7–20.

Tribe, J. (2020). *The economics of recreation, leisure and tourism.* Routledge. doi:10.4324/9780429278259

Tricco, A. C., Lillie, E., Zarin, W., O'Brien, K. K., Colquhoun, H., Levac, D., Moher, D., Peters, M. D. J., Horsley, T., Weeks, L., Hempel, S., Akl, E. A., Chang, C., McGowan, J., Stewart, L., Hartling, L., Aldcroft, A., Wilson, M. G., Garritty, C., ... Straus, S. E. (2018). PRISMA extension for scoping reviews (PRISMA-ScR): Checklist and explanation. *Annals of Internal Medicine*, *169*(7), 467–473. doi:10.7326/M18-0850 PMID:30178033

Truby, J. (2018). Decarbonizing Bitcoin: Law and policy choices for reducing the energy consumption of Blockchain technologies and digital currencies. *Energy Research & Social Science*, *44*, 399–410. doi:10.1016/j.erss.2018.06.009

Tsaih, R.-H., & Hsu, C. (2018). Artificial Intelligence in Smart Tourism: A Conceptual Framework. *ICEB 2018 Proceedings*. https://aisel.aisnet.org/iceb2018/89/

Tucci, G., & Bonora, V. (2015). Geomatics and Management of At-Risk Cultural Heritage. *Rendiconti Lincei. Scienze Fisiche e Naturali*, *26*(S1), 105–114. Advance online publication. doi:10.1007/s12210-015-0427-0

Tulloch, J. (2000). Terrorism, "killing events," and their audience: Fear of crime at the 2000 Olympics. In K. Schaffer & S. Sidonie (Eds.), *The Olympics at the millennium: Power, politics and the games* (pp. 224–242). Rutgers University Press.

Turek, J., Ocicka, B., Rogowski, W., & Jefmański, B. (2023). The role of Industry 4.0 technologies in driving the financial importance of sustainability risk management. *Equilibrium. Quarterly Journal of Economics and Economic Policy*, *18*(4), 1009–1044. doi:10.24136/eq.2023.032

Tussyadiah, I. P., & Zach, F. J. (2012). The Role of Geo-Based Technology in Place Experiences. *Annals of Tourism Research, 39*(2), 780–800. Advance online publication. doi:10.1016/j.annals.2011.10.003

Tussyadiah, I., Jung, T., & Kim, J. (2018). The effects of virtual reality (VR) on destination image, perceived authenticity, and purchase intention. *Journal of Travel Research, 57*(2), 233–247.

Tyan, I., Yagüe, M. I., & Guevara-Plaza, A. (2021). Blockchain technology's potential for sustainable tourism. In *Information and Communication Technologies in Tourism 2021: Proceedings of the ENTER 2021 eTourism Conference, January 19–22, 2021* (pp. 17-29). Springer International Publishing.

Ugur, N. G., & Akbıyık, A. (2020). Impacts of COVID-19 on global tourism industry: A cross-regional comparison. *Tourism Management Perspectives, 36*, 189–194. doi:10.1016/j.tmp.2020.100744 PMID:32923356

Ulusoy, E. (2016). Experiential responsible consumption. *Journal of Business Research, 69*(1), 284–297. doi:10.1016/j.jbusres.2015.07.041

Um, S., & Crompton, J. L. (1992). The roles of perceived inhibitors and facilitators in pleasure travel destination decisions. *Journal of Travel Research, 30*(3), 18–25. doi:10.1177/004728759203000303

UNESCO. (n.d.). *Medina of Fez*. UNESCO World Heritage Centre. Retrieved March 31, 2024, from https://whc.unesco.org/en/list/170/

United Nations. (2020). Beyond the Year of Return: Africa and the diaspora must forge closer ties. Author.

UNWTO (World Tourism Organization). (2017). *World Tourism Organization millennial report*. https://www.e-unwto.org/doi/pdf/10.18111/9789284417162

UNWTO. (2017). Discussion Paper on the Occasion of the International Year of Sustainable Tourism for Development 2017. *United Nations World Tourism Organization.*

UNWTO. (2020). *Impact assessment of the COVID-19 outbreak on international tourism*. www.unwto.org/impact-assessment-of-the-covid-19-outbreakon-international-tourism

Uslu, A., & Karabulut, A. N. (2018). *Touristic destinations' perceived risk and perceived value as indicators of E-WOM and revisit intentions*. Academic Press.

Utari, B. A., Arsallia, S., Ramdani, M. A., Rahmafitria, F., Belgiawan, P. F., Dirgahayani, P., & Nasution, R. A. (2024). Tourist destination choice on five priority destinations of Indonesia during health crisis. *Journal of Destination Marketing & Management, 32*, 100880. doi:10.1016/j.jdmm.2024.100880

Valeri, M., & Baggio, R. (2021). Social network analysis: Organizational implications in tourism management. *The International Journal of Organizational Analysis, 29*(2), 342–353. doi:10.1108/IJOA-12-2019-1971

van der Spek, S., van Schaick, J., de Bois, P., & de Haan, R. (2009). Sensing Human Activity: GPS Tracking. *Sensors (Basel), 9*(4), 3033–3055. Advance online publication. doi:10.3390/s90403033 PMID:22574061

Van Middelkoop, M., Borgers, A., & Timmermans, H. (2003). Inducing heuristic principles of tourist choice of travel mode: A rule-based approach. *Journal of Travel Research, 42*(1), 75–83. doi:10.1177/0047287503254116

Van Nuenen, T. (2019). Algorithmic authenticity: Sociotechnical authentication processes on online travel platforms. *Tourist Studies, 19*(3), 378–403. doi:10.1177/1468797619832314

Van Teijlingen, E., & Hundley, V. (2002). *Pilot Testing for Feasibility in a Study of Student Retention and Attrition in Online Undergraduate Programs*. Academic Press.

Vassakis, K., Petrakis, E., Kopanakis, I., Makridis, J., & Mastorakis, G. (2019). Location-based social network data for tourism destinations. *Big data and innovation in tourism, travel, and hospitality: Managerial approaches, techniques, and applications*, 105-114.

Vassakis, K., Petrakis, E., Kopanakis, I., & Makridis, J. (2021). Location intelligence for tourism destinations: A big data comparative analysis through location-based social networks. *International Journal of Tourism Policy*, *11*(3), 247–264. doi:10.1504/IJTP.2021.118664

Vazquez Melendez, E. I., Bergey, P., & Smith, B. (2024). Blockchain technology for supply chain provenance: Increasing supply chain efficiency and consumer trust. *Supply Chain Management*. Advance online publication. doi:10.1108/SCM-08-2023-0383

Vecchio, P. D., Mele, G., Ndou, V., & Secundo, G. (2018). Creating value from Social Big Data : Implications for Smart Tourism Destinations. *In (Big) Data we trust: Value creation in knowledge organizations, 54*(5), 847-860. doi:10.1016/j.ipm.2017.10.006

Vehbi, B. O. (2012). A model for assessing the level of tourism impacts and sustainability of coastal cities. *Strategies for tourism industry-Micro and Macro Perspectives*, 99-114.

Vermeir, I., & Verbeke, W. (2006). Sustainable Food Consumption : Exploring the Consumer "Attitude – Behavioral Intention" Gap. *Journal of Agricultural & Environmental Ethics*, *19*(2), 169–194. doi:10.1007/s10806-005-5485-3

Vermeir, I., & Verbeke, W. (2008). Sustainable food consumption among young adults in Belgium: Theory of planned behaviour and the role of confidence and values. *Ecological Economics*, *64*(3), 542–553. doi:10.1016/j.ecolecon.2007.03.007

Vigliotti, M., & Jones, H. (2020). *The Executive Guide to Blockchain: Using Smart Contracts and Digital Currencies in Your Business*. Springer Nature. doi:10.1007/978-3-030-21107-3

Vivek, S. D., Beatty, S. E., & Morgan, R. M. (2012). Customer engagement: Exploring customer relationships beyond purchase. *Journal of Marketing Theory and Practice*, *20*(2), 122–146. doi:10.2753/MTP1069-6679200201

Vu, H. Q., Li, G., Law, R., & Ye, B. H. (2015). Exploring the Travel Behaviors of Inbound Tourists to Hong Kong Using Geotagged Photos. *Tourism Management*, *46*, 222–232. doi:10.1016/j.tourman.2014.07.003

Vujičić, M. D., Stankov, U., & Vasiljević, Đ. A. (2022). Tourism at a crossroads–ignoring, adopting, or embracing alternative pathways for more sustainable post-pandemic tourism development. In *Crisis management, destination recovery and sustainability* (pp. 11–21). Routledge. doi:10.4324/9781003295839-3

Wachyuni, S. S., & Kusumaningrum, D. A. (2020). The effect of COVID-19 pandemic: How are the future tourist behavior? *Journal of Education. Society and Behavioural Science*, *33*(4), 67–76.

Walters, G., & Tacon, R. (2010). Corporate social responsibility in sport: Stakeholder management in the UK football industry. *Journal of Management & Organization*, *16*(4), 566–586. doi:10.5172/jmo.2010.16.4.566

Wang, D., Park, S., & Fesenmaier, D. R. (2012). The role of smartphones in mediating the touristic experience. *Journal of Travel Research*, *51*(4), 371–387. doi:10.1177/0047287511426341

Wang, L. H., Yeh, S. S., Chen, K. Y., & Huan, T. C. (2022). Tourists' travel intention: Revisiting the TPB model with age and perceived risk as moderator and attitude as mediator. *Tourism Review*, *77*(3), 877–896. doi:10.1108/TR-07-2021-0334

Wang, Q., Li, J., & Li, Y. (2022). Understanding the adoption of virtual consumption: A social exchange theory perspective. *Sustainability*, *14*(18), 11835.

Wang, Y., He, J., Zhu, N., Yi, Y., Zhang, Q., Song, H., & Xue, R. (2021). Security enhancement technologies for smart contracts in the blockchain: A survey. *Transactions on Emerging Telecommunications Technologies*, *32*(12), e4341. doi:10.1002/ett.4341

Wang, Z., Li, M., Lu, J., & Cheng, X. (2022). Business Innovation based on artificial intelligence and Blockchain technology. *Information Processing & Management*, *59*(1), 102759. doi:10.1016/j.ipm.2021.102759

Wan, P., Huang, L., & Holtskog, H. (2020). Blockchain-enabled information sharing within a supply chain: A systematic literature review. *IEEE Access : Practical Innovations, Open Solutions*, *8*, 49645–49656. doi:10.1109/ACCESS.2020.2980142

Watson, A.M., Haraldsdottir, K., Biese, K., Goodavish, L., Stevens, B., & McGuine, T. (2021). The Association of COVID-19 Incidence with Sport and Face Mask use in United States High School Athletes. *Journal of Athletic Training*.

Wen, J., Aston, J., Liu, X., & Ying, T. (2020). Effects of misleading media coverage on public health crisis: A case of the 2019 novel coronavirus outbreak in China. *Anatolia*, *31*(2), 331–336. doi:10.1080/13032917.2020.1730621

White, K., Habib, R., & Hardisty, D. J. (2019). How to SHIFT consumer behaviours to be more sustainable: A literature review and guiding framework. *Journal of Marketing*, *83*(3), 22–49. doi:10.1177/0022242919825649

Whitson, D., & Horne, J. (2006). Underestimated costs and overestimated benefits? Comparing the outcomes of sports mega-events in Canada and Japan. *The Sociological Review*, *54*(2), 73–89. doi:10.1111/j.1467-954X.2006.00654.x

Wibawa, I., Astawa, I., Sukmawati, N., Arjana, I., & Sutarma, I. (2022). Digital Marketing and Sustainable Tourism for Tourist Villages in Bangli Regency. *International Journal of Glocal Tourism*, *3*(2), 88–99. doi:10.58982/injogt.v3i2.209

Wider, W., Gao, Y., Chan, C. K., Lin, J., Li, J., Tanucan, J. C. M., & Fauzi, M. A. (2023). Unveiling trends in digital tourism research : A bibliometric analysis of co-citation and co-word analysis. *Environmental and Sustainability Indicators*, *100308*, 100308. Advance online publication. doi:10.1016/j.indic.2023.100308

Williams, A. M., & Hobson, J. (1995). Virtual reality in tourism: A potential application to heritage interpretation. *Journal of Travel Research*, *33*(1), 13–19.

Wired. (n.d.). https://www.wired.com/gallery/best-language-learning-apps/

Women on the road. (2021). https://www.women-on-the-road.com/best-travel-forums.html

Wong, E., Law, R., & Li, G. (2017). Reviewing Geotagging Research in Tourism. In Information and Communication Technologies in Tourism 2017. doi:10.1007/978-3-319-51168-9_4

Wong, I. A., Wan, Y. K. P., Huang, G. I., & Qi, S. (2022). Green event directed pro-environmental behaviours: An application of goal systems theory. In *Events and Sustainability* (pp. 210–231). Routledge. doi:10.4324/9781003314295-12

WordPress. (n.d.). https://es.wordpress.org/themes/the-travel-booking/

World Tourism Organization and Organization of American States. (2018). *Tourism and the Sustainable Development Goals – Good Practices in the Americas*. UNWTO.

Worldometer. (2020). *COVID-19 coronavirus / cases*. Retrieved January 3, 2021, from https://www.worldometers.info/coronavirus/coronavirus-cases/#case-distribution

Wu, X., Huang, Z., Peng, X., Chen, Y., & Liu, Y. (2018). Building a Spatially-Embedded Network of Tourism Hotspots from Geotagged Social Media Data. *IEEE Access : Practical Innovations, Open Solutions*, *6*, 21945–21955. Advance online publication. doi:10.1109/ACCESS.2018.2828032

Xiang, Z., & Fesenmaier, D. R. (2017). Big data analytics, tourism design and smart tourism. *Analytics in smart tourism design: concepts and methods*, 299-307.

Xiang, Z. (2018). From digitization to the age of acceleration: On information technology and tourism. *Tourism Management Perspectives*, *25*, 147–150. doi:10.1016/j.tmp.2017.11.023

Xiang, Z., Gretzel, U., Gong, X., & Li, Y. (2020). Tourists' experiences and satisfaction with virtual reality tours. *Journal of Hospitality & Tourism Research (Washington, D.C.)*, *44*(6), 881–903.

Xiang, Z., Wöber, K., & Fesenmaier, D. R. (2008). Representation of the online tourism domain in search engines. *Journal of Travel Research*, *47*(2), 137–150. doi:10.1177/0047287508321193

Xiao, W., Mills, J., Guidi, G., Rodríguez-Gonzálvez, P., Barsanti, S. G., & González-Aguilera, D. (2018). Geoinformatics for the Conservation and Promotion of Cultural Heritage in Support of the UN Sustainable Development Goals. *ISPRS Journal of Photogrammetry and Remote Sensing*, *142*, 389–406. Advance online publication. doi:10.1016/j.isprsjprs.2018.01.001

Xie, C., Zhang, J., & Huang, S. (2023). Effect of risk message framing on tourists' travel intention: Roles of resilience and impulsivity. *Journal of Travel Research*, *62*(4), 802–819. doi:10.1177/00472875221095212

Xie, K., & Mao, Z. (2017). The impacts of quality and quantity attributes of Airbnb hosts on listing performance. *International Journal of Contemporary Hospitality Management*, *29*(9), 2240–2260. doi:10.1108/IJCHM-07-2016-0345

Xin, L. X., & Law, R. (2020). *Network analysis of big data research in tourism*. Academic Press.

Xu, C., Li, H., & Li, Y. (2014). How does store environment influence brand experience and customer loyalty? A conceptual framework and empirical evidence. *Journal of Retailing and Consumer Services*, *21*(4), 673–681.

Yaegashi, K., & Yanai, K. (2009). Can Geotags Help Image Recognition? In. Lecture Notes in Computer Science: Vol. 5414. *Including Subseries Lecture Notes in Artificial Intelligence and Lecture Notes in Bioinformatics)*. LNCS. doi:10.1007/978-3-540-92957-4_32

Yallop, A., & Seraphin, H. (2020). Big data and analytics in tourism and hospitality: Opportunities and risks. *Journal of Tourism Futures*, *6*(3), 257–262. doi:10.1108/JTF-10-2019-0108

Yamagishi, K., Canayong, D., Domingo, M., Maneja, K., Montolo, A., & Siton, A. (2023). User-generated content on Gen Z tourist visit intention: a stimulus-organism-response approach. *Journal of Hospitality and Tourism Insights*. doi:10.1108/JHTI-02-2023-0091

Yan, A., & Jia, W. (2021). The influence of eliciting awe on pro-environmental behavior of tourist in religious tourism. *Journal of Hospitality and Tourism Management*, *48*, 55–65. doi:10.1016/j.jhtm.2021.05.007

Yang, H., & Han, S. Y. (2021). Understanding virtual reality continuance: An extended perspective of perceived value. *Online Information Review*, *45*(2), 422–439. doi:10.1108/OIR-02-2020-0058

Yang, T. T., Ruan, W. Q., Li, Y. Q., & Zhang, S. N. (2023). Virtual tourist motivation: The differences between virtual tourism and on-site tourism. *Tourism Review*, *78*(5), 1280–1297. doi:10.1108/TR-07-2022-0331

Yang, Y., & Han, K. (2021). The effects of virtual reality destination experiences on perceived authenticity, destination image, and revisit intention. *Journal of Travel Research*, *60*(7), 1517–1533.

Ye, B. H., Ye, H., & Law, R. (2020). Systematic review of smart tourism research. *Sustainability (Basel)*, *12*(8), 3401. doi:10.3390/su12083401

Yigitcanlar, T., Corchado, J. M., Mehmood, R., Li, R. Y. M., Mossberger, K., & Desouza, K. (2021). Responsible Urban Innovation with Local Government Artificial Intelligence (AI): A Conceptual Framework and Research Agenda. *Journal of Open Innovation*, *7*(1), 71. doi:10.3390/joitmc7010071

Yigitcanlar, T., Desouza, K. C., Butler, L., & Roozkhosh, F. (2020). Contributions and Risks of Artificial Intelligence (AI) in Building Smarter Cities: Insights from a Systematic Review of the Literature. *Energies*, *13*(6), 1473. doi:10.3390/en13061473

Yilmaz, E. (2020). The effects on consumer behavior of hotel related comments on the tripadvisor website: an istanbul case. *Advances in hospitality and tourism research-ahtr, 8*(1), 1-29. doi:10.30519/ahtr.536303

Yilmaz, E. S., & Aytekin, M. (2018). The determinants of tourist preferences of the mediterranean region in Turkey. *Administrative Sciences*, *8*(4), 81. Advance online publication. doi:10.3390/admsci8040081

Yin, D., Li, M., & Qiu, H. (2023). Do customers exhibit engagement behaviors in AI environments? The role of psychological benefits and technology readiness. *Tourism Management*, *97*, 104745. doi:10.1016/j.tourman.2023.104745

Yong, K., & Hassan, R. A. (2019). The relationships between social media marketing and entrepreneurial success: A conceptual study on homestay business in Sabah, Malaysia. *Review of Integrative Business and Economics Research*, *8*, 25–33.

Yoo, C., Kwon, S., Na, H., & Chang, B. (2017). Factors affecting the adoption of gamified smart tourism applications : An integrative approach. *Sustainability (Basel)*, *9*(12), 2162. doi:10.3390/su9122162

Young, G. (1973). *Tourism: Blessing or blight?* Penguin.

Yuan, Y., & Medel, M. (2016). Characterizing International Travel Behavior from Geotagged Photos: A Case Study of Flickr. *PLoS One*, *11*(5), e0154885. Advance online publication. doi:10.1371/journal.pone.0154885 PMID:27159195

Yung, R., & Khoo-Lattimore, C. (2019). New realities: A systematic literature review on virtual reality and augmented reality in tourism research. *Current Issues in Tourism*, *22*(17), 2056–2081. doi:10.1080/13683500.2017.1417359

Yu, Y., Li, X., & Jai, T. M. (2017). The impact of green experience on customer satisfaction: Evidence from TripAdvisor. *International Journal of Contemporary Hospitality Management*, *29*(5), 1340–1361. doi:10.1108/IJCHM-07-2015-0371

Zafar, S. Z., Zhilin, Q., Mabrouk, F., Ramirez-Asis, E., Alzoubi, H. M., Hishan, S. S., & Michel, M. (2023). Empirical linkages between ICT, tourism, and trade towards sustainable environment : Evidence from BRICS countries. *Ekonomska Istrazivanja*, *36*(2), 2127417. Advance online publication. doi:10.1080/1331677X.2022.2127417

Zaichkowsky, J. (1985). Effects of moderator variables on the use of arousal-activation and uncertainty in developing product-related emotions. *JMR, Journal of Marketing Research*, *22*(2), 132–141.

Zaifri, M., Azough, A., & El Alaoui, S. O. (2018). Experimentation of visual augmented reality for visiting the historical monuments of the medina of Fez. *2018 International Conference on Intelligent Systems and Computer Vision (ISCV)*, 1–4. 10.1109/ISACV.2018.8354051

Zaifri, M., Khalloufi, H., Kaghat, F. Z., Benlahbib, A., Azough, A., & Zidani, K. A. (2023). Enhancing Tourist Experiences in Crowded Destinations through Mobile Augmented Reality: A Comparative Field Study. *International Journal of Interactive Mobile Technologies*, *17*(20), 20. Advance online publication. doi:10.3991/ijim.v17i20.42273

Zajadacz, A., & Minkwitz, A. (2020). Using social media data to plan for tourism. *Quaestiones Geographicae*, *39*(3), 125–138. doi:10.2478/quageo-2020-0027

Zakrisson, I., & Zillinger, M. (2012). Emotions in Motion: Tourist Experiences in Time and Space. *Current Issues in Tourism*, *15*(6), 505–523. Advance online publication. doi:10.1080/13683500.2011.615391

Zarrin, J., Wen Phang, H., Babu Saheer, L., & Zarrin, B. (2021). Blockchain for decentralization of internet: Prospects, trends, and challenges. *Cluster Computing*, *24*(4), 2841–2866. doi:10.1007/s10586-021-03301-8 PMID:34025209

Zarrouk, I., & El Aidouni, A. (2023). The Digital Marketing Practiced by the Moroccan Tourism Sector : National-scale Communication Actions and Environmental Management. EDP Sciences. doi:10.1051/e3sconf/202341201035

Zeng, D., Tim, Y., Yu, J., & Liu, W. (2020). Actualizing big data analytics for smart cities : A cascading affordance study. *International Journal of Information Management*, *54*, 102156. doi:10.1016/j.ijinfomgt.2020.102156

Zeng, Z., Zhang, R., Liu, X., Guo, X., & Sun, H. (2012). Generating Tourism Path from Trajectories and Geo-Photos. In. Lecture Notes in Computer Science: Vol. 7651. *Including Subseries Lecture Notes in Artificial Intelligence and Lecture Notes in Bioinformatics)*. LNCS. doi:10.1007/978-3-642-35063-4_15

Zepeda, L., & Deal, D. (2009). Organic and local food consumer behaviour: Alphabet theory. *International Journal of Consumer Studies*, *33*(6), 697–705. doi:10.1111/j.1470-6431.2009.00814.x

Zhang, Y., & Hiltz, S. R. (2003). Factors that influence online relationship development in a knowledge sharing community. *AMCIS 2003 Proceedings*, 53.

Zhang, D., Wen, S., Chen, F., Li, Z., & Zhao, L. (2020). Spatial Crowdsourcing Based on Web Mapping Services. *World Wide Web (Bussum)*, *23*(1), 631–648. Advance online publication. doi:10.1007/s11280-019-00708-7

Zhang, S., Song, M., & He, H. (2020). Achieving the Success of Sustainability Development Projects through Big Data Analytics and Artificial Intelligence Capability. *Sustainability (Basel)*, *12*(3), 949. doi:10.3390/su12030949

Zhang, X., Xi, W., Xu, F. Z., & Wu, R. (2023). Deconstructing consumers' low-carbon tourism promotion preference and its consequences : A heuristic-systematic model. *Journal of Hospitality and Tourism Management*, *57*, 48–60. doi:10.1016/j.jhtm.2023.08.020

Zhang, Y., Xu, S., Zhang, L., & Yang, M. (2021). Big data and human resource management research: An integrative review and new directions for future research. *Journal of Business Research*, *133*, 34–50. doi:10.1016/j.jbusres.2021.04.019

Zheng, Y., & Wu, Y. (2023). An investigation of how perceived smart tourism technologies affect tourists' well-being in marine tourism. *PLoS One*, *18*(8), e0290539. doi:10.1371/journal.pone.0290539 PMID:37624780

Zhu, J. J., Liu, Z., Huang, T., & Guo, X. S. (2023). Roboethics of tourism and hospitality industry. *Systematic Reviews*, *18*(6), e0287439–e0287439. doi:10.1371/journal.pone.0287439 PMID:37390063

About the Contributors

Youssef El Archi is a Ph.D. candidate and research assistant at the National School of Business and Management of Tangier, Abdelmalek Essaadi University in Morocco, as well as the Institute of Rural Development and Sustainable Economy, Hungarian University of Agriculture and Life Sciences, Hungary. He serves as a research collaborator at the Centre of Tourism Research, Development, and Innovation (CiTUR) hosted by the Polytechnic Institute of Leiria in Portugal and is a member of prestigious associations dedicated to promoting tourism and ICT. Additionally, he has authored numerous papers on tourism, which have been published in top-tier journals.

Brahim Benbba holds the position of a full professor in marketing at the National School of Business and Management of Tangier, affiliated with Abdelmalek Essaadi University in Morocco. He actively participates in esteemed organizations and associations committed to advancing the fields of tourism and marketing research. Furthermore, he has made significant contributions to the literature on tourism, with multiple publications in high-impact journals.

Lóránt Dénes Dávid is an extremely versatile Hungarian scholar. He has become an internationally recognized researcher and professor in the fields of geography and tourism, and holds full-professor positions in Poland, Kazakhstan, Ukraine, Slovakia and Hungary. His extensive research has resulted in many prestigious journal articles, studies, papers and books. Furthermore, he has developed an exemplary international professional cooperation network, has served as a guest professor at many universities in Europe and overseas (in all continents in a lot of countries, including USA, Russia, Japan). As an awardee of the Jean Monnet Professorship (EU Commission), he designed and implemented a major educational program in tourism and regional development. Moreover, he has been very active in editing scholarly journals. He is an ordinary member of the European Academy of Sciences and Arts (EASA) in Austria.

Lucília Cardoso holds a background in Management and Marketing, complemented by academic studies in applying Management to Tourism. With 5 years in accounting and a 7-year tenure in management consultancy at Valpaços City Council, she handled diverse tasks, including Human Resources and Project Management. Engaging in International Relations, she collaborated with the Ministry of Foreign Affairs. In tourism, she contributed to the establishment of the Interactive Tourism Shop and Wine Museum. After earning her doctoral degree in 2010, she pursued an academic career focusing on Tourism Destination Image, applying principles from management, psychology, neuroscience, and neuromarketing. From 2010 to 2019, she served as a University Professor in Management, Marketing, and Tourism at four Portuguese institutions, transitioning to a research-focused career in 2020.

* * *

Mourad Aarabe is a PhD student at the National School of Business and Management of Fez, Sidi Mohamed de Ben Abdellah University. His research focuses on tourism, marketing, digital marketing, and management.

Munir Ahmad is a seasoned professional in the realm of Spatial Data Infrastructure (SDI), Geo-Information Productions, Information Systems, and Information Governance, boasting over 25 years of dedicated experience in the field. With a PhD in Computer Science, Dr. Ahmad's expertise spans Spatial Data Production, Management, Processing, Analysis, Visualization, and Quality Control. Throughout his career, Dr. Ahmad has been deeply involved in the development and deployment of SDI systems specially in the context of Pakistan, leveraging his proficiency in Spatial Database Design, Web, Mobile & Desktop GIS, and Geo Web Services Architecture. His contributions to Volunteered Geographic Information (VGI) and Open Source Geoportal & Metadata Portal have significantly enriched the geospatial community. As a trainer and researcher, Dr. Ahmad has authored over 50 publications, advancing the industry's knowledge base and fostering innovation in Geo-Tech, Data Governance, and Information Infrastructure, and Emerging Technologies. His commitment to Research and Development (R&D) is evident in his role as a dedicated educator and mentor in the field.

Lhoussaine Alla is a research professor at the National School of Applied Sciences of Fez, Sidi Mohamed University of Ben Abdellah (Morocco). His research focuses on reflexivity and applications in integrating intelligent solutions, resulting from digitization, AI and data engineering, in various areas of management of companies and territories, including tourism, marketing, logistics, governance, etc. Professor Alla is the author of numerous scientific articles and book publishers. He also coordinates many international scientific events on different aspects of this research axis.

Kandappan Balasubramanian (Dr Kandy) is an Associate Professor from the School of Hospitality, Tourism and Events;Faculty of Social Sciences and Leisure Management (FSLM), Taylor's University and Director for the Sustainable Tourism Impact Lab and Associate Director-Human Capital Development of the Centre for Research and Innovation in Tourism (CRiT), Taylor's University, Malaysia who has embraced technology as part of his teaching and learning pedagogy and continuously empowers to innovate himself with the latest developments of technology in order to meet the changing needs of today's learners. He has taken the mandate and made it as personal aspiration on, striving to broaden his ability in technology integration in the classroom. Dr Kandappan innovation and transformation in teaching and learning during the pandemic is well-recognised by awarding him as one of the "Taylor's Distinguished e-Learning Educator (DeE) Class of 2021".

Nouhaila Ben Khizzou is a PhD student at the National School of Business and Management of Fez, Sidi Mohamed de Ben Abdellah University.

Ahmed Benjelloun is a teacher-researcher at the National School of Business and Management of Fez, Sidi Mohamed de Ben Abdellah University.

Lazey (Doma) Bhutia, a Programme Leader and Lecturer at the School of Global Hospitality & Tourism, ignites a passion for tourism in students. Leveraging their expertise, Dr. Bhutia shapes future hospitality leaders by combining academic knowledge with real-world industry experience.

Naoual Bouhtati, PhD student at the National School of Business and Management of Fez, University Sidi Mohamed de Ben Abdellah (Morocco). His research focuses on marketing, logistics and performance management.

Sakhhi Chhabra is an assistant professor of marketing at XLRI Delhi NCR. She is an FPM from MDI- Gurgaon and an MSc—marketing student at Manchester Business School, UK.

Ikram Ed-Daakouri, a student researcher in digital marketing at Sidi Mohamed University of Ben Abdellah (Morocco).

Jose Antonio Fraiz is Professor of Marketing (University of Vigo). PhD in Business Administration.

Kevin Fuchs is currently affiliated with the Faculty of Hospitality and Tourism, Prince of Songkla University in Phuket, Thailand. He was educated in Sweden, Finland and Malaysia and developed a keen interest on topics that revolve around sustainable development.

Tamara Gajić was born on June 2, 1977 in Sarajevo. She finished elementary school "Simon Bolivar" in her hometown, and high school "Svetozar Marković" in Novi Sad. In 1996, she enrolled in studies at the Department of Geography, Tourism and Hotel Management, Faculty of Science, University of Novi Sad. She graduated in 2001. He completed his master's studies in tourism management in 2007 (average grade 9.11), and after defending his master's thesis, in the same year, enrolled in doctoral studies at the Department of Geography, Tourism and Hotel Management, Faculty of Science, University of Novi Sad. He completed his doctoral studies in 2010 on the geosciences study program, module: tourism (average grade during doctoral studies: 9.83). The doctoral dissertation, under the title "Management of tourism development in the Juznobacka district", was publicly defended on June 4, 2010. From 2008 to 2021, she was employed as a professor of vocational studies in Novi Sad. She received the title of scientific associate at the Geographical Institute, Jovan Cvijić, SANU, on December 23, 2015, with re-election on February 24, 2021. years. Since October 2019, she has been engaged as a senior scientific researcher (distance studies) at the Institute of Sports, Tourism and Services of the South Ural State University in Chelyabinsk (Russia). Since 2018, she has been employed part-time as an assistant professor at the University of Business Studies, Department of Trade, Tourism and Hotel Management in Banja Luka. In the period of 2021, she worked at Singidunum University, in Belgrade, and from September of the same year, she moved to work at the Geographical Institute, Jovan Cvijić, SASA, in Belgrade, as a research associate. Since February 2022, she has been employed in supplementary work (30%), at the Faculty of Hospitality and Tourism in Vrnjačka Banja, University of Kragujevac. She has participated in a large number of international scientific meetings, and until now she has published manuscripts in international and domestic journals, as well as proceedings of scientific meetings. Tamara is the author of the monograph entitled State and perspectives of the development of rural tourism in the Southbacka district, financed in its entirety by the Ministry of Education, Science and Technological Development of the Republic of Serbia. Her research is focused on broader aspects of tourism, hotel industry, gastro-tourism, geography, and especially on rural development, agritourism, local economy, assessment of the impact of tourism development, the role of tourism in the life of local communities, spatial aspects of tourism, management aspects of tourism business and quality determinants in hospitality sector. In the period from 2013 to 2016, she worked on the project TEMPUS MHTSPS 544543 (Modernization and Harmonization of Tourism Study Programs in Serbia), as a member of the Steering Committee of the project (544543-TEMPUS-1-2013-1-RS-TEMPUS-JPCR) and coordinator for the Business School of Vocational Studies. Also, she was part of the committee, as well as the mentor of several specialist and final theses at the Higher Business School of Vocational Studies, as well as a member of the committee and mentor of several graduate theses at the University of Business Studies in Banja Luka. She was a member of the commission for the equivalence of study programs, as well as the commission for the defense and evaluation of the final thesis for basic professional studies. She is also part of the committee (commentary) for writing and defending doctoral dissertations for two candidates at the Faculty of Hotel Management and Tourism in Vrnjačka Banja. Since 2017, she has been a permanent reviewer of the magazine Economic of Agriculture, as well as the magazine School of Business (new name - International Journal of Economic Practice and Policy. She has done several reviews for magazines in the category. She is a guest editor of the magazine Frontiers in Psychology (special edition of the journal category, for May 2022, entitled: The Paradigm of Creating a New Environment for Travel Industry Consumers During the Crisis and Instability in the World). She is also a member of the International Board of the CTTH conference (International Scientific Board) - International Conference Contemporary Trends in Tourism and Hospitality, 17-19, November 2022, University of Novi Sad, Serbia. She was also appointed as a member of the accreditation committee of the Master's study at the Faculty of Hotel Management and Tourism in Vrnjačka Banja. She is a member of the Scientific Council of the Geographical of the "Jovan Cvijić" Institute, Serbian Academy of Sciences and Arts, Belgrade (since 2021). She is a member of the Special Expert Committee of the Agency for Qualifications, Government of the Republic of Serbia, (sector 10 Services: subsector 1013 Hotels, restaurants and catering, subsector 1015 Travel, tourism and leisure). She is the mother of four adult children.

Rupam Konar is a senior lecturer at Taylor's University, Malaysia under the School of Hospitality, Tourism, and Events. His specialization area includes service innovation, service delivery, and service design in the hospitality industry. Within his research interest, he has published over 25 international refereed journals and has successfully complemented many national and university level research projects. Dr. Rupam is a certified Microsoft Innovative Educator and also has been awarded the gold medal at the e-learning carnival 2019. Currently, he is also the managing editor for the Asia Pacific Journal of Innovation in Hospitality and Tourism and an active member for the centre for research in tourism (CRiT) under Taylor's University.

Froilan Mobo is a Doctor of Public Administration graduate from the Urdaneta City University Class of 2016 and a graduate of the 2nd Doctorate Degree (Ph.D.) in Development Education program at the Central Luzon State University, Nueva Ecija, Philippines, Class of 2022. On March 11, 2024, Dr. Mobo was accredited and reclassified by the Commission on Higher Education (CHED) to the position of Professor II in the Philippine Merchant Marine Academy (PMMA), and this allowed him to work with different international research institutions, such as the Director and Research Consultant of the IKSAD Research Institute, Turkey. At present, he is in the process of finishing his 3rd master's degree, leading to social studies education at Bicol University. Recently, Dr. Mobo passed Batch 3—Certified Research Professional—and ranked in the top 5 in the National Examination. He was appointed Editor-in-Chief of the International Journal of Multidisciplinary: Applied Business and Education Research, Malang, Indonesia.

Muhammad Khizar Hayat Naeem is a distinguished Ph.D. candidate affiliated with the School of Management at Harbin Institute of Technology in Harbin, China. His notable contributions to the academic realm are underscored by a prolific publication record in both national and international journals, reflecting his dedication to advancing knowledge in his field of management.

Mario Ossorio is an Assistant Professor of Management at University of Campania "Luigi Vanvitelli", where he earned his Ph. D. in Innovation and Entrepreneurship. His last works are published in Journals such as Management Decision, Journal of Family Business Strategy, Global Business Review, European Journal of Innovation Management. He regularly serves as Chair and Discussant at international conferences of management.

Ramakrushna Padhy has received his PhD from the Indian Institute of Technology, Kharagpur. He also has a B.Tech in Mechanical engineering and PG diploma in Industrial Engineering. He has 22 years of consulting and teaching experience in the area of process management, lean, quality management and six sigma. His research has been in understanding the implementation strategy and process of Business Transformation Initiatives, enterprise sustainability, and circular economy.

Bassam Samir Al-Romeedy is a motivated Associate Professor at the faculty of tourism and hotels, University of Sadat City - Egypt, and an Accredited external auditor at the National Authority for Quality Assurance of Education and Accreditation – Egypt. I also work as a market research consultant and statistical analyst. Dr. Al-Romeedy has published in premier peer-reviewed journals including Tourism Review, Journal of Human Resources in Hospitality & Tourism, International Journal of Tourism and Hospitality Management, Corporate Social Responsibility and Environmental Management, and others.

Index

T

U

V

W

www.ingramcontent.com/pod-product-compliance
Lightning Source LLC
LaVergne TN
LVHW060423300926
849162LV00006B/1513
* 9 7 9 8 3 6 9 3 3 2 8 6 3 *